Australian Signpost MATHS

4

Alan McSeveny
Alan Parker
Diane McSeveny-Foster
Rachel McSeveny
Erika Johnson
Bob Collard

Short Contents

Sections

Pearson Australia
a division of Pearson Australia Group Pty Ltd)
459–471 Church Street, Level 1, Building B, Richmond, Victoria 3121
PO Box 23360, Melbourne, Victoria 8012
www.pearson.com.au

First published 2018 by Pearson Australia
2025 2024 2023 2022
10 9 8 7 6 5 4 3 2 1

Content Learning Specialist: Julian Lumb
Project Manager: Shelly Wang
Production Manager: Elizabeth Gosman & Aptara
Editor: Aptara
Designer: Anne Donald & Jennifer Johnston
Cover Designer: Jennifer Johnston
Desktop operator: Aptara
Rights & Permissions Editor: Samantha Russell-Tulip
Senior Publishing Services Analyst: Rob Curulli
Cover art: Michael Barter
Cover background: rf_vector/shutterstock.com
Illustrators: Michael Barter, Wendy Gorton, Nives Porcellato, Andrew Craig, Bruce Rankin
Printed in Malaysia (CTP-VVP)

ISBN 978 1 4886 2187 1
Pearson Australia Group Pty Ltd ABN 40 004 245 943

Acknowledgements
All material identified by ACARA is material subject to copyright under the *Copyright Act 1968* and is owned by the Australian Curriculum, Assessment and Reporting Authority 2018. ACARA neither endorses nor verifies the accuracy of the information provided and accepts no responsibility for incomplete or inaccurate information. In particular, ACARA does not endorse or verify that:

- The content descriptions are solely for a particular year and subject;
- All the content descriptions for that year and subject have been used; and
- The author's material aligns with the Australian Curriculum content descriptions for the relevant year and subject.

You can find the unaltered and most up to date version of this material at http://www.australiancurriculum.edu.au/ This material is reproduced with the permission of ACARA.

Every effort has been made to trace and acknowledge copyright. However, if any infringement has occurred, the publishers tender their apologies and invite the copyright holders to contact them.

What is Australian Signpost Maths?

Australian Signpost Maths is a mathematics activity book series for students from Foundation to Year 6. The series has been written to meet the requirements of the Australian Curriculum.

The components of the series include Student Books, Teacher's Books, Mentals Books and an interactive Website. Teachers can select an appropriate program for every student from the rich and varied material provided.

The content has been carefully sequenced within each year level and across the series to take into account students' likely mathematical development.

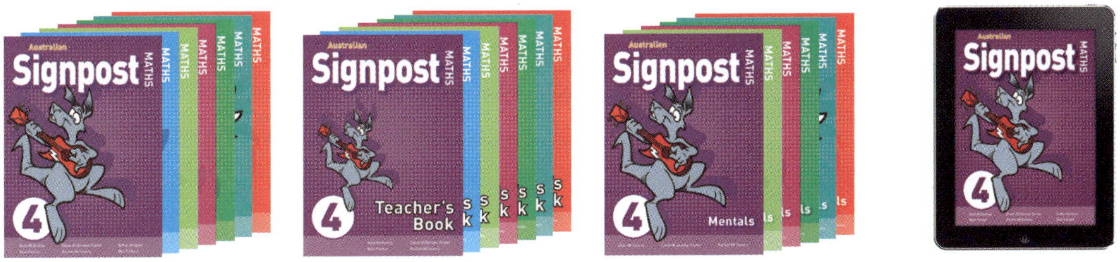

Student Books Teacher's Books Mentals Books Website

The Structure of Australian Signpost Maths

Australian Signpost Maths emphasises the curriculum's syllabus content as well as problem-solving strategies, language development and the use of technology.

The syllabus is organised into three content strands and four proficiency strands:

Content Strands

- Number and Algebra
- Measurement and Geometry
- Statistics and Probability

Proficiency Strands (see page v)

- Understanding
- Fluency
- Problem Solving
- Reasoning

The curriculum's **general capabilities** are developed throughout the Australian Signpost Maths program. These are:

- literacy
- numeracy
- information and communication technologies (ICT)
- critical and creative thinking.

Australian Signpost Maths also provides opportunities to develop other general capabilities, such as personal and social competence and intercultural understanding.

The cross-curriculum dimensions of the syllabus – 'Aboriginal and Torres Strait Islander histories and cultures', 'Asia and Australia's engagement with Asia' and 'Sustainability' – are embedded in the program.

To maximise the benefits of the program, the Student Book, Teacher's Book, Mentals Book and Website should be used together.

The structure of the **Student Book** allows teachers to determine both the order and the extent of content covered. Strands are organised separately so that the teacher, not the Student Book, decides the content of the next lesson. However, a suggested term program (see page x of this book) and a detailed program (see the Teacher's Book and Website) are also provided.

The **Teacher's Book** also provides lesson plans for each page of the Student Book and blackline masters to assist teachers in implementing the program.

The **Mentals Book** mixes examples from all strands. It revises the content covered in the Student Book. Each content strand is thoroughly covered, with the proficiency strands incorporated within each section. A special feature woven throughout the Mentals Book is the tables program in the four operations.

The innovative **Website** help teachers to bring mathematics alive with technology. The website provides interactive maths tools, games and practice opportunities as well as relevant resource masters and worksheets for all year levels. These can be used for whole-class, small-group and individual learning. The website also includes **Concept Check-In**, a new diagnostic screener.

Student Book pages are colour-coded by section.

Number and Algebra A	Measurement and Geometry A	Statistics and Probability
Number and Algebra B	Measurement and Geometry B	Answers

Australian Curriculum Proficiency Strands

The proficiency strands of the Australian Curriculum describe how content is explored or developed – that is, the 'thinking and doing' of mathematics.

Understanding

Learning the concepts

*Students build a robust knowledge of adaptable and transferable mathematical concepts. They make connections between related concepts and progressively apply the familiar to develop new ideas. They develop an understanding of the relationship between the 'why' and the 'how' of mathematics.**

Conceptual understanding of maths ideas includes the explanation of a concept using text and diagrams. This occurs throughout Australian Signpost Maths at the top of many pages and is indicated by the Concepts icon.

Fluency

Using the concepts

*Students develop skills in choosing appropriate procedures, carrying out procedures flexibly, accurately, efficiently and appropriately, and recalling factual knowledge and concepts readily.**

The practice of maths skills to build fluency occurs on every page of Australian Signpost Maths.

Problem Solving

Applying concepts and strategies to develop solutions to problems

*Students develop the ability to make choices, interpret, formulate, model and investigate problem situations, and communicate solutions effectively.**

Problem solving provides opportunities for students to use strategies and skills such as investigating and questioning, to collaborate with others and to communicate their findings to different audiences. Such activities are often indicated throughout Australian Signpost Maths by the Activity and Investigation icons.

Reasoning

Coherent and logical thought

*Students develop an increasingly sophisticated capacity for logical thought and actions, such as analysing, proving, evaluating, explaining, inferring, justifying and generalising.**

Students require opportunities to explain their mathematical thinking and can do so through both diagrams and written explanations. Reasoning questions are located throughout Australian Signpost Maths.

* The Australian Curriculum: Mathematics, v1.2 – Content structure

Special Features of Australian Signpost Maths

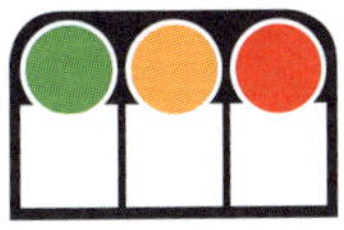

- **Traffic Light** system allows students to reflect on their work and highlight any units that they are having trouble understanding. They tick the red for units they feel they still don't understand, and green for those they feel they understand fully.
- Exercises are **well graded**. New work is reinforced in the Mentals Book.
- **Answers** are supplied in the back of this book as well as in the Teacher's Book.
- **Concept Check-In** diagnostic screener (on the Website) provides a snapshot of the class' conceptual understandings to aid in classroom management. It also allows teachers to measure progress over time.
- The eight **Diagnostic Tests** (now also in the back of this book) allow the teacher to discover each student's strengths and weaknesses, and the cross-references direct students to the pages where that work is treated. Answers are supplied in the Teacher's Book.
- The **Dictionary** at the beginning of this Student Book will help students to learn the language of mathematics.
- **ID Cards** (in the Mentals Book, Teacher's Book and Website) review the language of mathematics by asking students to identify common terms, shapes and symbols.
- Important **rules and concepts** are clearly highlighted.
- **Worked examples** and explanations are given throughout the Student Book where new concepts are introduced.
- The use of **colour** makes emphasis clear and is highly motivating.
- **Cartoons** give instruction and friendly advice.
- **Interactive activities** are provided on the website for whole-class, small-group and individual learning.

Australian Signpost Icons

Signpost icons are used throughout the book as cues to the essential nature of exercises and activities, and as a guide to ways of engaging with them. These icons often indicate alternative or more concrete approaches to dealing with concepts.

This icon highlights **important rules and concepts** occurring throughout the book. It often appears with worked examples.

Activities provide **applications and enrichment**. These activities usually involve the use of concrete materials and partner or group work.

These enjoyable activities are used to **motivate and involve** students in mathematical pursuits. They usually involve games and puzzles.

Investigations allow students to **explore and discover** maths concepts.

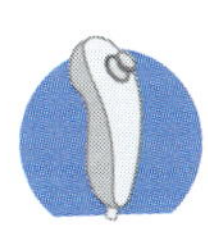

This icon indicates the use of computers, calculators or other **information and communications technology**.

4 Contents and Syllabus Overview

KEY

Blue	Number & Algebra
Green	Measurement & Geometry
Orange	Statistics & Probability

Number and Algebra A

Page	Unit	Title	Sub-strand	Number and place value	Fractions and decimals	Patterns and algebra	Content	Counting and numeration	Place value	Fractions	Decimals	Number patterns	Suggested Progress
1	1:01	Fractions			■					●			Term 1
2	1:02	Hundredths			■					●			
3	1:03	Decimals			■					●	●		
4	1:04	Numbers to 9999		■				●	●				
5	1:05	Numbers to 9999		■		■		●				●	
6	1:06	Solving Problems with Place Value		■				●	●				
7	1:07	Place Value to 10 000		■				●	●				T1, T2*
8	1:08	Rounding Off		■				●	●				
9	1:09	Expanded Notation		■				●	●				
10	1:10	Comparing Fractions			■					●			Term 2
11	1:11	Equivalent Fractions			■					●			
12	1:12	Equivalent Fractions			■					●			
13	1:13	Equivalent Fractions			■	■				●		●	
14	1:14	Improper Fractions and Mixed Numbers			■	■				●		●	T3, T4*
15	1:15	Mixed Numbers			■	■				●		●	
16	1:16	Numbers to 99 999		■				●	●				
17	1:17	Numbers to 99 999		■				●	●				
18	1:18	Equivalent Fractions			■					●			
19	1:19	Fractions and the Number Line			■	■				●		●	
20	1:20	Place Value in Decimals			■				●	●	●		Term 3
21	1:21	Tenths			■	■				●	●	●	
22	1:22	Comparing Decimals			■						●		
23	1:23	Place Value in Decimals			■				●		●		
24	1:24	Ordering Numbers to 99 999		■				●	●				T5, T6*
25	1:25	Reading and Writing Numbers		■				●	●				
26	1:26	Place Value to Hundredths		■	■				●		●		
27	1:27	Reading and Writing Decimals		■	■						●		
28	1:28	Reading and Writing Numbers		■				●	●				Term 4
29	1:29	Numbers to 999 999		■				●	●				
30	1:30	Place Value		■				●	●				
31	1:31	Rounding Off		■				●	●				T7, T8*
32	1:32	Fractions			■					●			
33	1:33	Fraction Patterns			■	■				●		●	
34	1:34	One Million		■				●	●				

* Suggested progress for Diagnostic Tests 1 to 8 is found in the Teacher's Book. The first of each pair of tests covers the first half of the period.

Number and Algebra B

Page	Unit	Title	Sub-strand	Number and place value	Money and financial mathematics	Patterns and algebra	Content	Addition	Subtraction	Multiplication	Division	Place value	Number patterns	Suggested Progress
35	2:01	Number Sequences		■		■				●			●	Term 1
36	2:02	Multiplication Tables Revision		■						●				
37	2:03	Multiplication Using Pictures		■						●				
38	2:04	Number Facts, × 2, × 4		■						●				
39	2:05	Number Facts, × 2, × 4		■						●				
40	2:06	Addition to 99, No Trading		■				●				●		
41	2:07	Addition, No Trading		■				●				●		
42	2:08	Addition to 99 with Trading		■				●				●		
43	2:09	Addition to 99 with Trading		■				●				●		
44	2:10	Addition with 2-Digit Numbers		■				●				●		T1, T2*
45	2:11	Addition Involving Hundreds		■				●				●		
46	2:12	Number Facts, × 8		■						●				Term 2
47	2:13	Number Facts, × 2, × 4, × 8		■		■				●			●	
48	2:14	Subtraction, No Trading		■					●					
49	2:15	Subtraction with Trading		■					●			●		
50	2:16	Subtraction with Trading to 99		■					●			●		
51	2:17	Subtraction with Trading to 99		■					●			●		
52	2:18	Subtraction with Trading to 99		■					●			●		
53	2:19	Addition Problems to 99		■				●				●		
54	2:20	Number Facts, × 3, × 6		■						●			●	
55	2:21	Number Facts, × 3, × 6		■		■				●				T3, T4*
56	2:22	Number Facts, × 9		■						●			●	
57	2:23	Number Facts, × 9		■						●			●	
58	2:24	Number Facts, × 7		■						●				
59	2:25	Multiplication Tables Review		■						●				
60	2:26	Addition to 999		■				●				●		
61	2:27	Addition to 999		■				●				●		
62	2:28	Writing the Addition Algorithm		■				●				●		
63	2:29	Finding Missing Numbers				■		●	●	●	●			
64	2:30	Subtraction without Trading to 999		■					●			●		Term 3
65	2:31	Subtraction with Trading to 999		■					●			●		
66	2:32	Subtraction with Trading to 999		■					●			●		
67	2:33	Subtraction with Trading to 999		■					●			●		
68	2:34	Subtraction from Hundreds		■					●			●		
69	2:35	Subtraction from Hundreds Strategy		■					●			●		
70	2:36	Division as Repeated Subtraction		■							●			
71	2:37	Understanding Division		■							●			T5, T6*
72	2:38	Division Facts		■						●	●			
73	2:39	Division Facts		■						●	●			

Number and Algebra B

Page	Unit	Title	Sub-strand: Number and place value	Sub-strand: Money and financial mathematics	Sub-strand: Patterns and algebra	Content: Addition	Content: Subtraction	Content: Multiplication	Content: Division	Content: Place value	Content: Number patterns	Suggested Progress
74	2:40	Division Facts	■					●	●			
75	2:41	Division Using the Multiplication Grid	■					●	●			
76	2:42	Money		■		●	●					
77	2:43	Counting Change		■		●	●					Term 4
78	2:44	Factors and Multiples			■			●	●		●	
79	2:45	Products and Factors			■			●	●		●	
80	2:46	Problem Solving			■	●	●	●	●			
81	2:47	Problem Solving			■	●	●	●	●			
82	2:48	Working with Numbers			■			●	●		●	T7, T8*
83	2:49	Number Patterns			■	●	●				●	
84	2:50	Using Odd and Even Numbers	■			●	●	●	●			
85	2:51	Rounding Off Money		■		●	●					
86	2:52	What's the Rule?			■	●	●	●	●		●	
87	2:53	Multiplication Using Place-Value Blocks	■					●		●		

Measurement and Geometry A

Page	Unit	Title	Sub-strand: Using units of measurement	Sub-strand: Shape	Content: Length	Content: Area	Content: Volume	Content: Capacity	Content: Mass	Content: Time	Content: Temperature	Suggested Progress
88	3:01	Analogue Time	■							●		Term 1
89	3:02	Analogue Time	■							●		
90	3:03	The Calendar	■							●		
91	3:04	The Calendar	■							●		
92	3:05	Using Centimetres and Millimetres	■		●							
93	3:06	Using Millimetres	■		●							
94	3:07	Recording Length	■		●							
95	3:08	Using Measurement Scales	■		●				●			T1, T2*
96	3:09	Temperature	■								●	Term 2
97	3:10	Recording Temperature	■								●	
98	3:11	Using Millilitres	■					●				
99	3:12	Using Millilitres	■					●				
100	3:13	Area Using Informal Units	■	■		●						
101	3:14	The Square Centimetre	■	■		●						T3, T4*
102	3:15	Measuring Mass	■						●			
103	3:16	Measuring Mass	■						●			
104	3:17	The Square Centimetre	■	■		●						
105	3:18	The Square Centimetre	■	■		●						Term 3
106	3:19	The Square Metre	■	■		●						
107	3:20	The Square Metre	■	■		●						
108	3:21	Finding Area	■	■		●						
109	3:22	Finding Area	■	■		●						
110	3:23	Analogue and Digital Time	■							●		
111	3:24	Time	■							●		

Measurement and Geometry A

Page	Unit	Title	Sub-strand	Using units of measurement	Shape	Content	Length	Area	Volume	Capacity	Mass	Time	Temperature	Suggested Progress
112	3:25	Volume		■					●	●				T5, T6*
113	3:26	Finding Volume		■					●					
114	3:27	The Cubic Centimetre		■					●					
115	3:28	Timetables		■								●		
116	3:29	Comparing Measurements		■			●							Term 4
117	3:30	Personal Benchmarks		■			●			●	●			
118	3:31	Analogue and Digital Time		■								●		
119	3:32	am and pm Time		■								●		
120	3:33	Recording Length		■			●							T7, T8*
121	3:34	Using Millilitres		■						●				
122	3:35	Using Grams		■							●			

Measurement and Geometry B

Page	Unit	Title	Sub-strand	Shape	Location and transformation	Geometric reasoning	Content	2D Space	3D Space	Location	Transformations	Angles	Suggested Progress
123	4:01	Flip, Slide and Turn		■	■			●			●		Term 1
124	4:02	Angles and Plane Figures		■		■		●				●	
125	4:03	Comparing Angles				■						●	
126	4:04	3D Objects		■				●	●				
127	4:05	Prisms and Pyramids		■				●	●				T1, T2*
128	4:06	Surfaces of Prisms and Pyramids		■				●	●				
129	4:07	Drawing Prisms and Pyramids		■					●				Term 2
130	4:08	Drawing Angles				■						●	
131	4:09	Angles as Half and Quarter Turns				■						●	T3, T4*
132	4:10	Investigating Polygons		■		■		●				●	
133	4:11	Visualising Shapes		■		■		●					
134	4:12	Maps			■					●			
135	4:13	Creating a Map			■					●			
136	4:14	Cones, Cylinders and Spheres		■					●				Term 3
137	4:15	Views of 3D Objects		■				●	●				
138	4:16	Compass Directions			■					●			
139	4:17	Compass Directions			■				●	●			T5, T6*
140	4:18	Describing Position			■			●	●	●			
141	4:19	Using Position in Maps			■			●		●			
142	4:20	Visualising Shapes		■		■		●					
143	4:21	Acute and Obtuse Angles				■						●	Term 4
144	4:22	Horizontal and Vertical				■		●	●			●	T7, T8*
145	4:23	Tessellations			■			●			●		
146	4:24	Tangrams			■			●			●		

* Suggested progress for Diagnostic Tests 1 to 8 is found in the Teacher's Book.
The first of each pair of tests covers the first half of the period.

Page	Unit	Title	Sub-strand	Chance	Data representation and interpretation	Content	Chance	Investigation	Data representation	Suggested Progress
147	5:01	Drawing Tables			■			●	●	Term 1
148	5:02	Chance		■			●			
149	5:03	Chance		■			●			
150	5:04	Using Graphs			■				●	T1, T2*
151	5:05	Reading Graphs			■				●	
152	5:06	Ordering Events		■			●	●		Term 2
153	5:07	Chance Used in Games		■			●			T3, T4*
154	5:08	Surveys			■			●		
155	5:09	Tally Marks			■			●	●	Term 3
156	5:10	Constructing Spinners		■	■		●	●		
157	5:11	Unequal Outcomes		■	■		●	●	●	T5, T6*
158	5:12	Collecting Information			■			●		
159	5:13	Graphing Data			■		●	●	●	Term 4
160	5:14	Chance Experiments		■	■		●	●	●	T7, T8*
161	5:15	Carry Out Your Own Survey			■			●	●	
162	5:16	Chance Experiments		■	■		●	●	●	

Table heading: **Statistics and Probability**

* Suggested progress for Diagnostic Tests 1 to 8 is found in the Teacher's Book.
The first of each pair of tests covers the first half of the period.

Suggested Program

It is assumed that there are 10 weeks in each term.

	Weeks 1–10	Weeks 11–20	Weeks 21–30	Weeks 31–end
Number and Algebra A	1:01–1:09	1:10–1:19	1:20–1:27	1:28–1:34
Number and Algebra B	2:01–2:11	2:12–2:29	2:30–2:42	2:43–2:53
Measurement and Geometry A	3:01–3:08	3:09–3:16	3:17–3:28	3:29–3:35
Measurement and Geometry B	4:01–4:06	4:07–4:13	4:14–4:20	4:21–4:24
Statistics and Probability	5:01–5:05	5:06–5:08	5:09–5:12	5:13–5:16

The eight Diagnostic Tests are found in the Teacher's Book.
See the Contents and Syllabus Overview on pages vi–x for suggested placement of each test.

Contents Cross-reference

Number and Algebra

1 Whole numbers	Pages	Australian Curriculum Reference
Four-, five- and six-digit numbers, and place value	4, 5, 6, 7, 8, 9, 16, 17, 24, 25, 28, 29, 30, 31, 34	Recognise, represent and order numbers to at least tens of thousands (ACMNA072); Apply place value to partition, rearrange and regroup numbers to at least tens of thousands to assist calculations and solve problems (ACMNA073)
Odd and even numbers	84	Investigate and use the properties of odd and even numbers (ACMNA071)
Rounding	8, 31, 60, 61, 85	Recognise, represent and order numbers to at least tens of thousands (ACMNA072); Apply place value to partition, rearrange and regroup numbers to at least tens of thousands to assist calculations and solve problems (ACMNA073)
'is less than', 'is greater than'	10, 11, 12, 25	Recognise, represent and order numbers to at least tens of thousands (ACMNA072); Investigate equivalent fractions used in contexts (ACMNA077); Use equivalent number sentences involving addition and subtraction to find unknown quantities (ACMNA083)
2 Addition		
Mental strategies	35, 63, 77	Apply place value to partition, rearrange and regroup numbers to at least tens of thousands to assist calculations and solve problems (ACMNA073); Investigate number sequences involving multiples of 3, 4, 6, 7, 8, and 9 (ACMNA074); Solve problems involving purchases and the calculation of change to the nearest five cents with and without digital technologies (ACMNA080); Use equivalent number sentences involving addition and subtraction to find unknown quantities (ACMNA083)
Written strategies	40, 41, 42, 43, 44, 45, 53, 60, 61, 62, 85	Apply place value to partition, rearrange and regroup numbers to at least tens of thousands to assist calculations and solve problems (ACMNA073)
Problem solving	40, 41, 43, 53, 62, 80, 81, 85	Apply place value to partition, rearrange and regroup numbers to at least tens of thousands to assist calculations and solve problems (ACMNA073); Use equivalent number sentences involving addition and subtraction to find unknown quantities (ACMNA083)
3 Subtraction		
Mental strategies	49, 50, 52, 69, 77	Apply place value to partition, rearrange and regroup numbers to at least tens of thousands to assist calculations and solve problems (ACMNA073); Solve problems involving purchases and the calculation of change to the nearest five cents with and without digital technologies (ACMNA080)
Written strategies	48, 49, 50, 51, 52, 64, 65, 66, 67, 68, 69	Apply place value to partition, rearrange and regroup numbers to at least tens of thousands to assist calculations and solve problems (ACMNA073); Use equivalent number sentences involving addition and subtraction to find unknown quantities (ACMNA083)
Problem solving	49, 52, 64, 68, 80, 81, 85	Apply place value to partition, rearrange and regroup numbers to at least tens of thousands to assist calculations and solve problems (ACMNA073); Use equivalent number sentences involving addition and subtraction to find unknown quantities (ACMNA083)
4 Multiplication		
Multiplication tables	36, 38, 39, 46, 47, 54, 55, 56, 57, 58, 59, 79	Investigate number sequences involving multiples of 3, 4, 6, 7, 8, and 9 (ACMNA074); Recall multiplication facts up to 10 × 10 and related division facts (ACMNA075); Explore and describe number patterns resulting from performing multiplication (ACMNA081)

Mental and written strategies	35, 37, 47, 54, 57, 58, 59, 72, 82, 87	Investigate number sequences involving multiples of 3, 4, 6, 7, 8, and 9 (ACMNA074); Recall multiplication facts up to 10 × 10 and related division facts (ACMNA075); Develop efficient mental and written strategies and use appropriate digital technologies for multiplication and for division where there is no remainder (ACMNA076); Explore and describe number patterns resulting from performing multiplication (ACMNA081); Solve word problems by using number sentences involving multiplication or division where there is no remainder (ACMNA082)
Multiples and factors	55, 78, 79	Recall multiplication facts up to 10 × 10 and related division facts (ACMNA075); Explore and describe number patterns resulting from performing multiplication (ACMNA081)
Problem solving	37, 80, 81, 87	Develop efficient mental and written strategies and use appropriate digital technologies for multiplication and for division where there is no remainder (ACMNA076); Solve word problems by using number sentences involving multiplication or division where there is no remainder (ACMNA082)
5 Division		
Mental and written strategies	70, 71, 72	Recall multiplication facts up to 10 × 10 and related division facts (ACMNA075); Develop efficient mental and written strategies and use appropriate digital technologies for multiplication and for division where there is no remainder (ACMNA076)
Division facts	72, 73, 74, 75	Recall multiplication facts up to 10 × 10 and related division facts (ACMNA075); Develop efficient mental and written strategies and use appropriate digital technologies for multiplication and for division where there is no remainder (ACMNA076)
Linking multiplication and division	72, 73, 74, 75, 82	Recall multiplication facts up to 10 × 10 and related division facts (ACMNA075); Develop efficient mental and written strategies and use appropriate digital technologies for multiplication and for division where there is no remainder (ACMNA076); Solve word problems by using number sentences involving multiplication or division where there is no remainder (ACMNA082)
Problem solving	70, 71, 72, 80, 81	Recall multiplication facts up to 10 × 10 and related division facts (ACMNA075); Develop efficient mental and written strategies and use appropriate digital technologies for multiplication and for division where there is no remainder (ACMNA076); Solve word problems by using number sentences involving multiplication or division where there is no remainder (ACMNA082)
6 Fractions and decimals		
Common fractions	1, 2, 3, 10, 11, 12, 18	Investigate equivalent fractions used in contexts (ACMNA077); Count by quarters halves and thirds, including with mixed numerals. Locate and represent these fractions on a number line (ACMNA078)
Mixed numbers and improper fractions	12, 13, 14, 15, 21, 32	Investigate equivalent fractions used in contexts (ACMNA077); Count by quarters halves and thirds, including with mixed numerals. Locate and represent these fractions on a number line (ACMNA078)
Comparing fractions	10, 11, 13, 18	Investigate equivalent fractions used in contexts (ACMNA077); Count by quarters halves and thirds, including with mixed numerals. Locate and represent these fractions on a number line (ACMNA078)
Equivalent fractions	11, 12, 13, 18, 19	Investigate equivalent fractions used in contexts (ACMNA077); Count by quarters halves and thirds, including with mixed numerals. Locate and represent these fractions on a number line (ACMNA078)
Counting using fractions	11, 12, 13, 14, 15, 18, 19, 21, 33	Investigate equivalent fractions used in contexts (ACMNA077); Count by quarters halves and thirds, including with mixed numerals. Locate and represent these fractions on a number line (ACMNA078)

Decimal notation	2, 3, 20, 21, 22, 23, 26, 27	Recognise that the place value system can be extended to tenths and hundredths. Make connections between fractions and decimal notation (ACMNA079)
Comparing decimals	22, 23	Recognise that the place value system can be extended to tenths and hundredths. Make connections between fractions and decimal notation (ACMNA079)
7 Money		
Calculating costs and giving change	76, 77, 85	Solve problems involving purchases and the calculation of change to the nearest five cents with and without digital technologies (ACMNA080)
8 Number patterns		
Patterns	5, 13, 15, 21, 33, 35, 55, 56, 57, 82, 83, 86	Recognise, represent and order numbers to at least tens of thousands (ACMNA072); Investigate number sequences involving multiples of 3, 4, 6, 7, 8, and 9 (ACMNA074); Investigate equivalent fractions used in contexts (ACMNA077); Count by quarters halves and thirds, including with mixed numerals. Locate and represent these fractions on a number line (ACMNA078); Explore and describe number patterns resulting from performing multiplication (ACMNA081)
Pattern rules	33, 86	Count by quarters halves and thirds, including with mixed numerals. Locate and represent these fractions on a number line (ACMNA078); Explore and describe number patterns resulting from performing multiplication (ACMNA081)
Number sentences with missing elements	37, 57, 59, 63, 71, 72, 73, 74, 75, 79	Recall multiplication facts up to 10 × 10 and related division facts (ACMNA075); Explore and describe number patterns resulting from performing multiplication (ACMNA081); Use equivalent number sentences involving addition and subtraction to find unknown quantities (ACMNA083)
9 Calculators		
Using a calculator	6, 29, 46, 51, 79	Investigate and use the properties of odd and even numbers (ACMNA071); Apply place value to partition, rearrange and regroup numbers to at least tens of thousands to assist calculations and solve problems (ACMNA073); Develop efficient mental and written strategies and use appropriate digital technologies for multiplication and for division where there is no remainder (ACMNA076)

Measurement and Geometry

1 Using units of measurement	Pages	Australian Curriculum Reference
Reading scales	30, 83, 95, 96, 97, 98, 116, 117, 120	Use scaled instruments to measure and compare lengths, masses, capacities and temperatures (ACMMG084)
Length	92, 93, 94, 95, 116, 117, 120	Use scaled instruments to measure and compare lengths, masses, capacities and temperatures (ACMMG084)
Area	100, 101, 104, 105, 106, 107, 108, 109	Compare objects using familiar metric units of area and volume (ACMMG290); Compare the areas of regular and irregular shapes by informal means (ACMMG087)
Volume	112, 113, 114	Compare objects using familiar metric units of area and volume (ACMMG290)
Capacity	83, 98, 99, 112, 117, 121	Use scaled instruments to measure and compare lengths, masses, capacities and temperatures (ACMMG084)

Mass	30, 95, 102, 103, 117, 122	Use scaled instruments to measure and compare lengths, masses, capacities and temperatures (ACMMG084)
Temperature	96, 97	Use scaled instruments to measure and compare lengths, masses, capacities and temperatures (ACMMG084)
Time	88, 89, 110, 111, 118, 119	Convert between units of time (ACMMG085); Use am and pm notation and solve simple time problems (ACMMG086)
Time problems	90, 91, 110, 111, 115	Convert between units of time (ACMMG085); Use am and pm notation and solve simple time problems (ACMMG086)
Calendars	90, 91	Convert between units of time (ACMMG085)
Timelines, timetables	90, 115	Convert between units of time (ACMMG085); Use am and pm notation and solve simple time problems (ACMMG086)
2 Shape		
2D shapes	132, 133, 142, 145, 146	Compare and describe two-dimensional shapes that result from combining and splitting common shapes, with and without the use of digital technologies (ACMMG088); Compare angles and classify them as equal to, greater than or less than a right angle (ACMMG089); Describe and draw two-dimensional shapes, with and without digital technologies (ACMMG042) *[Progression]*
3D objects	126, 127, 128, 129, 136, 137, 144	Make models of three-dimensional objects and describe key features (ACMMG063) *[Progression]*
Cones, cylinders and spheres	126, 136, 137	Make models of three-dimensional objects and describe key features (ACMMG063) *[Progression]*
Prisms and pyramids	126, 127, 128, 129, 137, 144	Make models of three-dimensional objects and describe key features (ACMMG063) *[Progression]*
Angles	124, 125, 130, 131, 143	Compare angles and classify them as equal to, greater than or less than a right angle (ACMMG089)
3 Location and transformation		
Maps, scales and legends	134, 135, 138, 139	Use simple scales, legends and directions to interpret information contained in basic maps (ACMMG090)
Compass directions and position	138, 139, 140, 141	Compare angles and classify them as equal to, greater than or less than a right angle (ACMMG089); Use simple scales, legends and directions to interpret information contained in basic maps (ACMMG090)
Using coordinates	140, 141	Use simple scales, legends and directions to interpret information contained in basic maps (ACMMG090)
Symmetrical patterns and pictures	123, 142, 145, 146	Create symmetrical patterns, pictures and shapes with and without digital technologies (ACMMG091)

Statistics and Probability

1 Data representation and interpretation	Pages	Australian Curriculum Reference
Data collection	154, 155, 156, 158, 161	Select and trial methods for data collection, including survey questions and recording sheets (ACMSP095)
Graphs	147, 150, 151, 159, 162	Construct suitable data displays, with and without the use of digital technologies, from given or collected data. Include tables, column graphs and picture graphs where one picture can represent many data values (ACMSP096); Evaluate the effectiveness of different displays in illustrating data features including variability (ACMSP097)
Tables	147, 155, 157, 158, 159, 160	Construct suitable data displays, with and without the use of digital technologies, from given or collected data. Include tables, column graphs and picture graphs where one picture can represent many data values (ACMSP096); Evaluate the effectiveness of different displays in illustrating data features including variability (ACMSP097)
2 Chance		
Chance experiments	152, 153, 155, 156, 157, 160, 162	Describe possible everyday events and order their chances of occurring (ACMSP092); Identify everyday events where one cannot happen if the other happens (ACMSP093); Identify events where the chance of one will not be affected by the occurrence of the other (ACMSP094)
Describing likelihood	148, 149, 152, 153, 156, 157, 160, 162	Describe possible everyday events and order their chances of occurring (ACMSP092); Identify everyday events where one cannot happen if the other happens (ACMSP093); Identify events where the chance of one will not be affected by the occurrence of the other (ACMSP094)

4 Dictionary

abacus

An instrument used for counting and calculating.

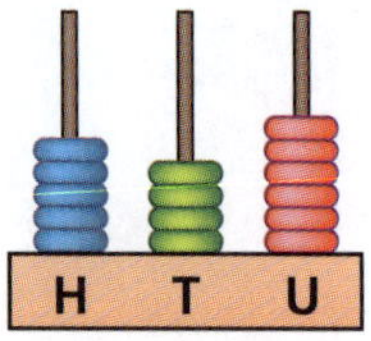

- The abacus shows 546.

am (ante meridiem)

Any time between midnight and midday.

- The time is 25 past 7 in the morning. It is 7:25 am.

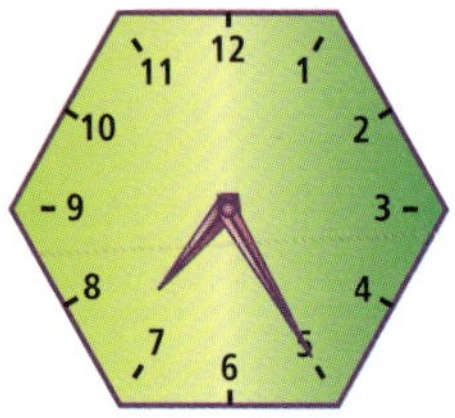

Morning

analogue time

The time shown on a clock face.

- 13 minutes to 6 is the time on this analogue clock.

angle

The amount of turning between two lines about a common point.

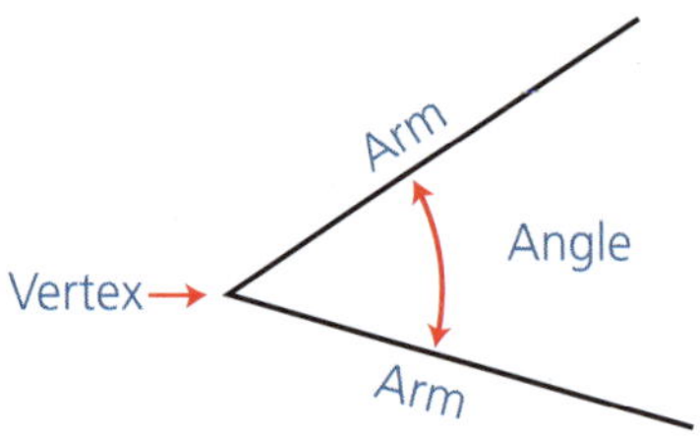

- Right Angle (Quarter Turn)

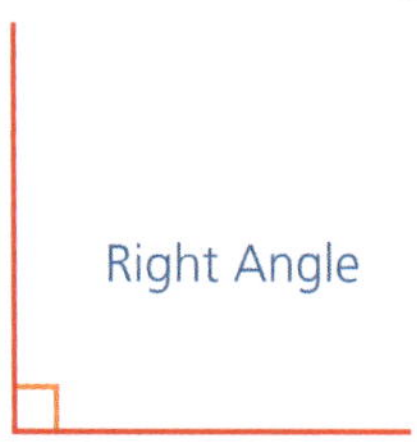

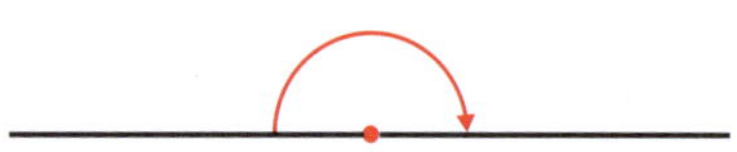

- Straight Angle (Half Turn)

anticlockwise and clockwise

The direction of a turn.

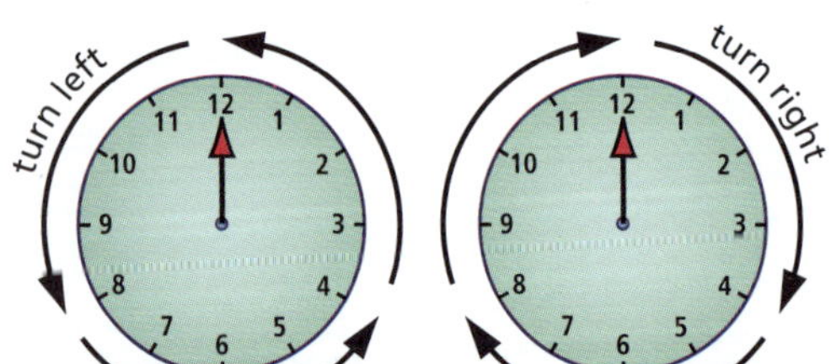

Anticlockwise

Clockwise (as on a clock)

area

The measure of the amount of surface.
Area is measured in square units:

- square centimetres cm^2
- square metres m^2

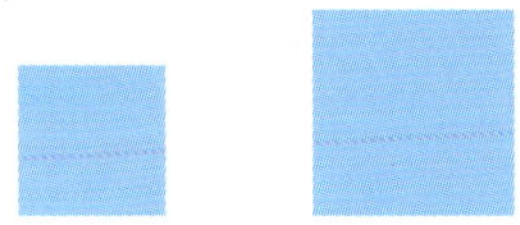

arrow cards

Arrow cards help us to understand the value of a number.

- This arrow card represents 638 419.

axis of symmetry

a line that divides a picture in half so that each half is the mirror image of the other part

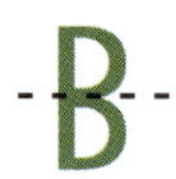

- One axis of symmetry

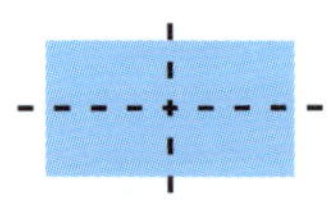

- Two axes of symmetry

The plural of *axis* is *axes*.

capacity

The amount that a container can hold.

- Less than 1 litre
- More than 1 litre

centimetre (cm)

A unit of length equal to one hundredth of a metre.

- 100 cm = 1 m

chance

The chance (or probability) of something happening is its likelihood of happening.

- If you toss a coin, there is an even chance of tossing a head.

See also *probability*.

cone

A three-dimensional object with a circular base that tapers to a point.

counting number

Any number used in counting.

- 1, 2, 3, 4, 5, …

  Zero is not a counting number.

cube

A three-dimensional object that has six equal square faces, eight vertices and twelve equal edges.

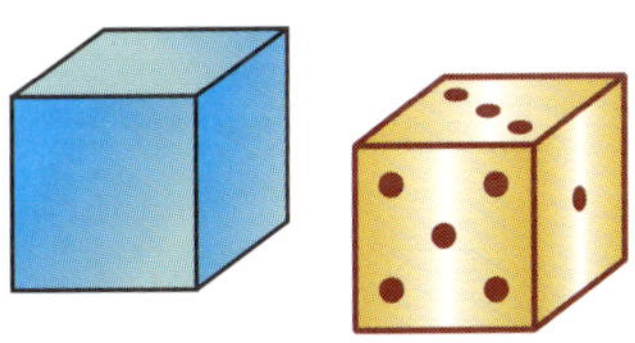

cubic centimetre (cm^3)

A unit of volume equal to the volume of a cube of side length 1 cm.

cylinder

A three-dimensional object with two equal circular faces and one curved surface.

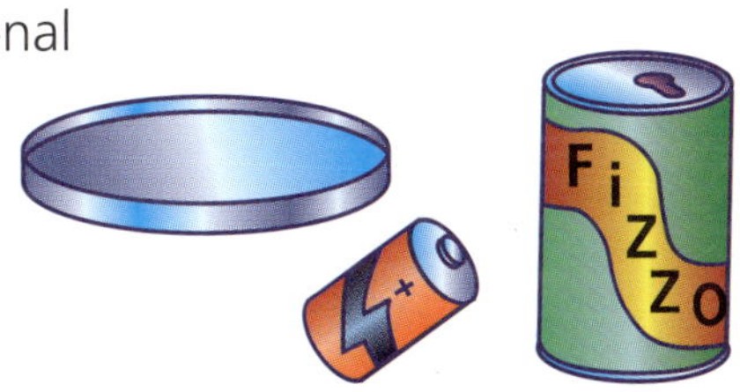

decrease

To make smaller.

diagonal

A line that joins any two non-adjacent corners of a polygon.

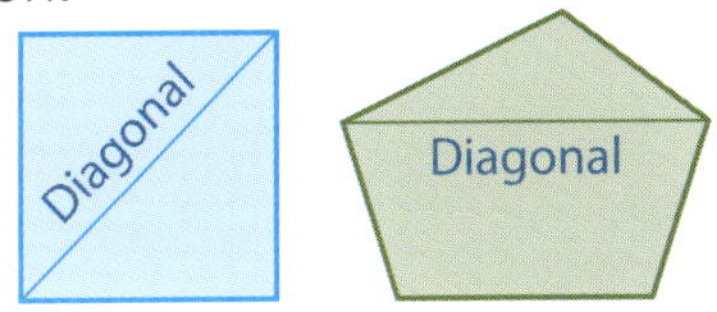

difference

The amount by which two numbers are separated.

- 9 – 6 = 3

  The difference between 9 and 6 is 3.

digital time

Time expressed using digits.

- This digital clock shows 24 minutes past 10.

digits

Symbols used to write a number.

- 6

  Six is a 1-digit number.
- 47

  Forty-seven is a 2-digit number.

division (÷)

Breaking up groups into equal parts.

- 6 ÷ 3

  a Sharing
  2 each

  b Grouping
  2 groups

double

Twice as much.

- Double 6 = 6 + 6 or 2 × 6

  12 is double 6.

edge

The intersection of two surfaces.

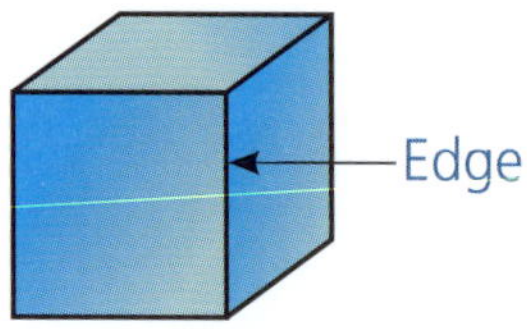

even number

Any number that is a multiple of two and can be grouped in twos.

- 16, 300, 4 394

expanded notation

A way of writing numerals to show the place value of each digit.

- 137 = (1 × 100) + (3 × 10) + 7

face

A flat surface of a three-dimensional object.

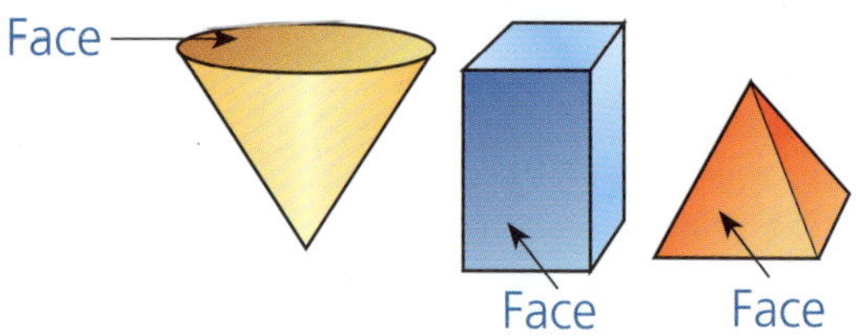

factor

Any whole number that divides another whole number exactly, leaving no remainder.

- The factors of 12 are 1,12, 2, 6, 3 and 4.

flip (reflection)

To turn over.

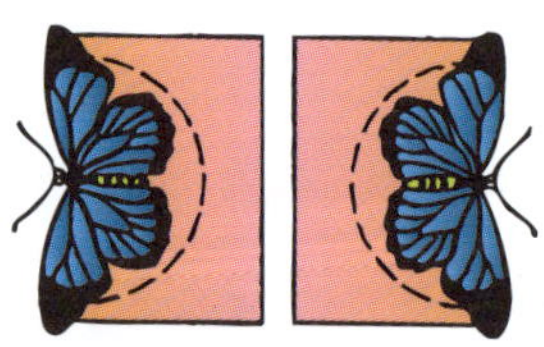

fraction

Any part of a whole, group or object.

- 2 out of 6 coloured

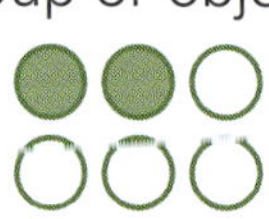

- $\frac{1}{4}$ is coloured

Equivalent fractions

Fractions of equal size.

- $\frac{1}{2} = \frac{5}{10} = \frac{7}{14} = \ldots$

Improper fraction

A fraction with a numerator that is bigger than the denominator.

- $\frac{9}{8}$

Mixed number

A number that has a whole number part and a fraction part.

- $1\frac{2}{3}$

gram (g)

A unit of mass used to measure how heavy something is.

- 1 kilogram = 1 000 grams

graphs

- Bar Graph

  A graph which uses horizontal bars to compare the size of groups.

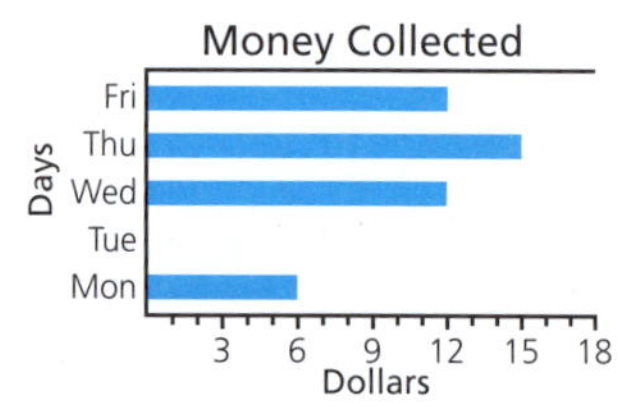

- Column Graph

  Groups are compared using the heights of columns.

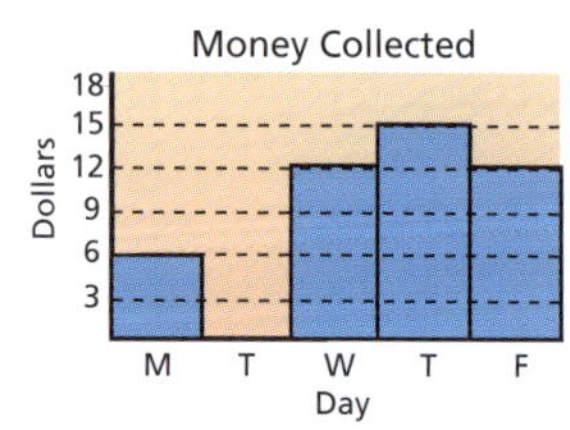

- Dot Plot

  A graph which uses dots to compare the size of groups.

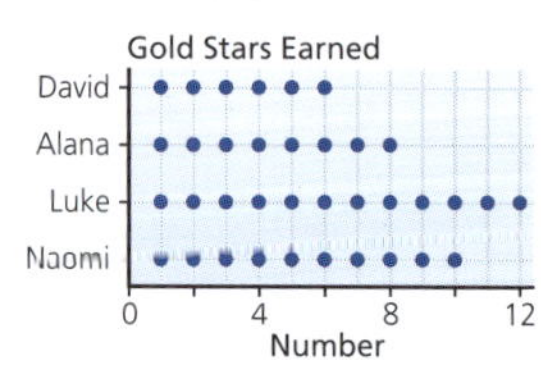

- Picture Graph

  A picture is used as a unit to show how many.

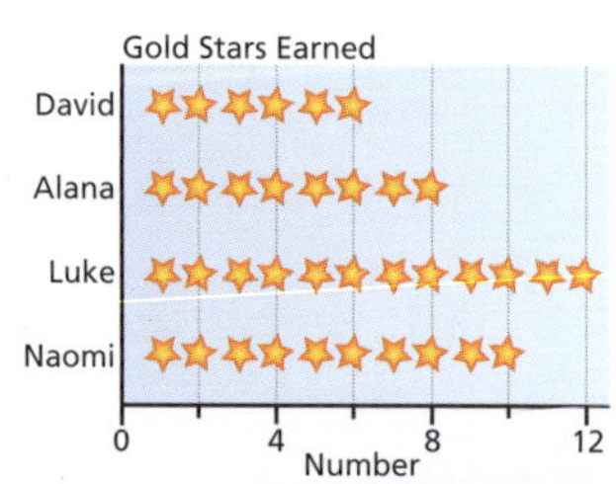

greater than (>)

A way of showing that a number is larger than another number.

- 7 > 3

See also *less than (<)*.

hexagon

A polygon with six sides.

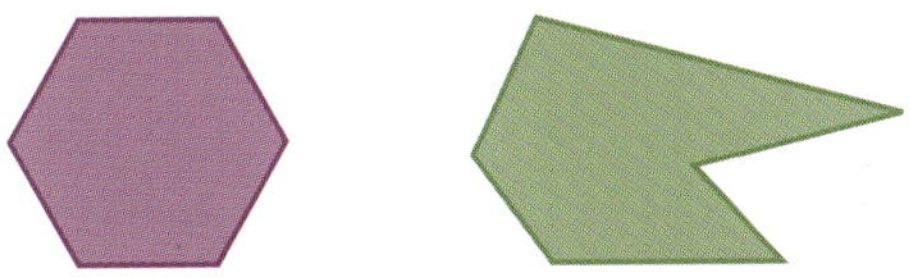

Regular Hexagon Irregular Hexagon

See also *polygon*.

horizontal

- parallel to the horizon
- level or flat
- any direction at right angles to the vertical.

increase

To make larger.

inverse operations

Adding 8 is the opposite (the inverse) of subtracting 8.

- 100 + 8 – 8 = 100

Multiplying by 2 is the opposite (the inverse) of dividing by 2.

- 4 × 2 ÷ 2 = 4

jump strategy

Adding or subtracting numbers, jumping by tens or ones.

- 52 – 14 = 38

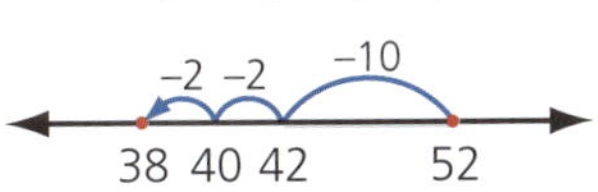

kilo (k)

Thousand (1 000).

kilogram (kg)

The basic unit of mass, equal to 1000 grams.

- 1 kg = 1 000 g

less than (<)

A way of showing that a number is smaller than another number.

- 3 < 7

See also *greater than (>)*.

line

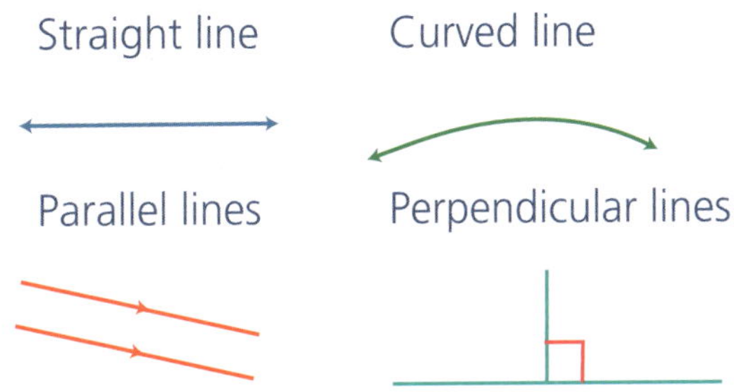

litre (L)

A unit of capacity (or volume) used for the measurement of liquids.

- 1 L = 1 000 mL

map or plan

A picture of an area viewed from above.

mass

The amount of matter in an object, a measure of how heavy it is.

metre (m)

The basic unit of length, equal to 100 centimetres.

- 1 m = 100 cm

millilitre (mL)

A unit of capacity equal to one thousandth of a litre.

- 1 000 mL = 1 L

millimetre (mm)

A unit of length equal to one tenth of a centimetre, or one thousandth of a metre.

- 10 mm = 1 cm
- 1 000 mm = 1 m

multiple

The result of multiplying a counting number by another counting number.

- The multiples of 5 are 5, 10, 15, 20, …

multiplication (×)

Combining equal groups.

- 3 × 5 = 15

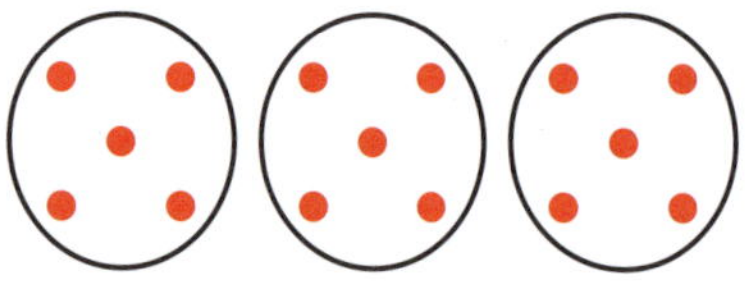

net

A flat shape that can be folded to make a three-dimensional shape.

Net of a Cube

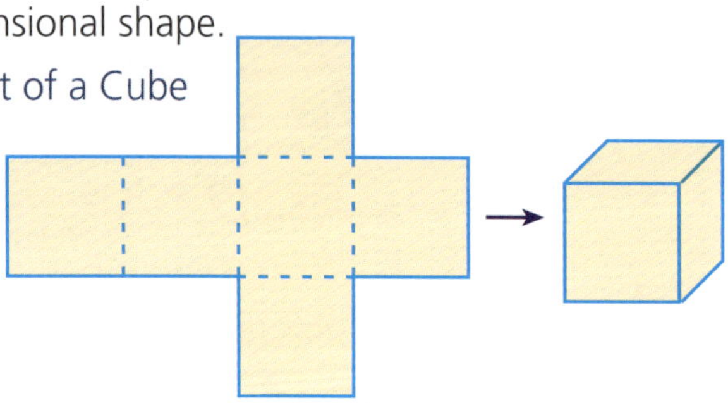

number line

A line on which numbers are marked. Number lines can be used to show operations.

- 2 + 2 + 2 = 6

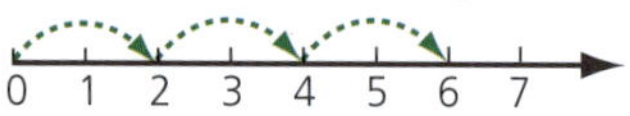

numeral expander

Breaks up a number into hundreds, tens and ones.

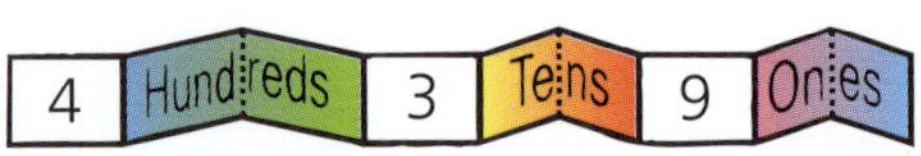

- This numeral expander shows 439.

object

The term used to describe a three-dimensional shape.

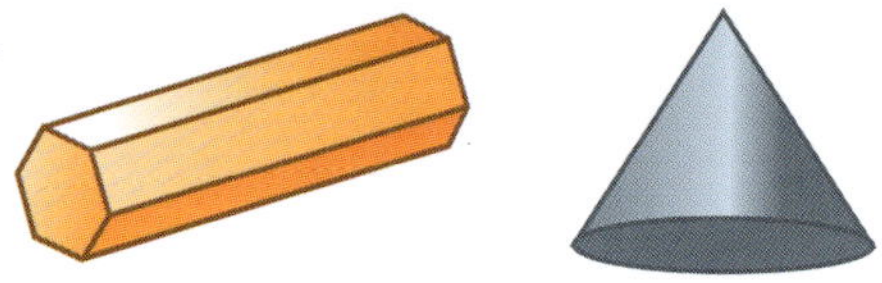

Hexagonal Prism   Cone

octagon

A polygon with eight sides.

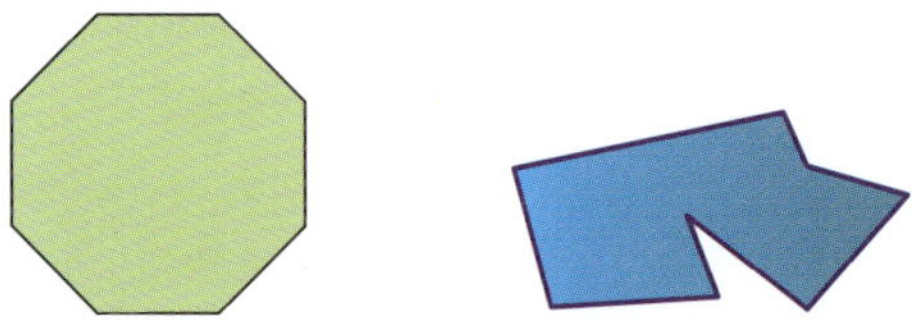

Regular Octagon   Irregular Octagon

See also *polygon*.

oval

A squashed circle.

parallel lines

Straight lines on the same flat surface that do not meet.

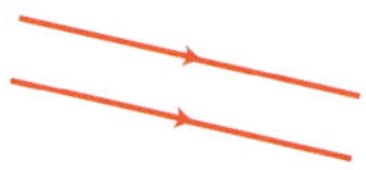

parallelogram

A shape with four sides such that the pairs of opposite sides are parallel and equal.

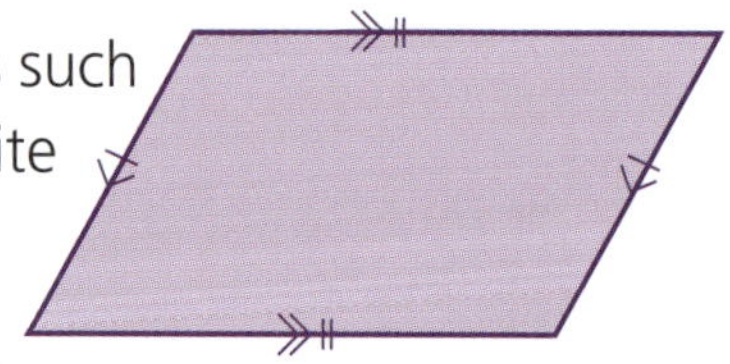

pentagon

A polygon with five sides.

Regular Pentagon   Irregular Pentagon

See also *polygon*.

perimeter

The distance around the outside of a shape; the boundary.

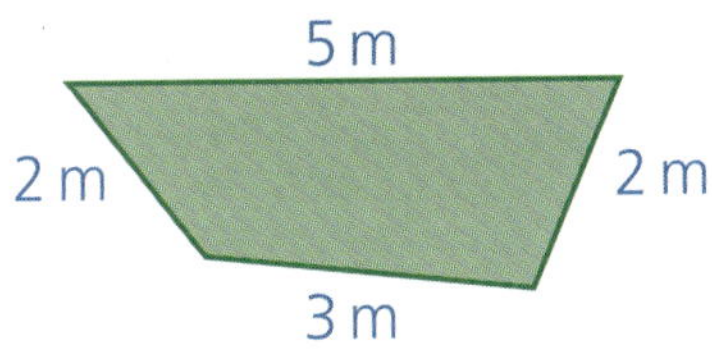

- Perimeter = 2 m + 3 m + 2 m + 5 m
  = 12 m

perpendicular lines

Lines that meet at right angles.

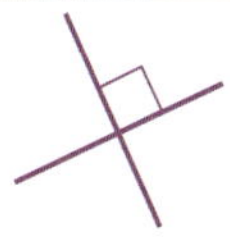

place value

The column value of a digit.

- 396 =

Hundreds	Tens	Ones
3	9	6

pm (post meridiem)

Any time between midday and midnight.

- The time is 20 past 1 in the afternoon. It is 1:20 pm.

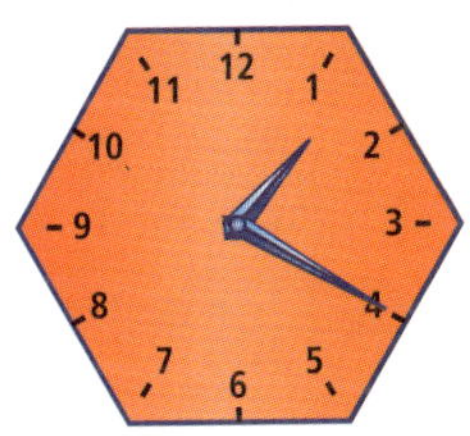

Afternoon

polygon

A two-dimensional shape with three or more straight sides, such as a triangle, quadrilateral, pentagon etc.

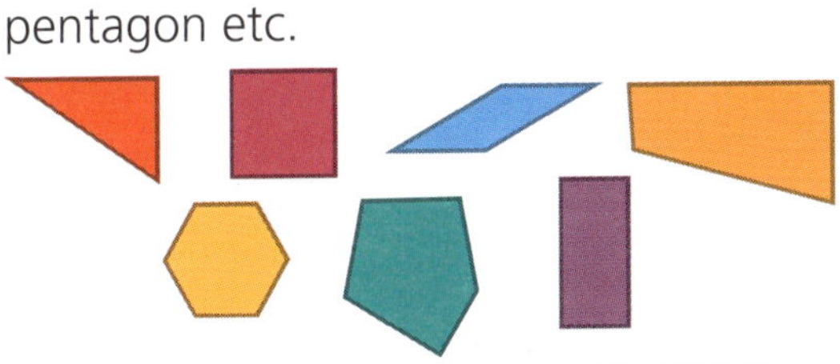

prism

A three-dimensional object with a uniform cross-section. The ends are identical shapes and all other faces are rectangles. Prisms are named by the shape of their ends.

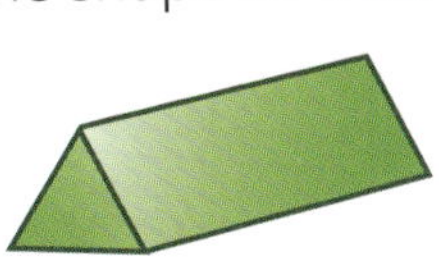

Triangular Prism

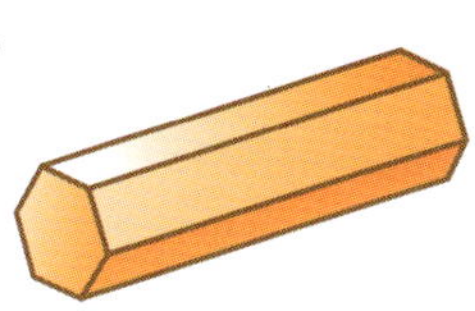

Hexagonal Prism

probability

The probability (or chance) of something happening is its likelihood of happening.

- The probability of tossing an even number on a dice is 50%.

pyramid

A three-dimensional object that has a polygon for a base and triangles for all other faces. Pyramids are named by the shape of their base.

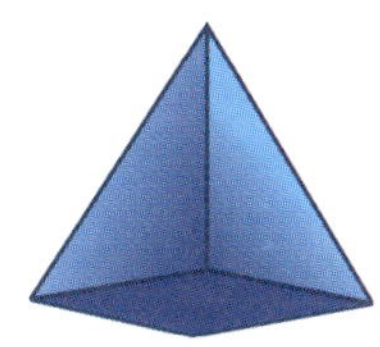

Square Pyramid

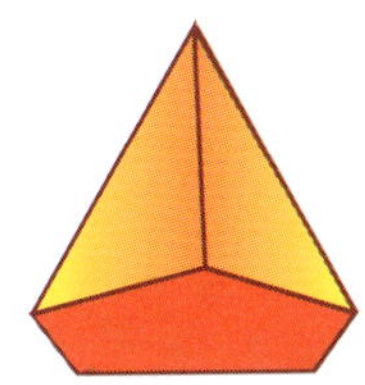

Pentagonal Pyramid

quadrilateral

A two-dimensional shape with four straight sides.

random selection

Choosing without looking.

Each item has an equal chance of being chosen.

rectangle

A quadrilateral with four right angles and two pairs of equal and parallel sides.

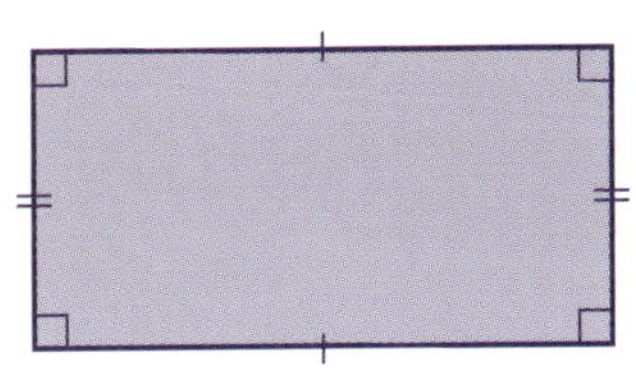

reflection

See *flip*.

regular and irregular shapes

Regular shapes have all sides and all angles equal. Irregular shapes do not.

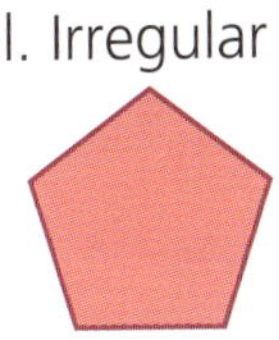

Regular Shape   Irregular Shape

rhombus

A shape with 4 sides such that the pairs of opposite sides are parallel and all sides are equal.

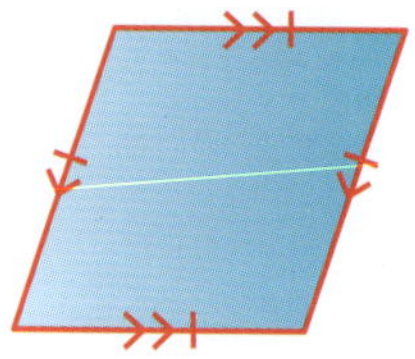

Roman numerals

A number system devised by the ancient Romans.

Roman numerals use letters for numbers:

I	V	X	L	C	D	M
1	5	10	50	100	500	1000

- XXVIII = 28

rotation

See *turn*.

rounding

Rewriting a number to the nearest 5, 10 or 100.

skip counting

Counting on, adding the same number each time.

- 5, 10, 15, 20, 25, … is skip counting by 5.

slide (translation)

To move a shape in any direction without changing its orientation.

- 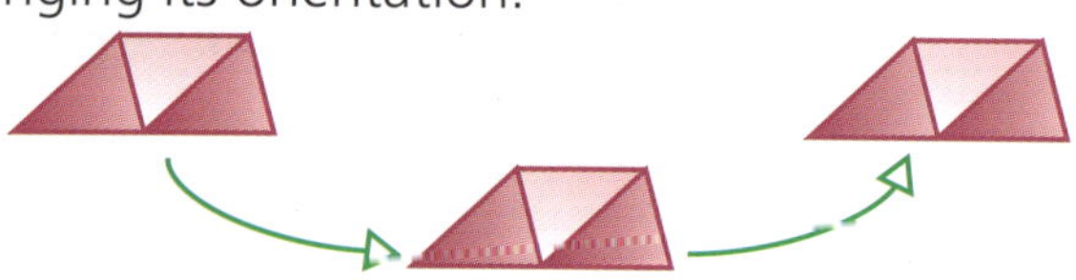

solid

A term used to describe a three-dimensional shape.

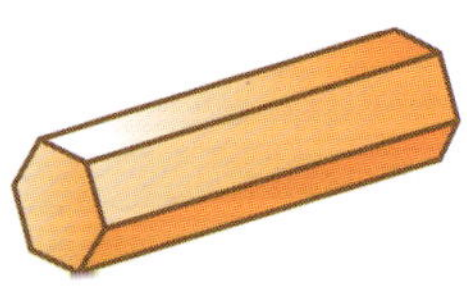

Hexagonal Prism

Cone

sphere

A three-dimensional object that is ball-shaped and round.

All points on the surface of a sphere are the same distance from its centre.

split strategy

Adding numbers by splitting them into their parts.

- $36 + 52 = 30 + 6 + 50 + 2$
  $= (30 + 50) + (6 + 2)$
  $= 80 + 8$
  $= 88$

square

A quadrilateral that has four equal sides and four right angles.

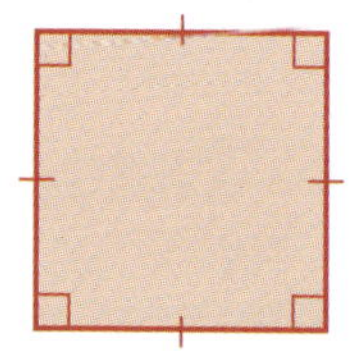

square centimetre (cm^2)

A unit for measuring area that is equal to a square with sides of 1 cm.

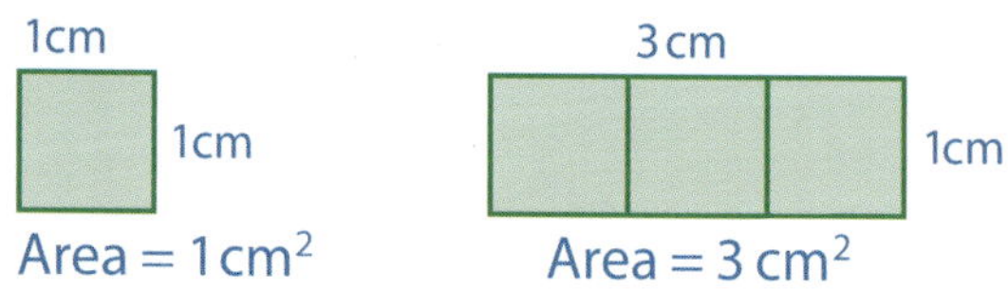

square metre (m^2)

A unit for measuring area that is equal to a square with sides of 1 m.

sum

The answer when you add numbers.

surface

The outside layer of a three-dimensional object. A surface can be flat or curved.

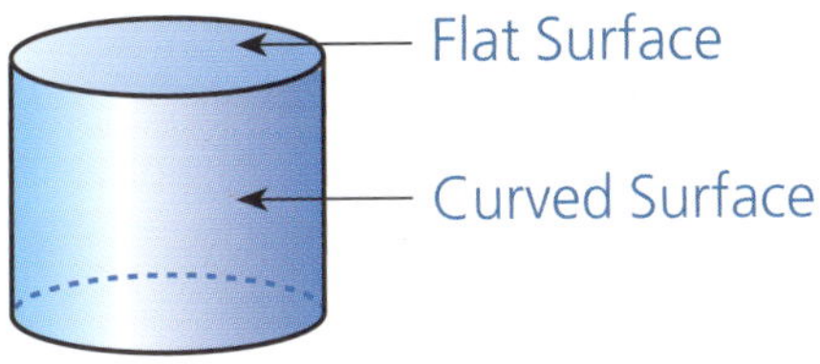

See also *face*.

survey

A list of questions used to discover information.

symmetry

If a figure has a line of symmetry, it can be folded so that the two halves exactly overlap.

Each half is a mirror image of the other.

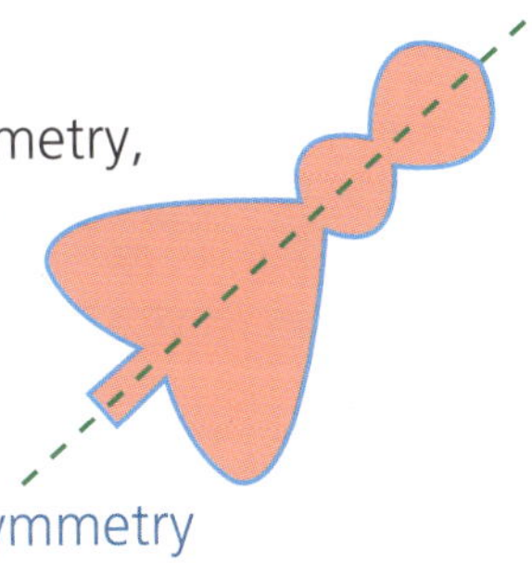

tally

To keep count by making a mark for each item. To make counting easy, the marks are drawn in groups of five with each fifth mark crossed over the other four marks.

-  = 18

tangram

A traditional Chinese puzzle. A square is cut into seven pieces that can be rearranged to make different pictures.

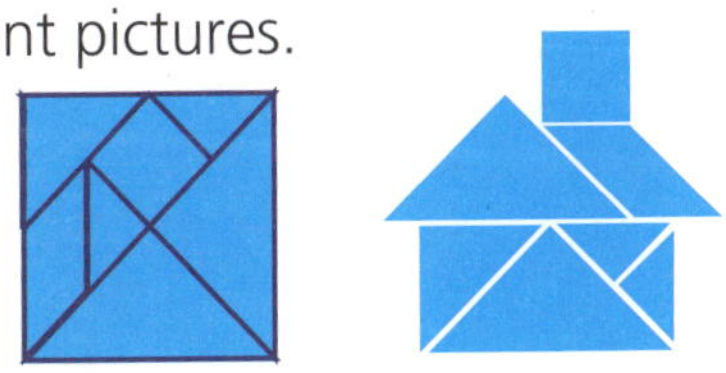

tessellation

A pattern of identical shapes that fit together without gaps or overlaps.

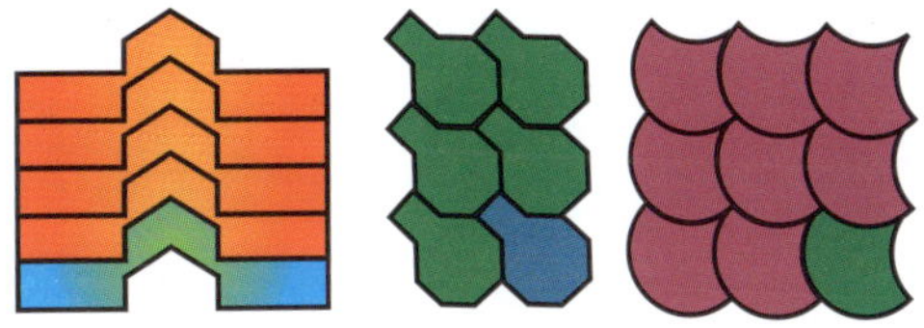

- These shapes tessellate.

three-dimensional (3D) objects

Objects are three-dimensional.

They have length, width and height.

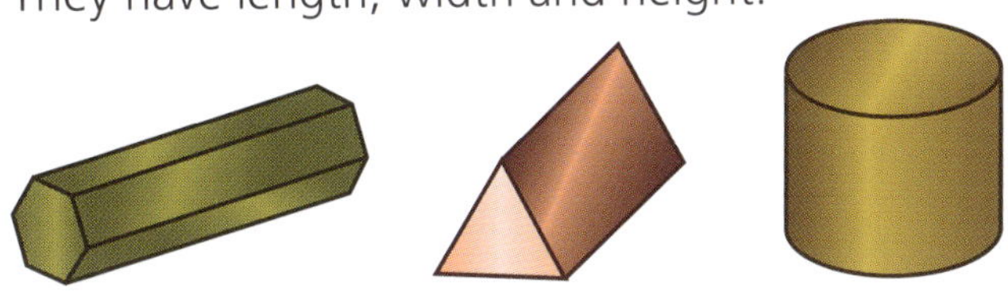

time

time line

Shows a sequence of events in time.

translation

See *slide*.

trapezium

A quadrilateral with one pair of parallel sides.

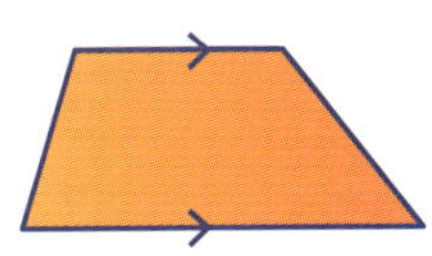

triangle

A polygon with three sides and three angles.

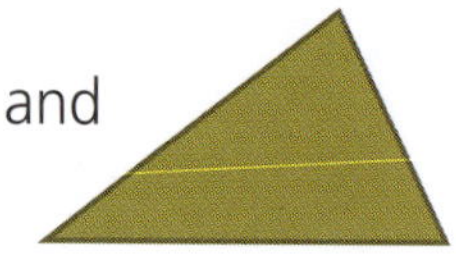

See also *polygon*.

turn (rotation)

To rotate a shape about a point.

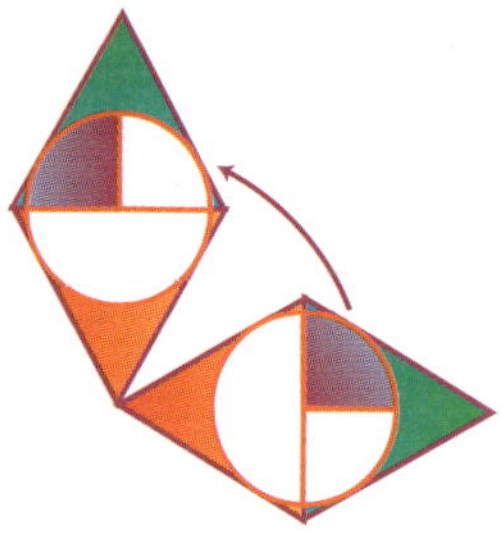

two-dimensional (2D) shapes

Plane shapes are two-dimensional. They have length and width.

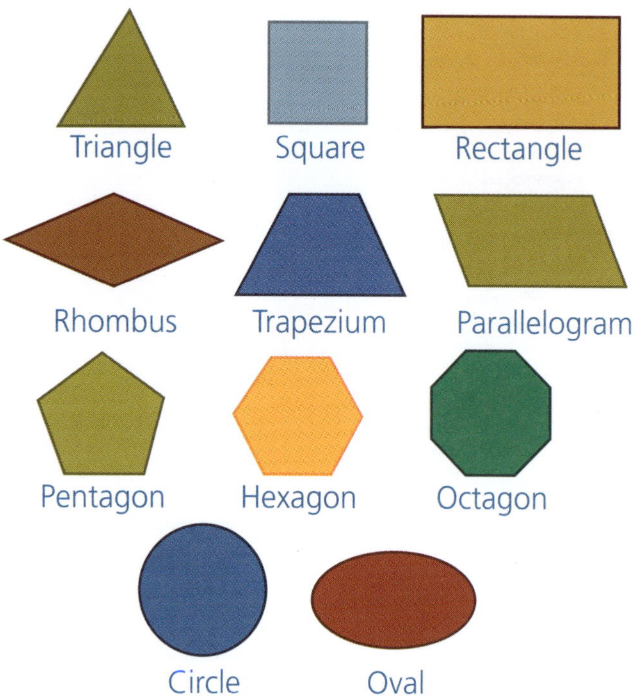

two-way tables

Tables of data that show categories down the side as well as across the top.

	Blue	Red	White
Bus	2	5	0
Car	8	10	21
Truck	1	6	2

vertex

A point at which two or more lines meet to form a corner on a plane or object.

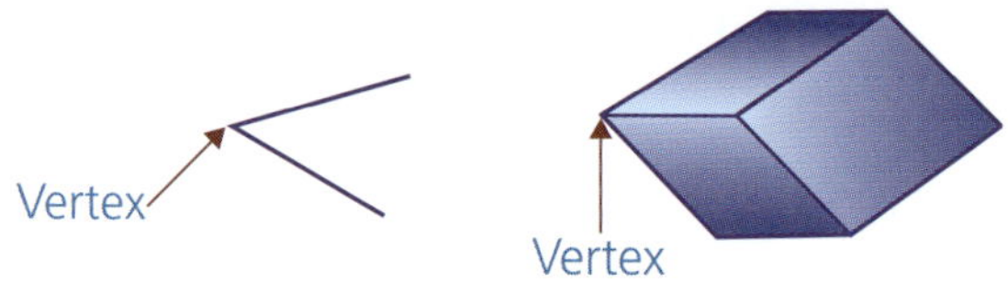

The plural of *vertex* is *vertices*.

vertical

- at right angles to the horizontal
- straight up and down
- the direction in which an object falls under gravity.

volume

The amount of space occupied by a three-dimensional object.

- Volume = 10 cubic units

1:01 Fractions

CONCEPT

Numerator → 5
Denominator → 12

$\frac{5}{12}$ is 5 of 12 equal parts.

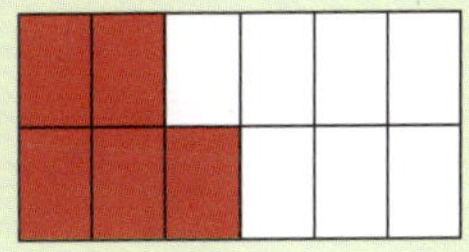

$\frac{5}{12}$

1 What part of each shape has been coloured?

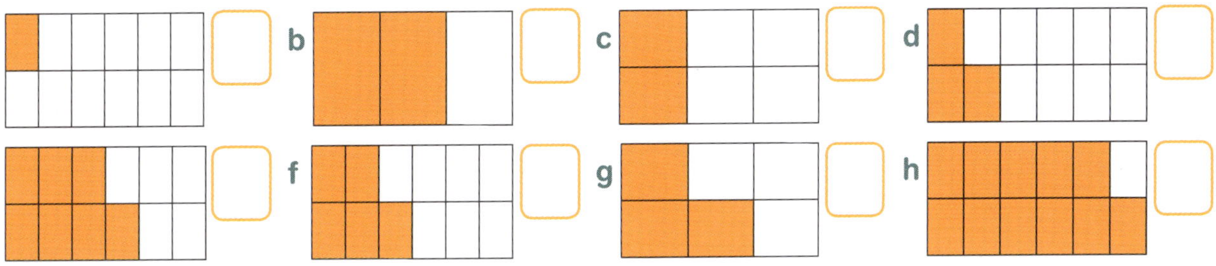

a ☐ b ☐ c ☐ d ☐ e ☐ f ☐ g ☐ h ☐

2 What part of each shape above has not been coloured?

a ☐ b ☐ c ☐ d ☐ e ☐ f ☐ g ☐ h ☐

3 Colour part of each shape to match the given fraction.

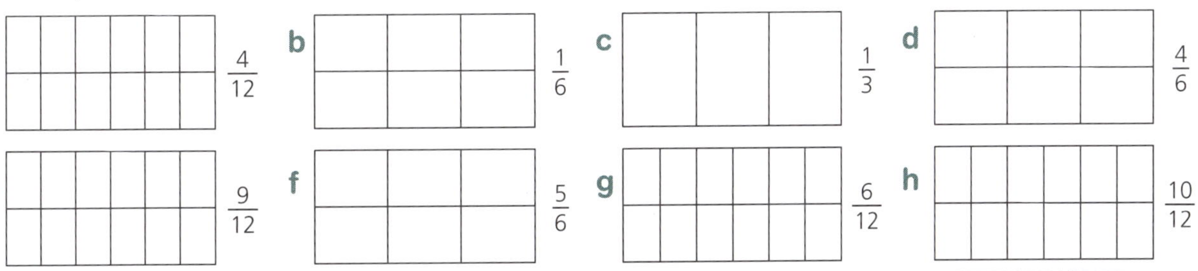

a $\frac{4}{12}$ b $\frac{1}{6}$ c $\frac{1}{3}$ d $\frac{4}{6}$

e $\frac{9}{12}$ f $\frac{5}{6}$ g $\frac{6}{12}$ h $\frac{10}{12}$

4 What part of each group has been coloured?

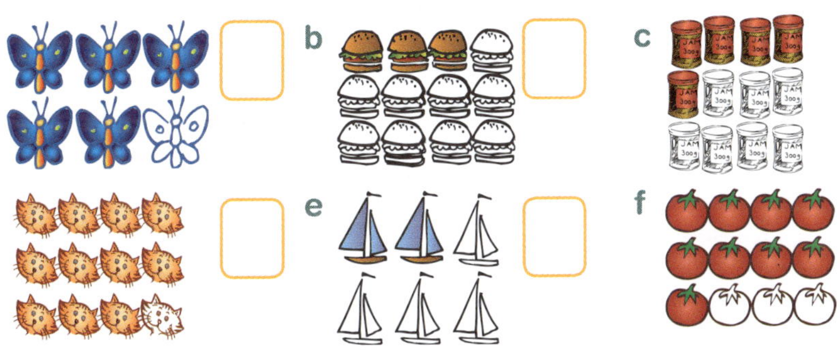

a ☐ b ☐ c ☐ d ☐ e ☐ f ☐

5 What part of each group above has not been coloured?

a ☐ b ☐ c ☐ d ☐ e ☐ f ☐

1:02 Hundredths

CONCEPT

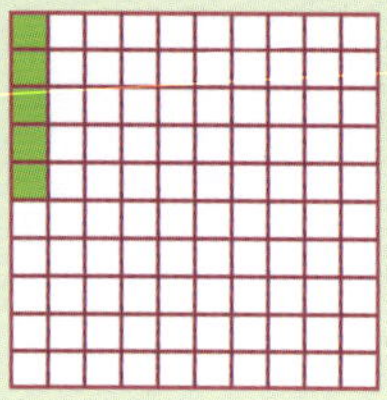

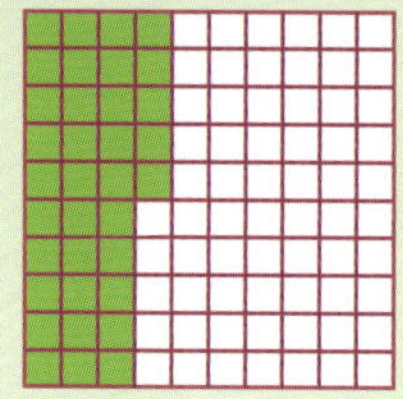

One hundred hundredths = 1 whole

$\frac{100}{100}$ = 1 whole

1 What part of each hundred square has been coloured?

a 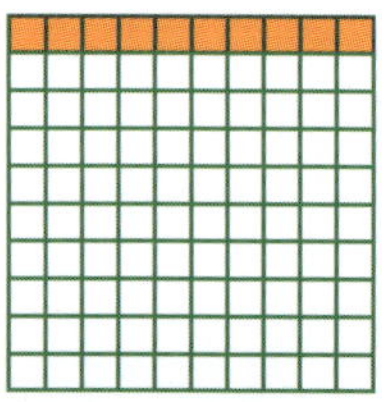 ☐

b 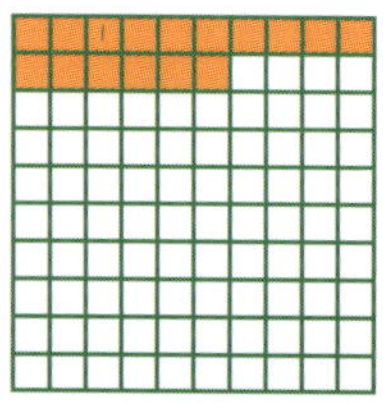 ☐

c 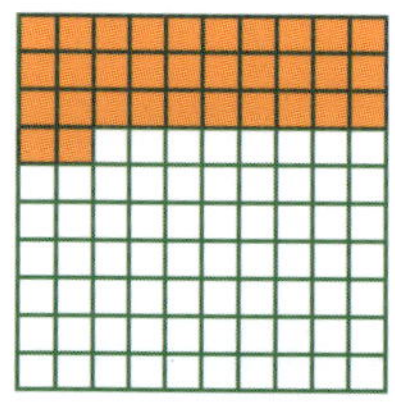 ☐

d  ☐

e 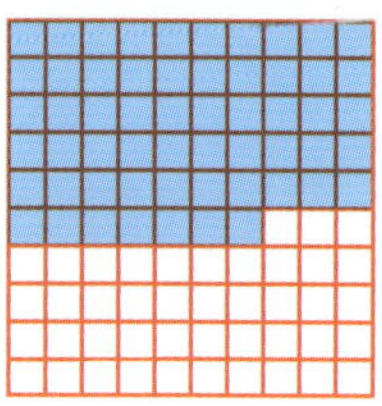 ☐

f 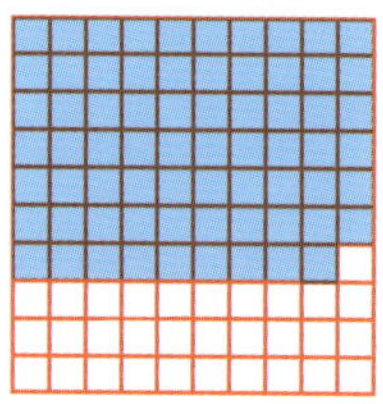 ☐

g 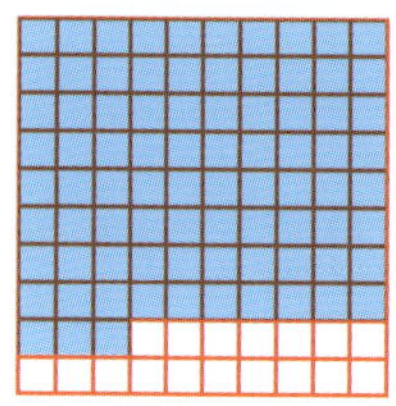 ☐

h 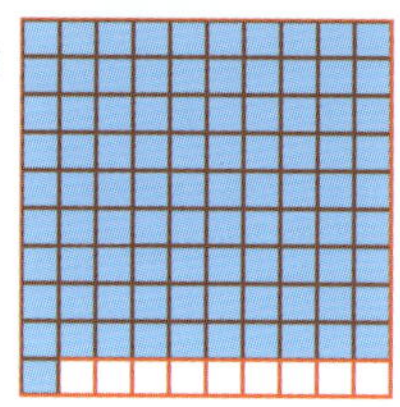 ☐

2 Colour part of each hundred square to match the given fraction.

a 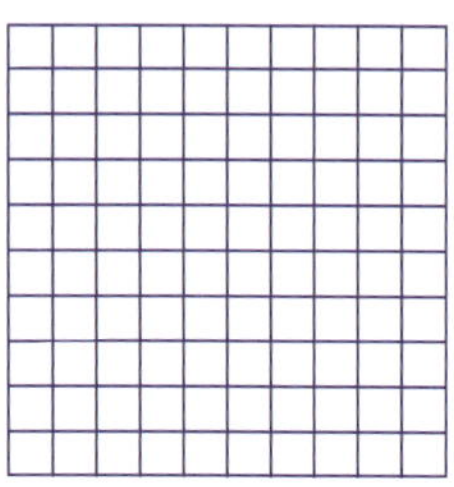
$\frac{25}{100}$

b 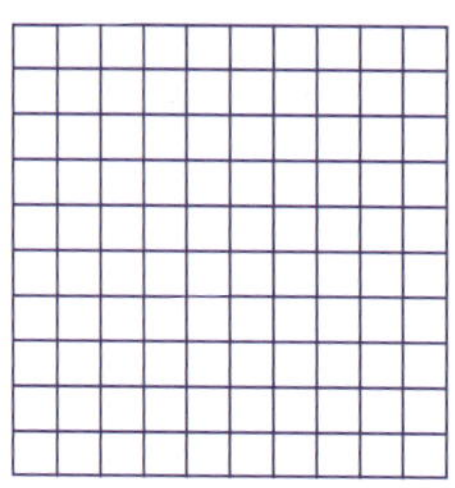
$\frac{43}{100}$

c 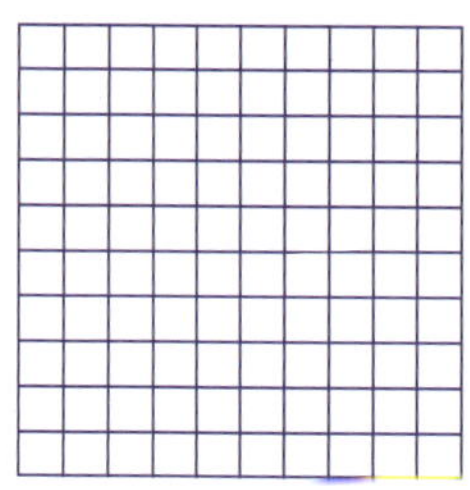
$\frac{19}{100}$

d 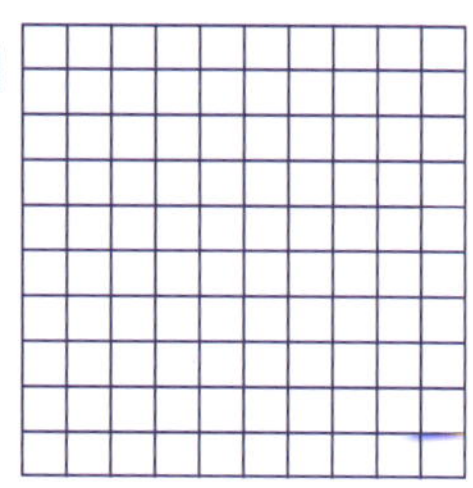
$\frac{36}{100}$

ACTIVITY

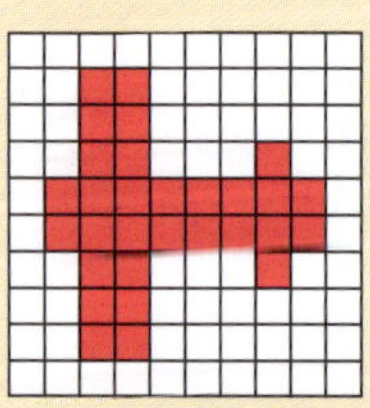

3 Estimate what part of this hundred square has been covered. Check by counting.

- Do this for pictures of your own, drawn on hundred squares.

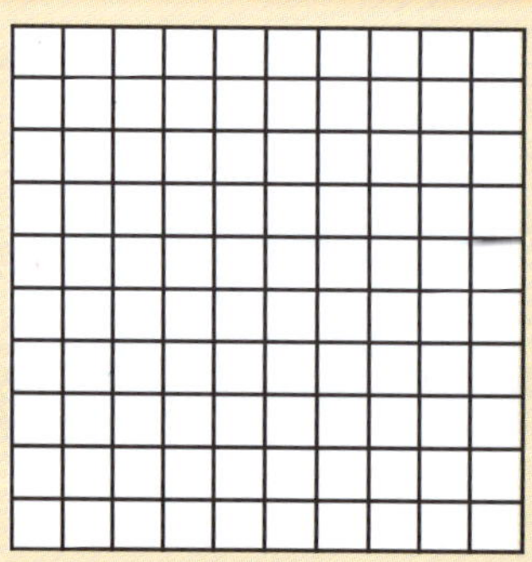

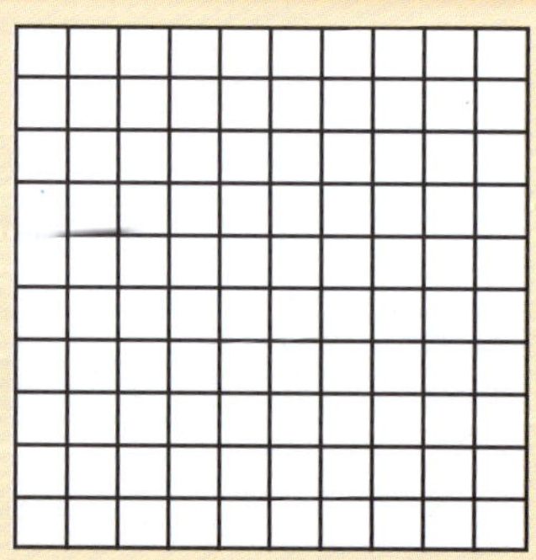

 ISBN 978 1 4886 2187 1

1:03 Decimals

CONCEPT

We can write fractions as decimals.

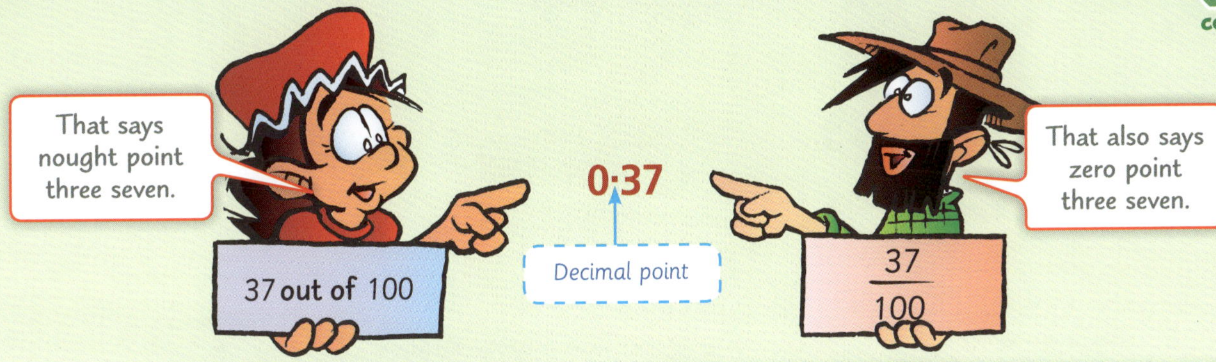

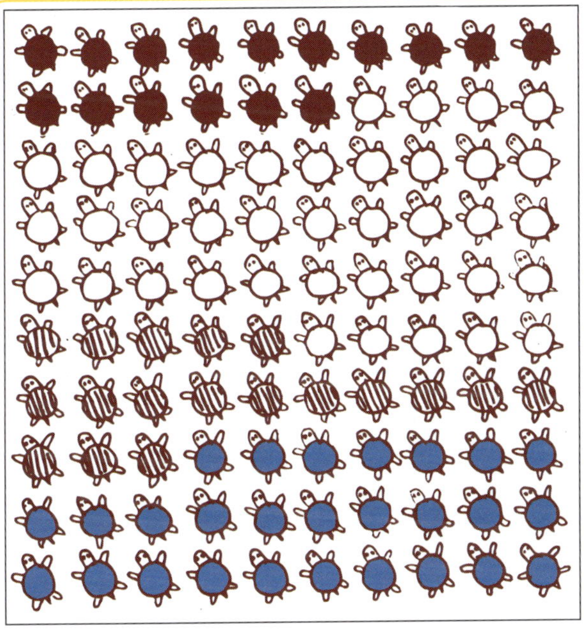

1 How many turtles are inside the box? ☐

2 What part of the group of turtles:

a is  ?

☐ out of 100   ☐/100   0·☐

b is  ?

☐ out of 100   ☐/100   0·☐

c is  ?

☐ out of 100   ☐/100   0·☐

3 What part of the group of turtles is  ? 0·☐

4 Write the decimal and fraction shown on each hundred square.

a 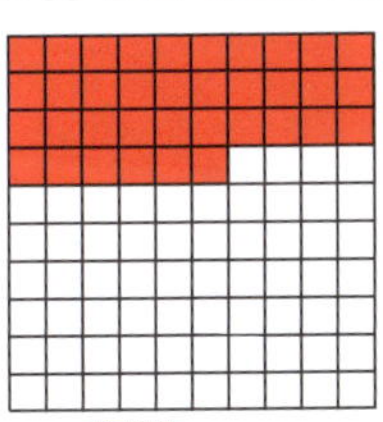 ☐/☐ 0·☐

b 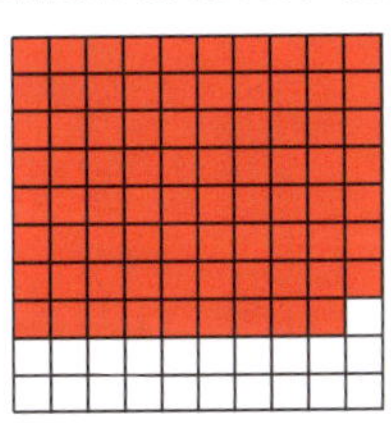 ☐/☐ 0·☐

c 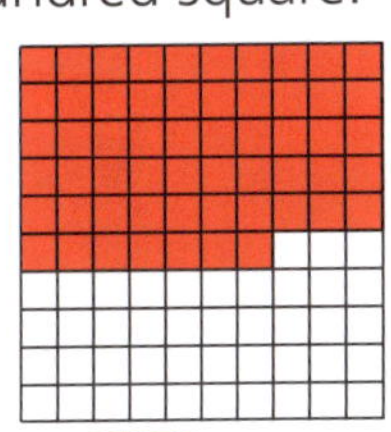 ☐/☐ 0·☐

d 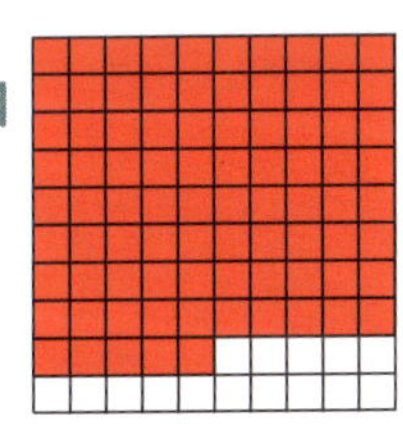 ☐/☐ 0·☐

e 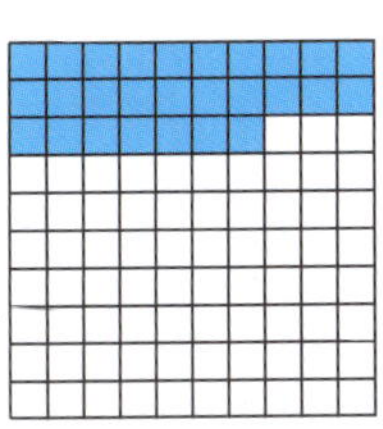 ☐/☐ 0·☐

f 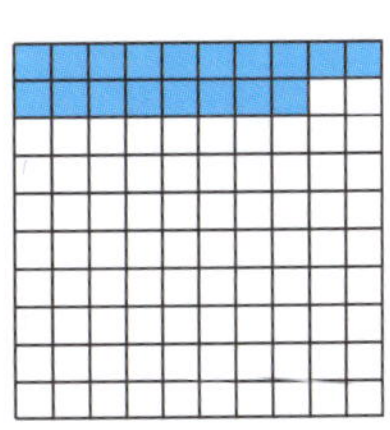 ☐/☐ 0·☐

g 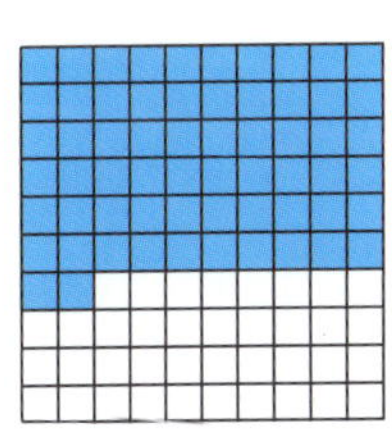 ☐/☐ 0·☐

h 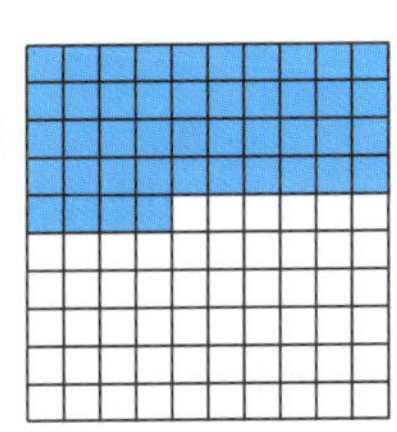 ☐/☐ 0·☐

1:04 Numbers to 9999

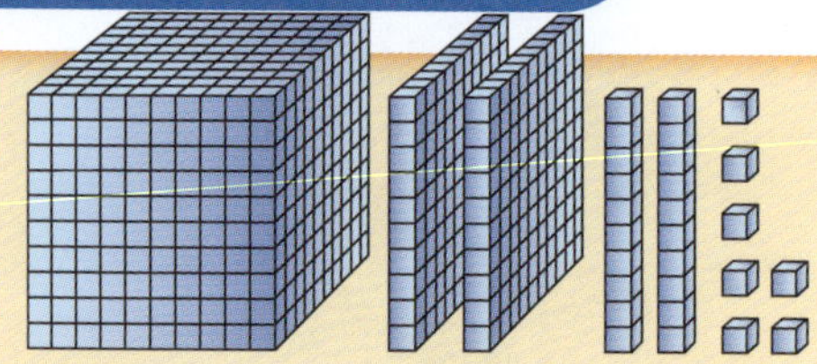

1	Thousands	2	Hundreds	2	Tens	7	Ones

one thousand, two hundred and twenty-seven

ACTIVITY

1 Write the numeral, fill in the numeral expander and write the number in words.

a

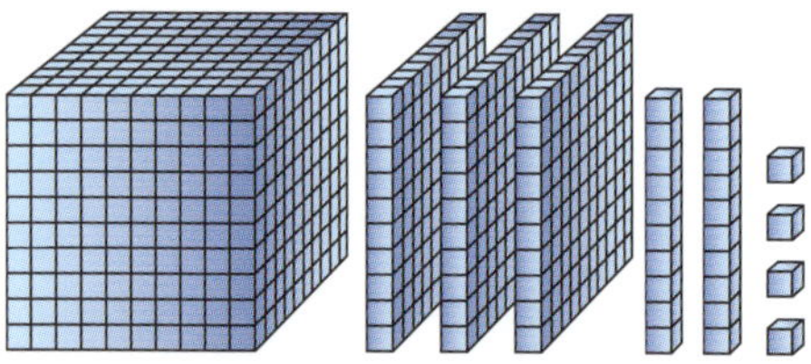

	Thousands		Hundreds		Tens		Ones

b

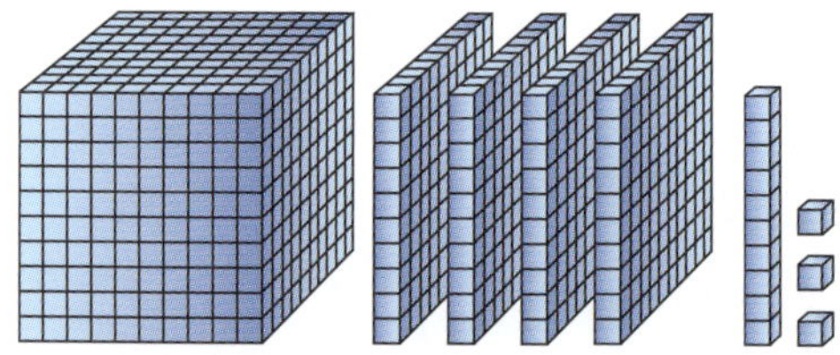

	Thousands		Hundreds		Tens		Ones

c

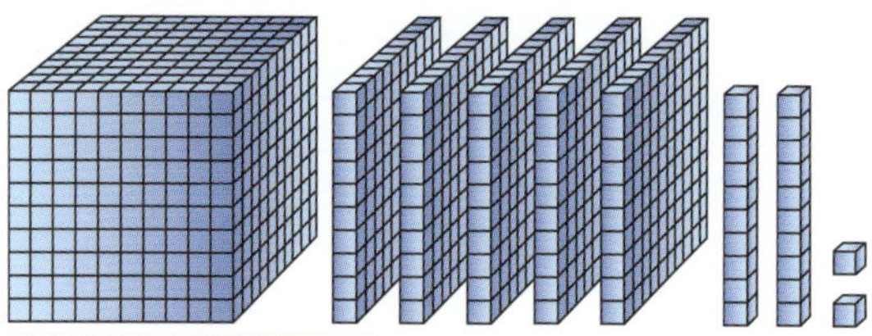

	Thousands		Hundreds		Tens		Ones

d

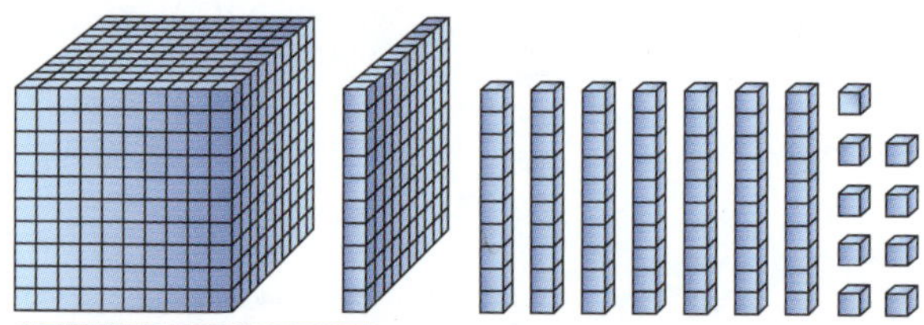

	Thousands		Hundreds		Tens		Ones

2 How many digits are there in each numeral?

a 639 ☐ **b** 1256 ☐ **c** 3970 ☐ **d** 567 ☐

e 4982 ☐ **f** 135 ☐ **g** 76 ☐ **h** 5200 ☐

Place value is built on the × and ÷ of tens.

3 Write these as numerals.

a two hundred and sixty ☐

b one hundred and fifty-two ☐

c one thousand and forty ☐

d seven thousand and eighteen ☐

e five thousand one hundred and seventy-nine ☐

f two thousand six hundred and thirty-four ☐

g eight thousand five hundred and sixty-eight ☐

1:05 Numbers to 9999

1 What number is:

a 2 more than 1000? ____ **b** 5 more than 1000? ____

c 7 more than 1020? ____ **d** 1 less than 1000? ____

e 10 less than 1030? ____ **f** 3 less than 1020? ____

2 Complete these number patterns.

a 1005, 1006, 1007, ____, ____ **b** 3002, 3004, 3006, ____, ____

c 5010, 5015, 5020, ____, ____ **d** 990, 1000, 1010, ____, ____

e 1020, 1019, 1018, ____, ____ **f** 7030, 7020, 7010, ____, ____

3 Write these numbers.

a one thousand and sixteen ____ **b** one thousand and twenty ____

c one thousand and twenty-seven ____ **d** five thousand and ten ____

e eight thousand and eighteen ____ **f** two thousand and twelve ____

g seven thousand and twenty-four ____ **h** six thousand and one ____

i nine thousand four hundred and thirty-six ____

j four thousand nine hundred and sixty-five ____

k three thousand six hundred and eighty-nine ____

> 10000
> = *ten thousands*
> = 10 × 1000

4 Write in words:

a 4023 ____

b 9030 ____

c 7500 ____

d 2901 ____

5 Draw a line to join each numeral and its name.

a

1007	one thousand and four
1017	one thousand and seventeen
1004	one thousand and seven

b

1003	one thousand and thirteen
1013	one thousand and five
1005	one thousand and three

 • *AUSTRALIAN SIGNPOST MATHS 4* • ISBN 978 1 4886 2187 1

1:06 Solving Problems with Place Value

CONCEPT

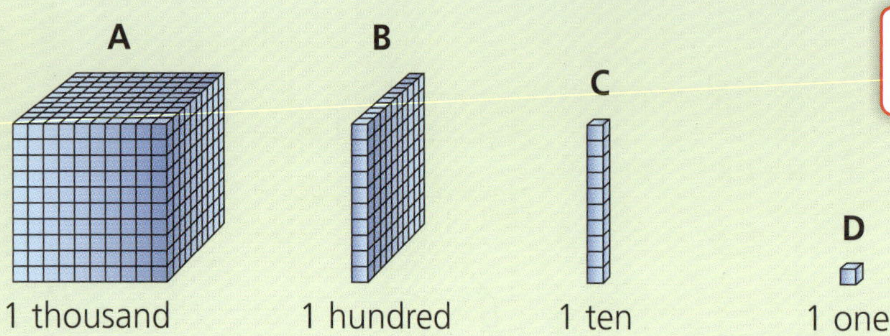

1 How many times as big is the number shown in:

a **A**, compared to the one shown in **B**?

b **B**, compared to the one shown in **C**?

c **C**, compared to the one shown in **D**?

d **A**, compared to the one shown in **C**?

e **B**, compared to the one shown in **D**?

f **A**, compared to the one shown in **D**?

2 Which number is larger:

a **A**: 60 000 + 7 000 + 600 + 80 + 1 or **B**: 60 000 + 900 + 90 + 9?

b **C**: 80 000 + 1 000 + 200 + 40 + 9 or **D**: 80 000 + 2 000 + 100 + 60 + 2?

c **E**: 20 000 + 5 000 + 700 + 10 + 8 or **F**: 20 000 + 5 000 + 800 + 80 + 1?

d **G**: 50 000 + 3 000 + 900 + 90 + 2 or **H**: 50 000 + 9 000 + 700 + 90 + 2?

3 **A** 74 186 **B** 79 146 **C** 60 715 **D** 40 207 **E** 97 364 **F** 98 170

a Which number has a 7 that stands for 7 000?

b Which numbers contain 6s that have the same value?

c Which numbers contain 9s that have the same value?

d Which numbers contain 7s that have the same value?

e How many times as big is the 7 in **B** compared to the 7 in **E**?

Wipe Out a Digit

ICT

- A student enters any 5-digit number into a calculator.
- A partner selects any digit to be 'wiped out', i.e. changed to zero.
- Only one operation can be entered into the calculator to wipe out a digit.
- Take turns and score one point for each successful wipe out.

 • *AUSTRALIAN SIGNPOST MATHS 4* • ISBN 978 1 4886 2187 1

1:07 Place Value to 10 000

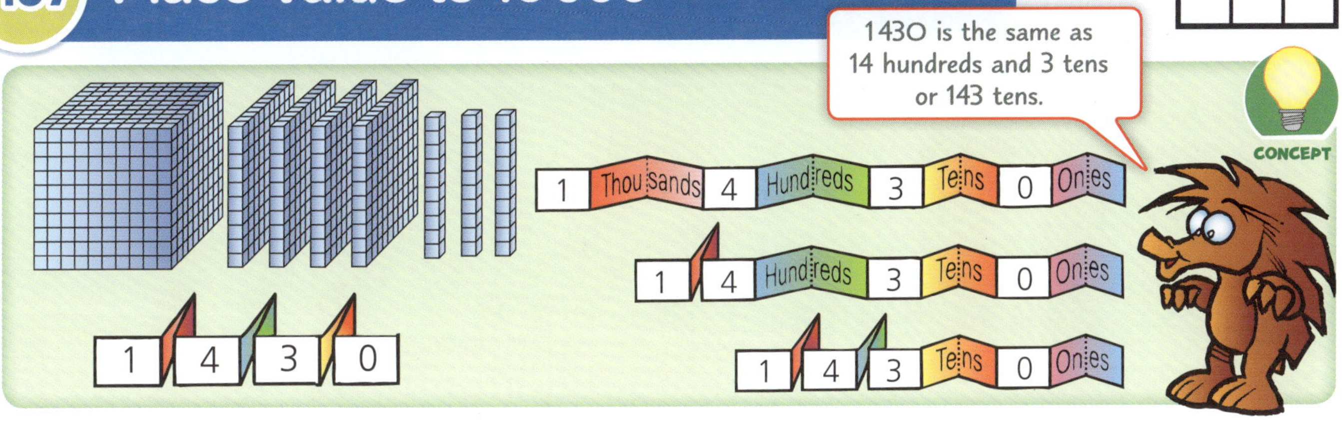

1. Complete the numeral expanders.

a 1 578

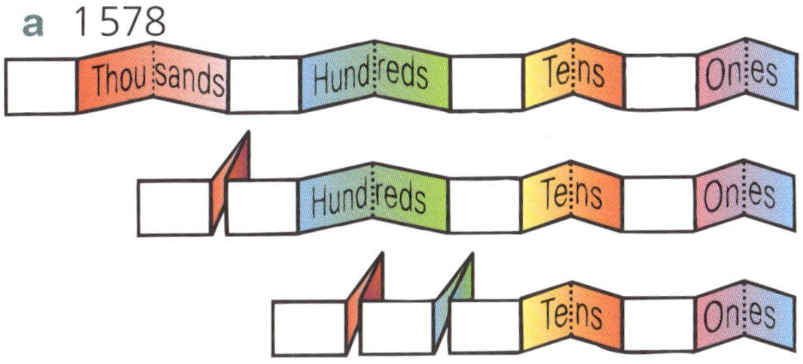

b 2 365

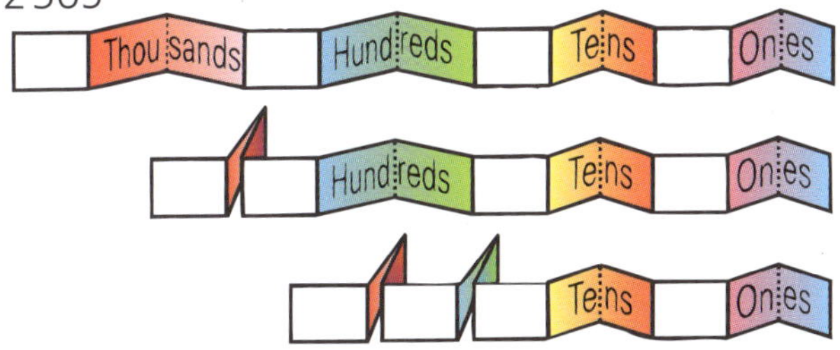

c 3 490

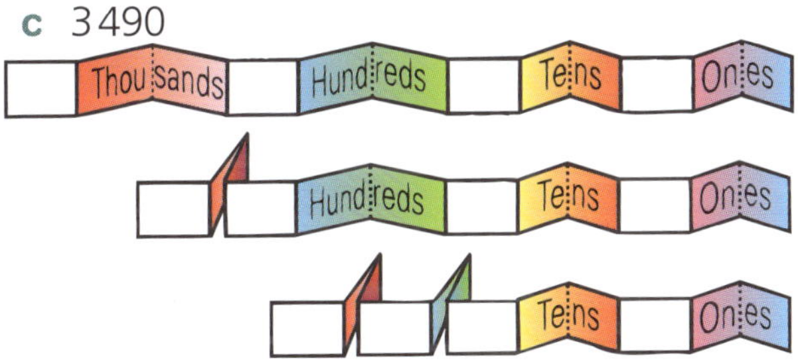

d 4 206

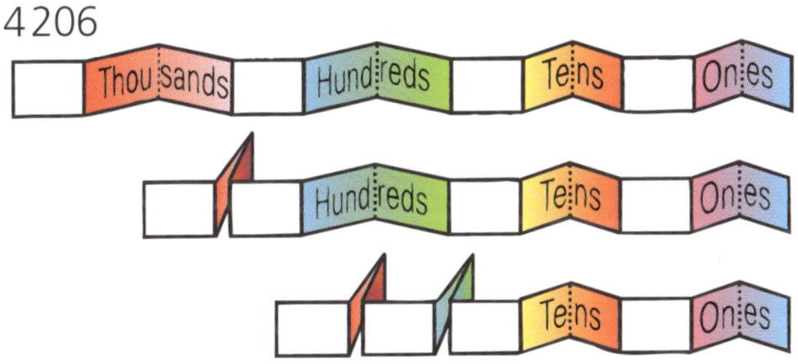

2. How many hundreds are in each number altogether?

a 1 760 ☐ b 1 934 ☐ c 1 503 ☐ d 2 641 ☐

e 3 981 ☐ f 6 275 ☐ g 9 634 ☐ h 7 190 ☐

3. How many tens are in each number altogether?

a 1 623 ☐ b 1 470 ☐ c 2 194 ☐ d 3 478 ☐

e 5 384 ☐ f 7 360 ☐ g 4 576 ☐ h 8 257 ☐

4. a 300 = ☐ tens b 800 = ☐ tens

c 3 000 = ☐ tens d 7 000 = ☐ tens

e 23 hundreds = ☐ thousands + ☐ hundreds

f 10 000 = ☐ ten thousand

 • *AUSTRALIAN SIGNPOST MATHS 4* • ISBN 978 1 4886 2187 1

1:08 Rounding Off

1 Round off these numbers to the nearest hundred.

a 3674 ☐ b 4237 ☐ c 1396 ☐ d 9271 ☐

e 6549 ☐ f 6704 ☐ g 8962 ☐ h 5854 ☐

2 Round off these numbers to the nearest thousand.

a 31569 ☐ b 82738 ☐ c 10846 ☐ d 57249 ☐

e 23496 ☐ f 52301 ☐ g 46972 ☐ h 69347 ☐

3 a Circle numbers that round off to 2700.

2694	2781	2639
2721	2675	2642
2765	2748	2683

b Circle numbers that round off to 53000.

53640	52967	52849
52621	52076	53297
53599	53346	52374

4 Answer **true** or **false** for each statement.

a 4639 rounds off to 4600. ☐

b 1854 rounds off to 1800. ☐

c 6341 rounds off to 6400. ☐

d 9782 rounds off to 9800. ☐

5 Complete these numeral expanders.

a

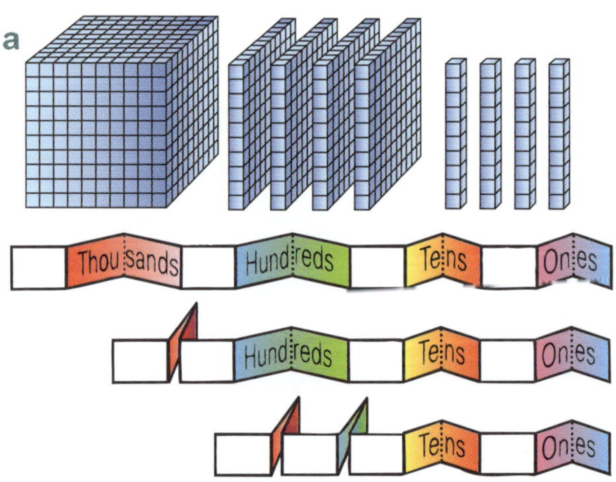

b

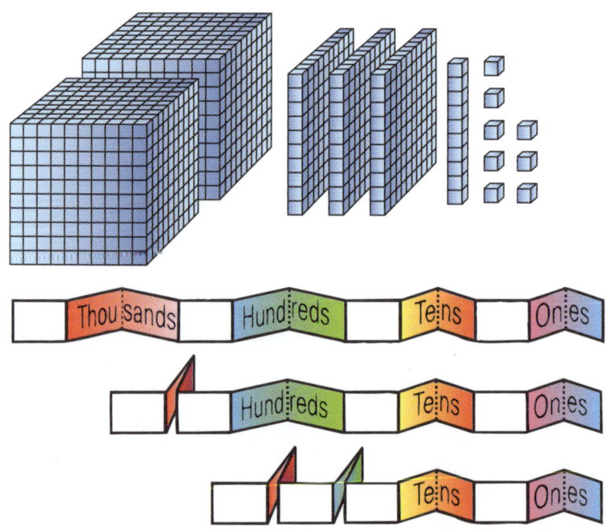

© PEARSON AUSTRALIA 2018 • *AUSTRALIAN SIGNPOST MATHS 4* • ISBN 978 1 4886 2187 1

1:09 Expanded Notation

CONCEPT

Thousands	Hundreds	Tens	Ones
3	9	6	2
3000 +	900 +	60 +	2

The expanded form of 3962 is 3000 + 900 + 60 + 2.

1 Write the numeral for:

a 2000 + 700 + 40 + 9

b 7000 + 100 + 20 + 4

c 4000 + 500 + 90 + 2

d 6000 + 400 + 30 + 2

e 30000 + 6000 + 800 + 30 + 6

f 80000 + 3000 + 500 + 70 + 5

g 10000 + 2000 + 500 + 70

h 50000 + 700 + 40 + 6

2 Write these numerals in expanded form.

a 4286

b 9317

c 3752

d 6489

e 71423

f 29460

3 Write the numeral for the number shown on each abacus.

a

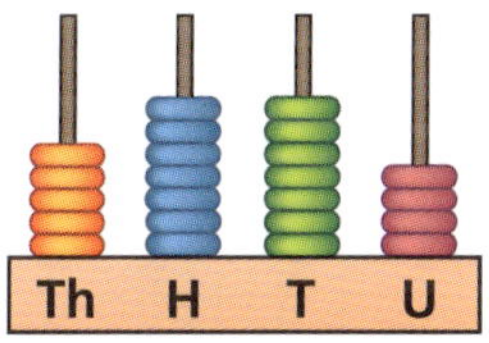

b

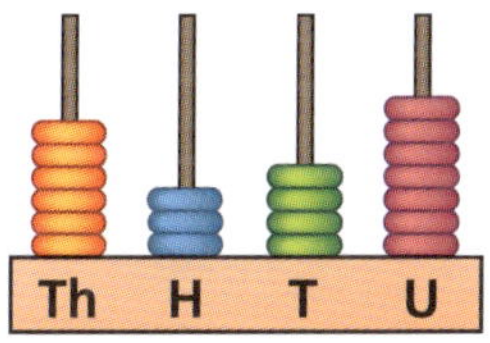

c

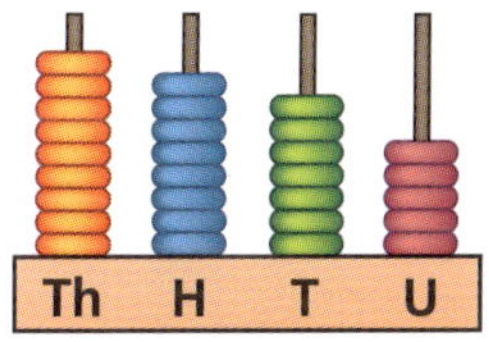

d

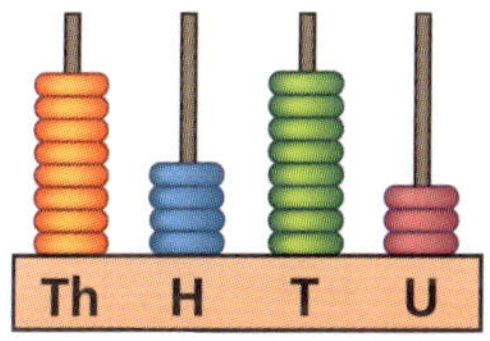

e

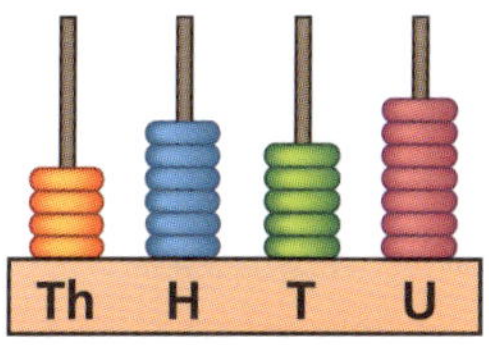

f

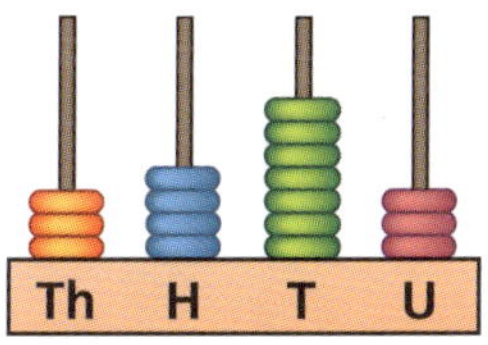

g

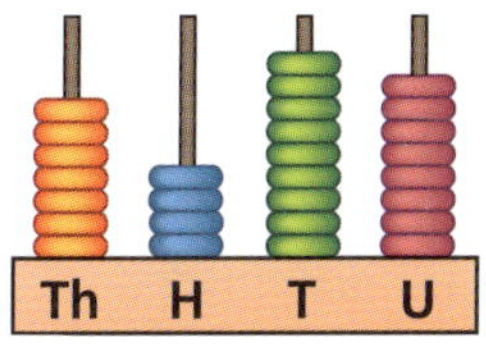

h

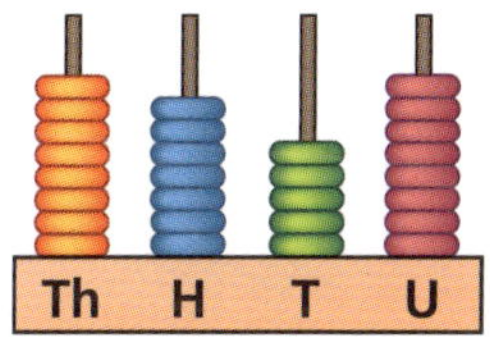

4 Write the value for each coloured digit in these numbers.

a 936**7**4

b 3186**2**

c 7**5**437

d 49**5**28

e 1**4**762

f 52**1**06

g 87**3**15

h 689**4**1

ACTIVITY

With a partner, use place-value blocks to demonstrate these numbers.

- 3681
- 1579
- 2532
- 2436
- 3659
- 1234
- 5215
- 4051
- 5478
- 1110

1:10 Comparing Fractions

CONCEPT

$\frac{1}{2}$ is bigger than $\frac{1}{5}$.

$\frac{1}{2}$ $\frac{1}{5}$

1 Circle the larger fraction.

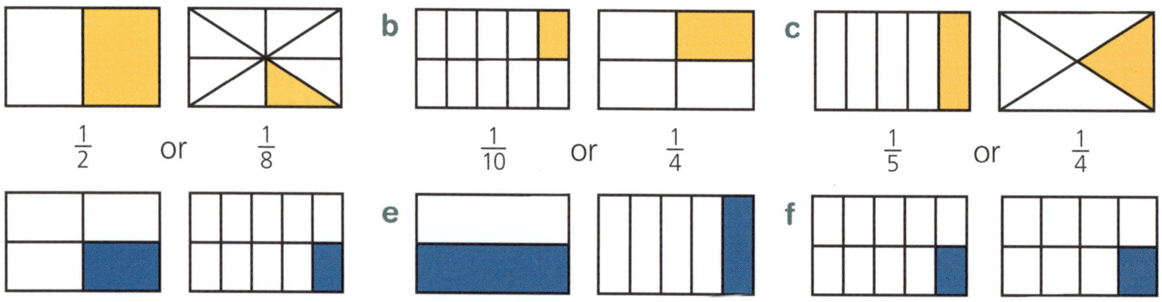

a $\frac{1}{2}$ or $\frac{1}{8}$ b $\frac{1}{10}$ or $\frac{1}{4}$ c $\frac{1}{5}$ or $\frac{1}{4}$

d $\frac{1}{4}$ or $\frac{1}{10}$ e $\frac{1}{2}$ or $\frac{1}{5}$ f $\frac{1}{10}$ or $\frac{1}{8}$

2 Colour part of each shape to match the given fraction.

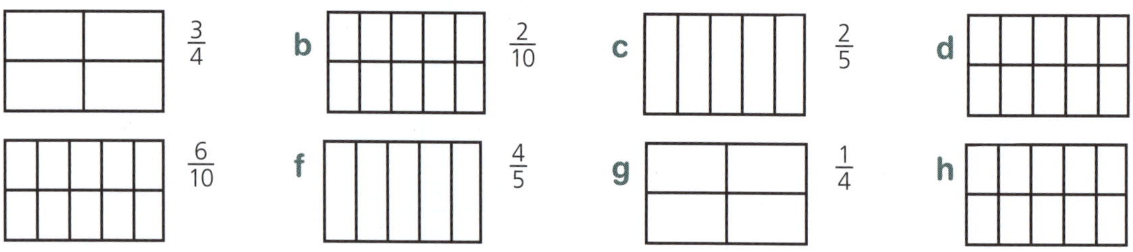

a $\frac{3}{4}$ b $\frac{2}{10}$ c $\frac{2}{5}$ d $\frac{3}{10}$

e $\frac{6}{10}$ f $\frac{4}{5}$ g $\frac{1}{4}$ h $\frac{7}{10}$

3 Write **true** or **false** for each statement.

a $\frac{2}{2} = 1$ ☐ b $\frac{4}{5} = 1$ ☐ c $\frac{8}{8} = 1$ ☐

d $1 = \frac{10}{10}$ ☐ e $1 = \frac{3}{8}$ ☐ f $1 = \frac{5}{5}$ ☐

> is greater than

< is less than

4 Find the coloured fraction.

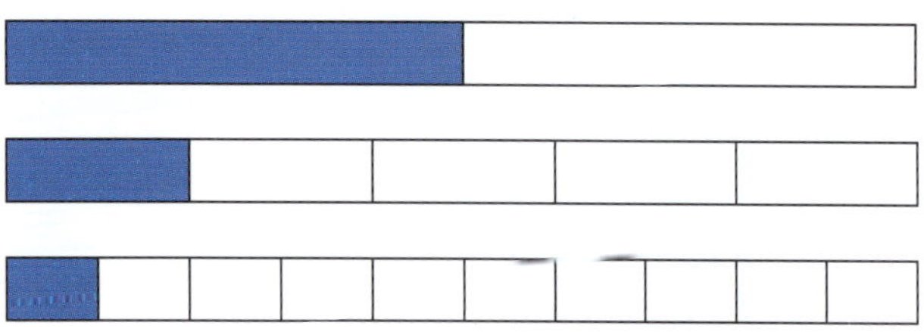

a Fraction coloured = ☐

b Fraction coloured = ☐

c Fraction coloured = ☐

Use > or < to make each number sentence true.

d $\frac{1}{10}$ ☐ $\frac{1}{5}$ e $\frac{1}{2}$ ☐ $\frac{1}{5}$ f $\frac{1}{5}$ ☐ $\frac{1}{2}$ g $\frac{1}{2}$ ☐ $\frac{3}{5}$

h $\frac{7}{10}$ ☐ $\frac{1}{2}$ i $\frac{2}{5}$ ☐ $\frac{1}{2}$ j $\frac{2}{5}$ ☐ $\frac{1}{10}$ k $\frac{3}{10}$ ☐ $\frac{1}{5}$

1:11 Equivalent Fractions

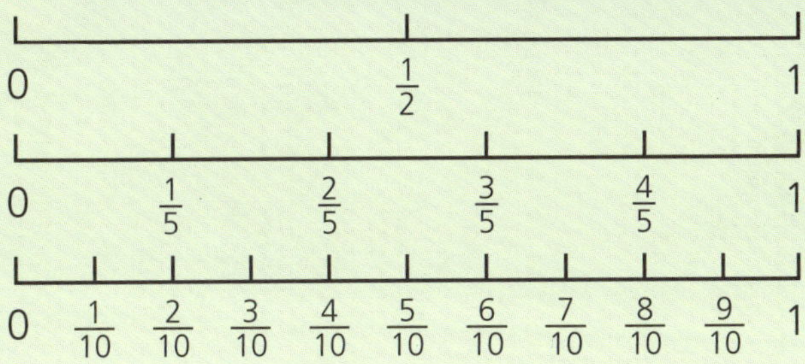

CONCEPT

$\frac{1}{2} = \frac{5}{10}$

$\frac{4}{5} = \frac{8}{10}$

1 Use the number lines to show an equivalent fraction for:

a $\frac{1}{5}$ = ☐ **b** $\frac{4}{10}$ = ☐ **c** $\frac{4}{5}$ = ☐ **d** $\frac{6}{10}$ = ☐

e $\frac{5}{10}$ = ☐ **f** $\frac{2}{5}$ = ☐ **g** $\frac{1}{2}$ = ☐ **h** $\frac{10}{10}$ = ☐

2 Use the number lines above to answer **true** or **false**.

a $\frac{1}{5} > \frac{1}{10}$ ☐ **b** $\frac{6}{10} > \frac{1}{2}$ ☐ **c** $\frac{8}{10} > \frac{3}{5}$ ☐ **d** $\frac{4}{5} < \frac{9}{10}$ ☐

3 Complete the number lines. Count forwards and backwards.

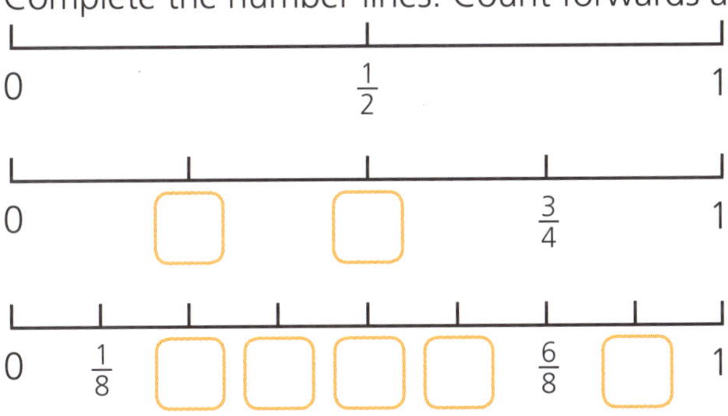

4 Use the number lines above to show an equivalent fraction for:

a $\frac{1}{4}$ = ☐ **b** $\frac{4}{8}$ = ☐ **c** $\frac{2}{4}$ = ☐

d $\frac{1}{2}$ = ☐ **e** $\frac{6}{8}$ = ☐ **f** 1 = ☐

5 Use the number lines above to answer **true** or **false**.

a $\frac{3}{4} > \frac{5}{8}$ ☐ **b** $\frac{3}{8} < 1$ ☐ **c** $\frac{1}{4} < \frac{1}{2}$ ☐ **d** $\frac{3}{4} > \frac{1}{2}$ ☐

ACTIVITY

- The number line below represents 1 unit.
- Write 8 fractions in their correct position below the number line.

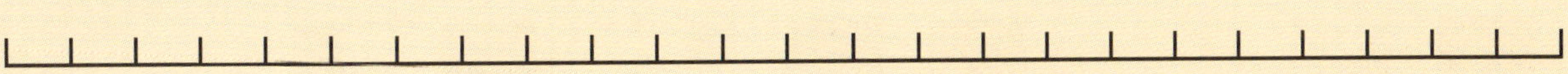

1:12 Equivalent Fractions

Equivalent fractions have the same value.

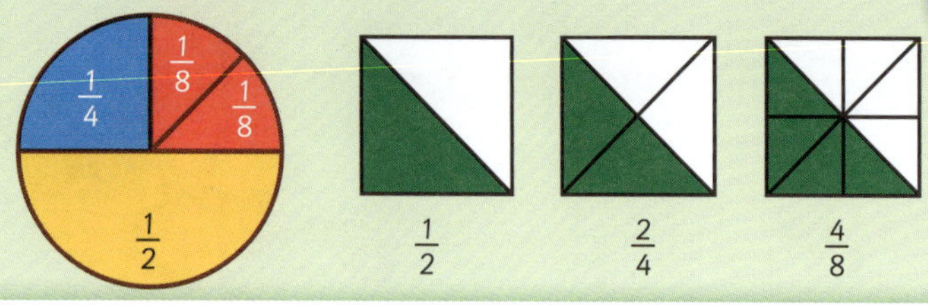

CONCEPT

1 Colour part of each shape to match the given fractions.

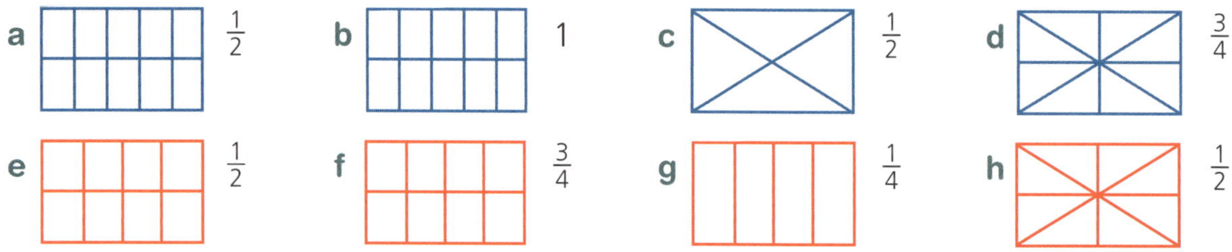

2 Use > or < to make each number sentence true.

a $\frac{1}{8}$ ☐ $\frac{3}{8}$ b $\frac{3}{10}$ ☐ $\frac{1}{10}$ c $\frac{3}{4}$ ☐ $\frac{1}{4}$ d $\frac{3}{10}$ ☐ $\frac{7}{10}$

e $\frac{6}{10}$ ☐ $\frac{9}{10}$ f $\frac{5}{8}$ ☐ $\frac{2}{8}$ g $\frac{4}{5}$ ☐ $\frac{3}{5}$ h $\frac{9}{10}$ ☐ $\frac{8}{10}$

3 Write the equivalent fraction for:

a $\frac{4}{8} = \frac{☐}{4}$ b $\frac{8}{10} = \frac{☐}{5}$ c $\frac{4}{8} = \frac{☐}{2}$ d $\frac{5}{10} = \frac{☐}{2}$

e $\frac{6}{10} = \frac{☐}{5}$ f $\frac{2}{8} = \frac{☐}{4}$ g $\frac{6}{8} = \frac{☐}{4}$ h $\frac{4}{10} = \frac{☐}{5}$

4 Complete each number line. Practise counting forwards and backwards.

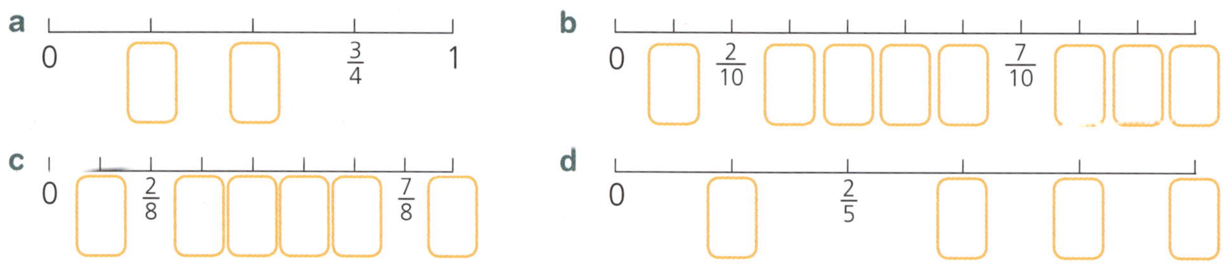

5 What fraction is shown? Write each answer as a mixed number.

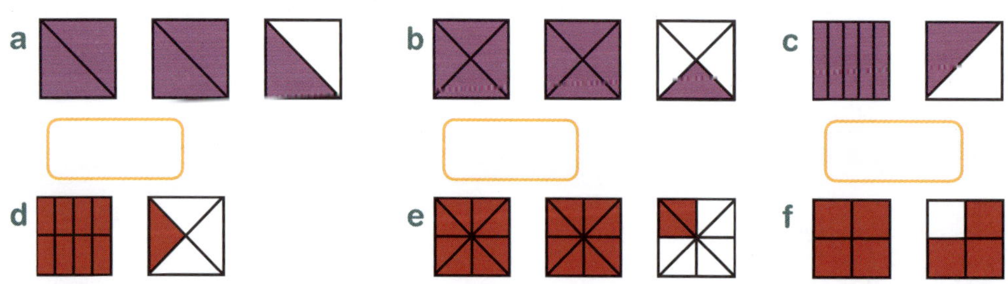

 • *AUSTRALIAN SIGNPOST MATHS 4* • ISBN 978 1 4886 2187 1

1:13 Equivalent Fractions

CONCEPT

1							
$\frac{1}{2}$				$\frac{1}{2}$			
$\frac{1}{4}$		$\frac{1}{4}$		$\frac{1}{4}$		$\frac{1}{4}$	
$\frac{1}{8}$	$\frac{1}{8}$	$\frac{1}{8}$	$\frac{1}{8}$	$\frac{1}{8}$	$\frac{1}{8}$	$\frac{1}{8}$	$\frac{1}{8}$

1					
$\frac{1}{2}$			$\frac{1}{2}$		
$\frac{1}{3}$		$\frac{1}{3}$		$\frac{1}{3}$	
$\frac{1}{6}$	$\frac{1}{6}$	$\frac{1}{6}$	$\frac{1}{6}$	$\frac{1}{6}$	$\frac{1}{6}$

1 What fraction is equal to:

a $\frac{2}{4}$? ☐ b $\frac{4}{8}$? ☐

c $\frac{6}{8}$? ☐ d $\frac{4}{4}$? ☐

e 1 ? ☐ f $\frac{1}{2}$? ☐

2 What fraction is equal to:

a $\frac{2}{6}$? ☐ b $\frac{4}{6}$? ☐

c $\frac{6}{6}$? ☐ d $\frac{3}{3}$? ☐

e $\frac{1}{3}$? ☐ f $\frac{2}{3}$? ☐

3 True (**T**) or false (**F**)?

a $\frac{1}{2} = \frac{2}{4}$ ☐ b $\frac{3}{4} = \frac{6}{8}$ ☐

c $\frac{4}{8} = \frac{2}{4}$ ☐ d $\frac{5}{8} = \frac{3}{4}$ ☐

e $\frac{8}{8} = \frac{4}{4}$ ☐ f $\frac{2}{2} = 1$ ☐

4 True (**T**) or false (**F**)?

a $\frac{2}{3} = \frac{4}{6}$ ☐ b $\frac{1}{3} = \frac{2}{6}$ ☐

c $\frac{3}{3} = \frac{6}{6}$ ☐ d $\frac{1}{6} = \frac{1}{3}$ ☐

e $\frac{3}{6} = \frac{2}{3}$ ☐ f $\frac{6}{6} = 1$ ☐

5 Complete each pattern.

a $\frac{1}{3}, \frac{2}{3}, \frac{3}{3},$ ☐, ☐, ☐, ☐

b $\frac{1}{4}, \frac{2}{4}, \frac{3}{4},$ ☐, ☐, ☐, ☐, ☐

c $\frac{1}{6}, \frac{2}{6}, \frac{3}{6},$ ☐, ☐, ☐, ☐, ☐, ☐

d $\frac{1}{8}, \frac{2}{8}, \frac{3}{8},$ ☐, ☐, ☐, ☐, ☐, ☐, ☐

FUN SPOT

- Cut 5 strips of paper the same length to construct your own fraction wall.
- Use one strip at a time to make separate strips folded into:
  - halves • quarters • eighths • thirds • sixths

1:14 Improper Fractions and Mixed Numbers

CONCEPT

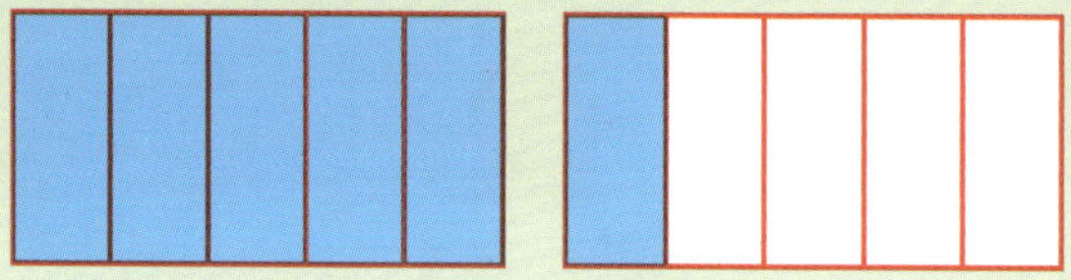

$\frac{6}{5}$ coloured or $1\frac{1}{5}$ coloured

1 Write the improper fraction for the parts coloured.

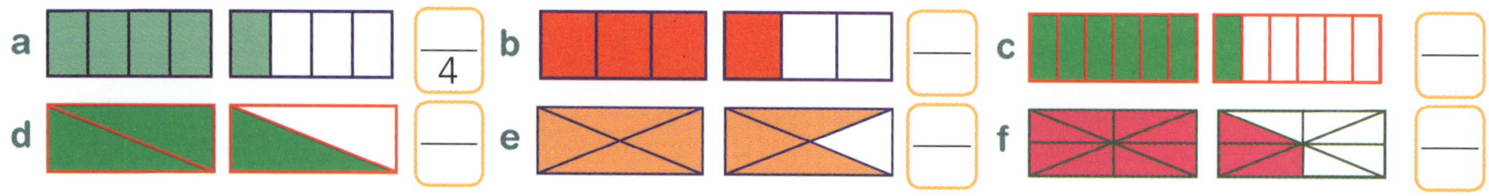

2 Write the mixed number for the parts coloured.

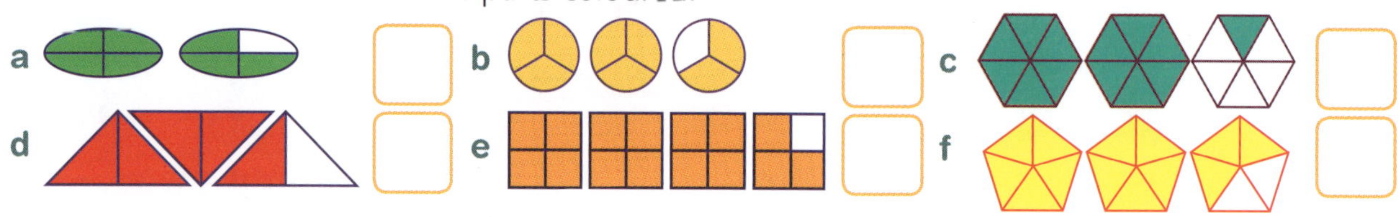

3 Complete the number line. Practise counting forwards and backwards.

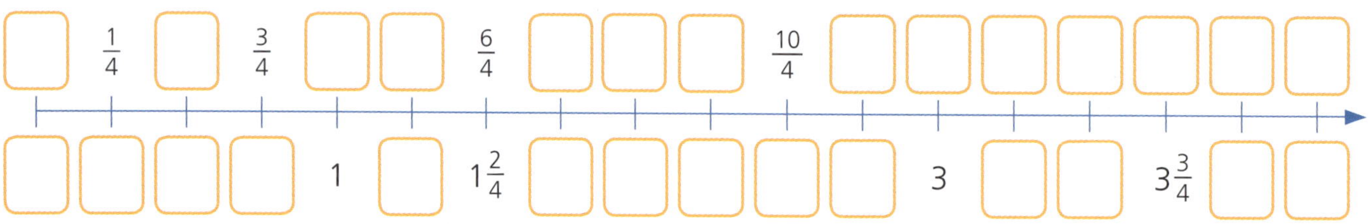

4 Write the mixed number for:

a $\frac{7}{4}$ b $\frac{9}{5}$ c $\frac{11}{4}$ d $\frac{11}{5}$

e $\frac{8}{5}$ f $\frac{13}{4}$ g $\frac{14}{5}$ h $\frac{17}{4}$

5 Write the improper fraction for:

a $1\frac{1}{2}$ b $2\frac{1}{5}$ c $3\frac{2}{4}$ d $1\frac{4}{5}$

e $2\frac{3}{4}$ f $2\frac{5}{4}$ g $4\frac{2}{3}$ h $2\frac{5}{8}$

- Explain to a friend how you got your answers to Questions 4 and 5.

 • *AUSTRALIAN SIGNPOST MATHS 4* • ISBN 978 1 4886 2187 1

1:15 Mixed Numbers

1 Write the mixed number for each part.

a 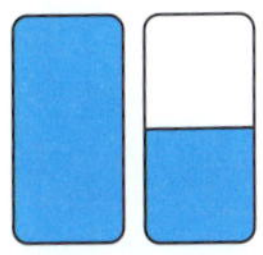 ☐

b 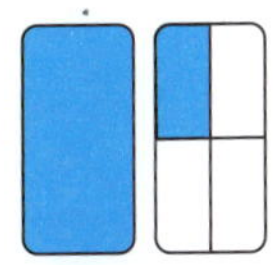 ☐

c 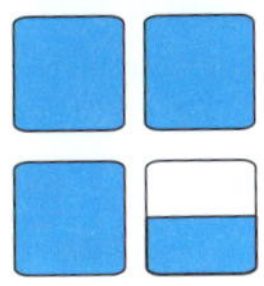 ☐

d 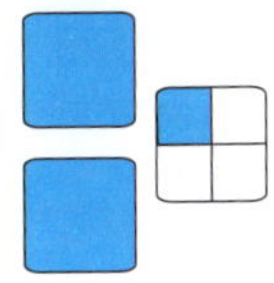 ☐

CONCEPT

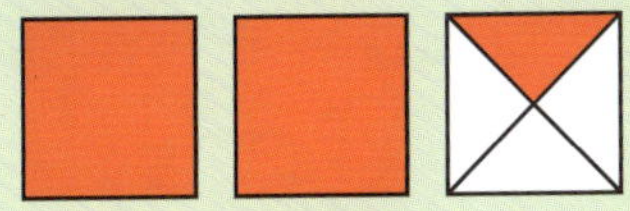

$2\frac{1}{4}$

When a number has a **whole number part** and a **fraction part**, it is called a mixed number.

2 Colour shapes to show each mixed number.

a $1\frac{1}{2}$ 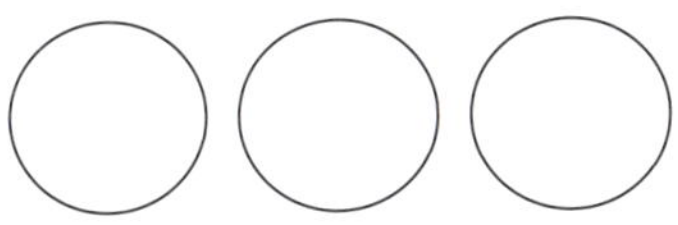

b $2\frac{1}{4}$ 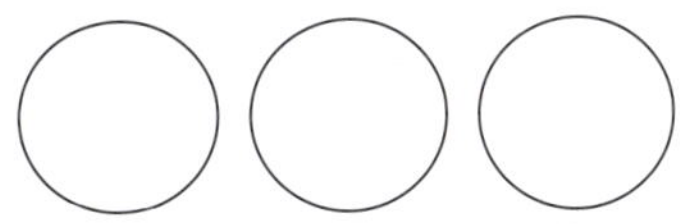

c $2\frac{1}{2}$ 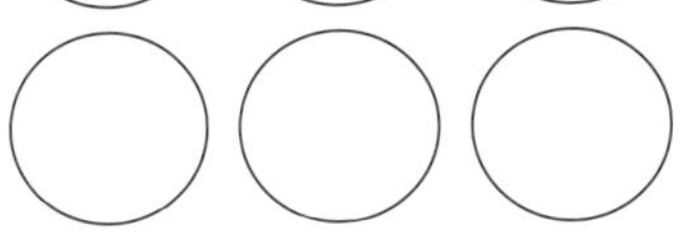

d $1\frac{1}{4}$ 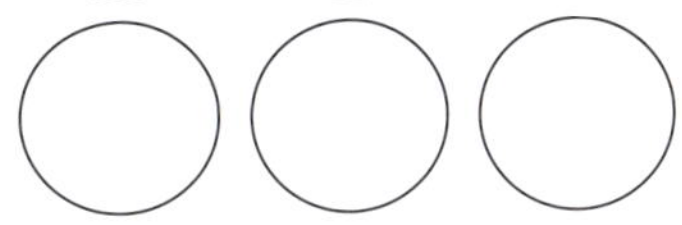

e $7\frac{1}{2}$ 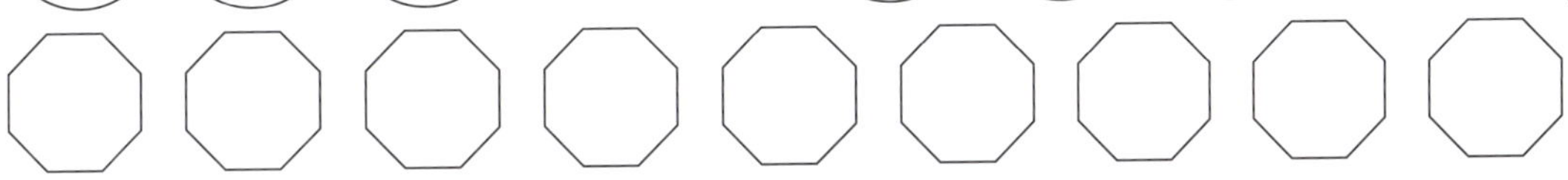

3 Continue each fraction pattern to describe the shape pattern. Discuss your answers.

a 

$\frac{1}{2}$, 1, $1\frac{1}{2}$, 2, $2\frac{1}{2}$, ☐, ☐, ☐, ☐

b 

$\frac{1}{4}$, $\frac{2}{4}$, $\frac{3}{4}$, $\frac{4}{4}$, $\frac{5}{4}$, ☐, ☐, ☐, ☐

4 Use the answers to Question 3 to count forwards and backwards.

ACTIVITY

- Draw and label mixed numbers of your own, on paper.

1:16 Numbers to 99 999

ACTIVITY

Ten Thousands	Thousands	Hundreds	Tens	Ones
7	5	2	4	8

Seventy-five thousand two hundred and forty-eight

75 248

Use counters and a place-value chart to model:

a twenty-five thousand six hundred and fifty-two

b sixty-one thousand five hundred and thirty-one

c fifty-six thousand eight hundred and forty-nine

d 93 471 e 72 345 f 19 862 g 21 978

h 47 532 i 66 234 j 23 348 k 97 643

1 Write the numeral for:

a 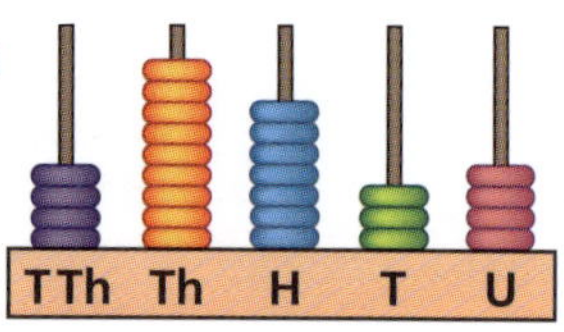

b 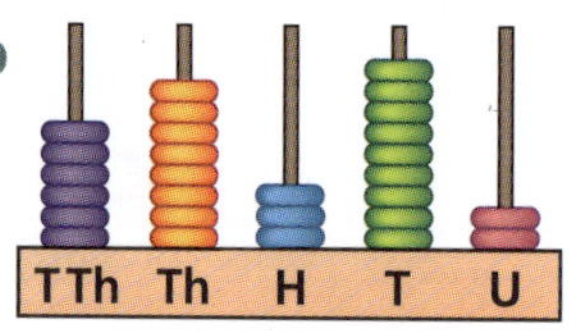

c 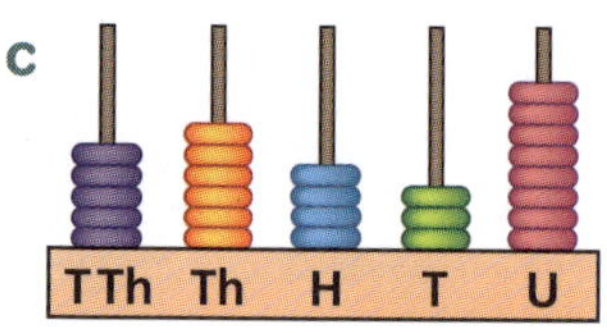

d 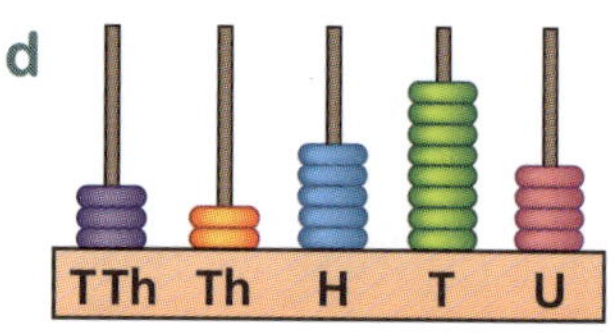

2 Read these numbers aloud and then write them in figures on the place-value chart .

a twenty-six thousand three hundred and twenty-four

b thirty-five thousand one hundred and sixty-two

c eighty-two thousand nine hundred and seventy

d fifty-two thousand eight hundred and fourteen

e seventy-four thousand two hundred and sixty

T Thous	Thous	Hund	Tens	Ones

3 Write the numeral for:

a 30 000 + 4 000 + 500 + 20 + 8

b 60 000 + 7 000 + 900 + 30 + 4

c 50 000 + 8 000 + 400 + 60 + 2

d 90 000 + 2 000 + 700 + 40 + 8

e 80 000 + 2 000 + 300 + 50 + 9

f 40 000 + 8 000 + 600 + 70 + 3

© PEARSON AUSTRALIA 2018 • *AUSTRALIAN SIGNPOST MATHS 4* • ISBN 978 1 4886 2187 1

1:17 Numbers to 99 999

1 Write the numeral for:

a
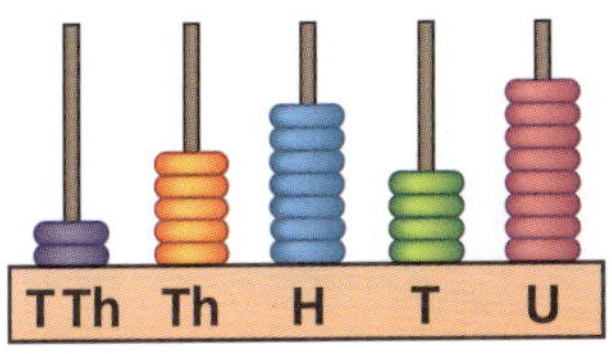

b
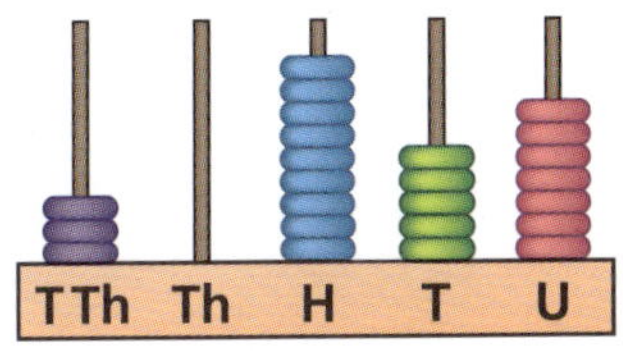

c
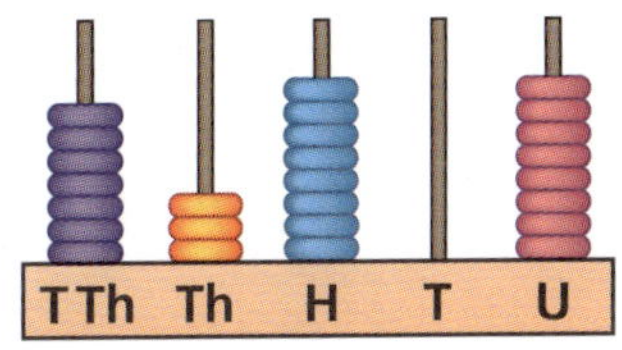

2 Read these numbers aloud and write them in the place-value chart.

a sixty-four thousand three hundred and seventy

b seventy thousand seven hundred and thirty-nine

c forty-one thousand six hundred and two

d eighty-five thousand and seventy-two

e thirty-nine thousand five hundred and eight

f fifty-three thousand and twenty-nine

T Thous	Thous	Hund	Tens	Ones

3 Write the numeral for:

a 90 000 + 6 000 + 200 + 40 + 8

b 70 000 + 9 000 + 500 + 30 + 1

c 40 000 + 7 000 + 90 + 3

d 60 000 + 2 000 + 800 + 50 + 6

e 20 000 + 8 000 + 400 + 60

f 10 000 + 5 000 + 700 + 4

4 Write the value for each coloured digit in these numbers.

a 48 603 (coloured digit: 8)

b 27 534 (coloured digit: 4)

c 91 738 (coloured digit: 1)

d 68 507 (coloured digit: 6)

e 32 480 (coloured digit: 8)

f 10 791 (coloured digit: 9)

g 80 965 (coloured digit: 9)

h 50 673 (coloured digit: 6)

Make the Number

- For this game you need a set of playing cards marked with digits 0 to 9.
- The dealer selects a number to be made, e.g. 'Make the number closest to 30 000'.
- Five cards are then dealt to each player. The winner is the player able to arrange the five cards closest to the chosen number.
- Each winner scores one point.

 • *AUSTRALIAN SIGNPOST MATHS 4* • ISBN 978 1 4886 2187 1

1:18 Equivalent Fractions

CONCEPT

Fractions can be positioned on number lines.

0 $\frac{1}{3}$ $\frac{2}{3}$ $\frac{3}{3}$

0 $\frac{1}{6}$ $\frac{2}{6}$ $\frac{3}{6}$ $\frac{4}{6}$ $\frac{5}{6}$ 1

$\frac{3}{3} = 1$

1 Use the number lines above to write a fraction equal to:

a $\frac{2}{6}$ ☐ **b** $\frac{4}{6}$ ☐ **c** $\frac{6}{6}$ ☐ **d** $\frac{2}{3}$ ☐

2 Complete the number lines.

a

0 $\frac{1}{2}$ 1

0 $\frac{1}{4}$ ☐ ☐ 1

0 $\frac{1}{8}$ ☐ $\frac{3}{8}$ $\frac{4}{8}$ ☐ $\frac{6}{8}$ ☐ 1

b

0 $\frac{1}{2}$ 1

0 ☐ ☐ 1

0 ☐ ☐ $\frac{3}{6}$ ☐ ☐ 1

3 Use the number lines above to answer **true** (**T**) or **false** (**F**).

a $\frac{4}{8} = \frac{1}{2}$ ☐ **b** $\frac{3}{4} = \frac{6}{8}$ ☐ **c** $\frac{1}{2} = \frac{3}{6}$ ☐ **d** $\frac{2}{6} = \frac{1}{2}$ ☐

e $\frac{7}{8} = \frac{3}{4}$ ☐ **f** $\frac{2}{8} = \frac{1}{4}$ ☐ **g** $\frac{4}{6} = \frac{2}{3}$ ☐ **h** $\frac{2}{6} = \frac{1}{3}$ ☐

i $\frac{3}{4} = \frac{5}{8}$ ☐ **j** $\frac{4}{4} = \frac{2}{8}$ ☐ **k** $\frac{5}{6} = \frac{1}{3}$ ☐ **l** $\frac{2}{2} = \frac{3}{3}$ ☐

ACTIVITY

4 Share 24 cards into these fractions.

a Quarters = ☐ in each share. **b** Eighths = ☐ in each share.

c Thirds = ☐ in each share. **d** Sixths = ☐ in each share.

- Use other concrete materials to demonstrate halves, quarters and eighths of an object or a collection of objects.
- Explain your answers to a partner.

 • *AUSTRALIAN SIGNPOST MATHS 4* • ISBN 978 1 4886 2187 1

1:19 Fractions and the Number Line

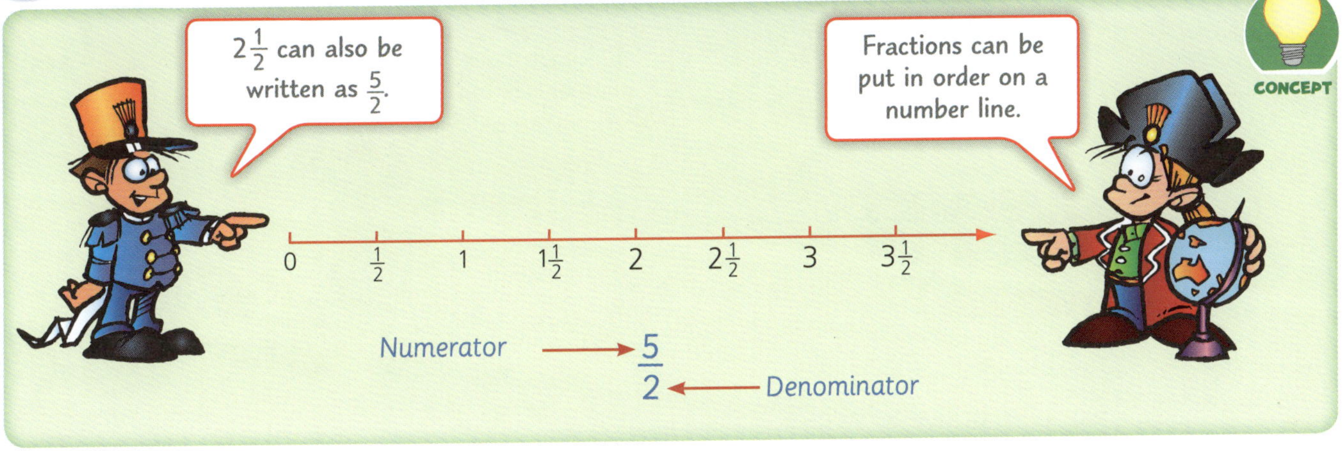

1 Complete the number lines.

a 0, $\frac{1}{4}$, □, □, 1

b 0, $\frac{1}{8}$, □, $\frac{3}{8}$, □, $\frac{5}{8}$, □, $\frac{7}{8}$, □

c 0, $\frac{1}{3}$, □, □, $1\frac{1}{3}$, □, □

d 0, $\frac{1}{6}$, □, $\frac{3}{6}$, □, $\frac{5}{6}$, □, $1\frac{1}{6}$, □, $1\frac{3}{6}$, □, $1\frac{5}{6}$, □

e 0, □, $\frac{2}{5}$, □, □, 1

f 0, $\frac{1}{10}$, □, □, □, □, □, $\frac{7}{10}$, $\frac{8}{10}$, $\frac{9}{10}$, □

2 Complete the number lines.

Top: 0, □, $\frac{2}{4}$, □, □, □, $\frac{6}{4}$, □, □, $\frac{9}{4}$, □, □, □

Bottom: 0, □, $\frac{2}{4}$, □, □, $1\frac{1}{4}$, □, □, 2, □, □, □, 3

3 Write a mixed number that is the same as:

a $\frac{6}{4}$ = □ **b** $\frac{7}{4}$ = □ **c** $\frac{12}{4}$ = □

© PEARSON AUSTRALIA 2018 • *AUSTRALIAN SIGNPOST MATHS 4* • ISBN 978 1 4886 2187 1

1:20 Place Value in Decimals

CONCEPT

27 out of 100

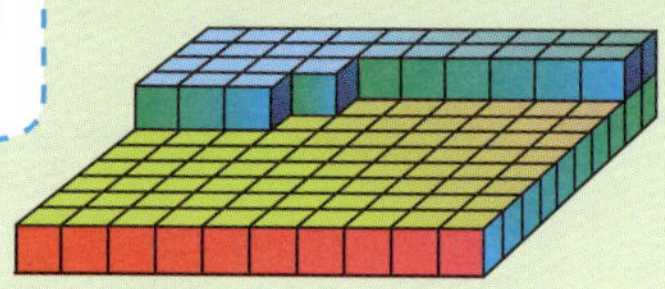

Ones	Tenths	Hundredths
0 ·	2	7

Decimals have place values too.

1 Complete the label for each model.

a

Ones	Tenths	Hundredths
0 ·		

b

Ones	Tenths	Hundredths
0 ·		

c

Ones	Tenths	Hundredths
·		

This is one whole and twenty-eight hundredths.

2 Complete the numeral expanders for each decimal.

a 0·31

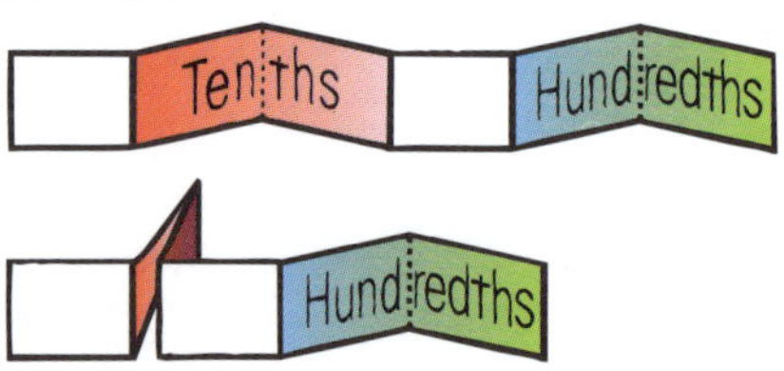

b 0·68

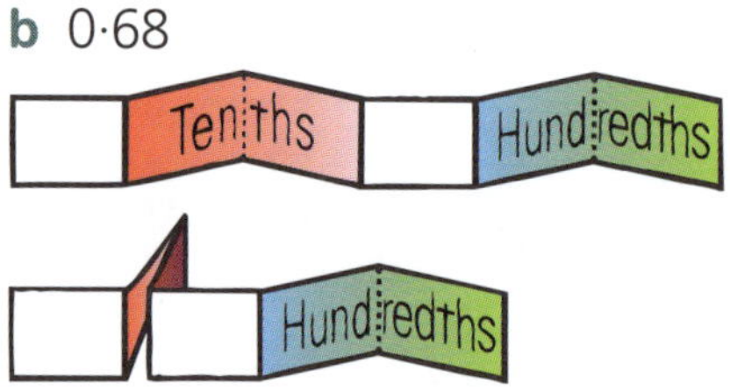

c 0·95

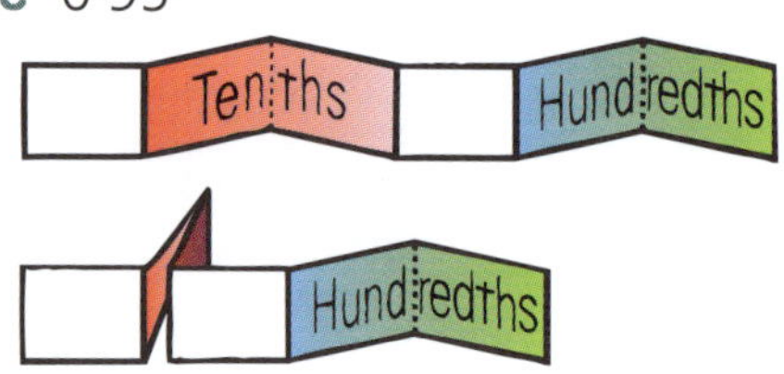

d 0·43

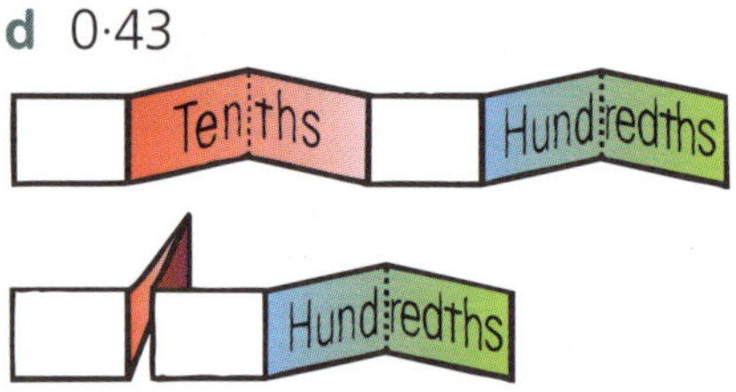

e 0·72

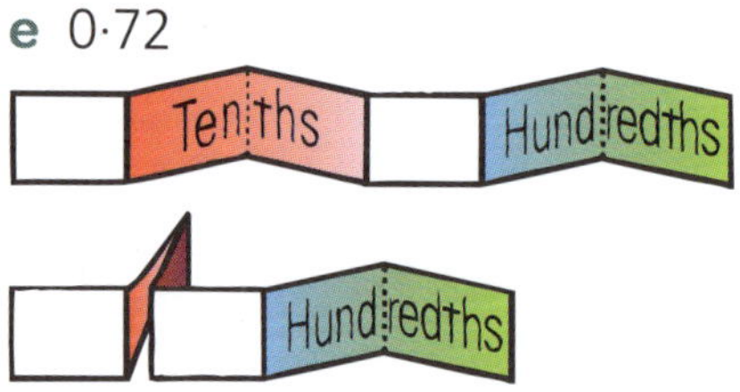

f 0·86

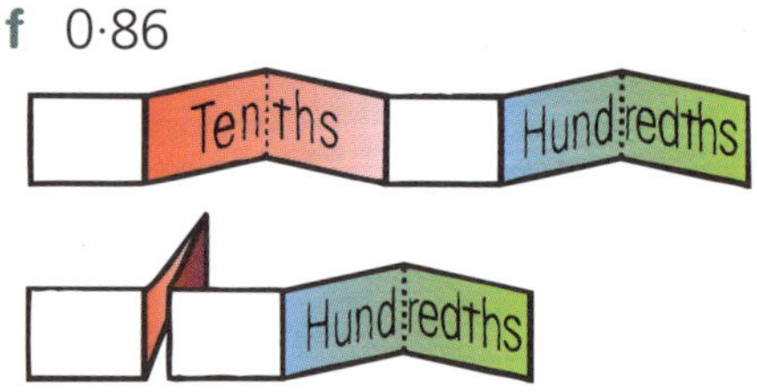

3 Draw a line from each decimal to the correct fraction.

a

$\frac{35}{100}$	0·59
$\frac{59}{100}$	1·22
$1\frac{22}{100}$	0·35

b

$\frac{41}{100}$	0·63
$2\frac{97}{100}$	0·41
$\frac{63}{100}$	2·97

c

$\frac{89}{100}$	3·38
$1\frac{56}{100}$	0·89
$3\frac{38}{100}$	1·56

4 Use place-value blocks to show your answers to Question 3.

 • *AUSTRALIAN SIGNPOST MATHS 4* • ISBN 978 1 4886 2187 1

1:21 Tenths

CONCEPT

1 Write the decimal for:

a $\frac{5}{10}$ ☐ b $\frac{3}{10}$ ☐ c $\frac{6}{10}$ ☐ d $\frac{1}{10}$ ☐

e $\frac{8}{10}$ ☐ f $\frac{9}{10}$ ☐ g $\frac{2}{10}$ ☐ h $\frac{4}{10}$ ☐

i $1\frac{1}{10}$ ☐ j $1\frac{3}{10}$ ☐ k $1\frac{9}{10}$ ☐ l $1\frac{8}{10}$ ☐

2 Match each fraction with the correct decimal.

a

$\frac{4}{10}$	0·7
$\frac{9}{10}$	0·4
$\frac{7}{10}$	0·9

b

$\frac{3}{10}$	0·2
$\frac{1}{10}$	0·3
$\frac{2}{10}$	0·1

c

$1\frac{9}{10}$	1·8
$1\frac{8}{10}$	1·9
$1\frac{6}{10}$	1·6

3 Use decimals to write:

a 5 tenths ☐ b 3 tenths ☐ c 9 tenths ☐ d 7 tenths ☐

e 4 tenths ☐ f 8 tenths ☐ g 6 tenths ☐ h 10 tenths ☐

i zero point one ☐ j zero point eight ☐ k zero point five ☐

l one point nine ☐ m one point three ☐ n one point zero ☐

4 Complete the number lines.

a

0 $\frac{1}{10}$ ☐ $\frac{3}{10}$ ☐ ☐ $\frac{6}{10}$ ☐ ☐ ☐ 1

0 0·1 ☐ ☐ ☐ 0·5 0·6 ☐ ☐ ☐ 1·0

b

0 $\frac{1}{10}$ $\frac{2}{10}$ $\frac{3}{10}$ ☐ ☐ ☐ ☐ ☐ ☐ ☐

0 0·1 0·2 0·3 ☐ ☐ ☐ ☐ ☐ ☐ ☐

1:22 Comparing Decimals

1 Write the label for each coloured part.

A B C D E

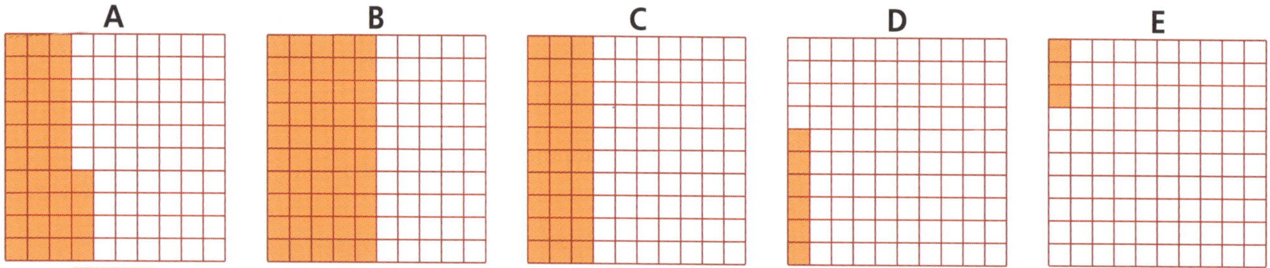

0· 0· 0· 0· 0·

2 Which hundred square shows the bigger coloured fraction:

a **D** or **C**? b **A** or **D**? c **A** or **C**? d **D** or **E**? e **E** or **B**?

f **B** or **D**? g **E** or **C**? h **A** or **B**? i **C** or **B**? j **A** or **E**?

3 Which hundred square shows the biggest coloured fraction:

a **C**, **D** or **E**? b **A**, **C** or **E**? c **B**, **C** or **D**? d **A**, **D** or **E**? e **A**, **B** or **E**?

4 Write the decimal fraction that follows:

a 0·26, 0·27, b 0·83, 0·84, c 0·47, 0·48, d 0·35, 0·36,

5 Complete the number line.

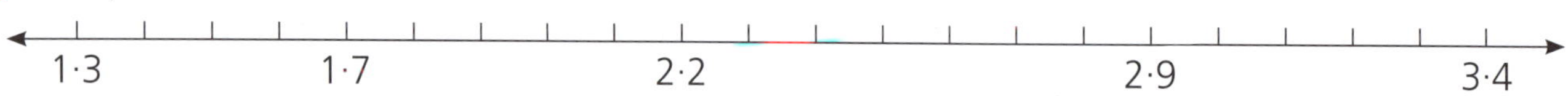

6 Complete the decimal table below, counting by one hundredths from 0·01 to 1·00.

0·01							0·08		
			0·34	0·35					
0·51									
							0·68		
									1·00

1:23 Place Value in Decimals

CONCEPT

Ones	Tenths	Hundredths
0 ·	6	0

6 tenths

Ones	Tenths	Hundredths
0 ·	0	6

6 hundredths

1. Match the part coloured on each hundred square with the correct label.

0·5 0·02 0·1 0·3 0·05 0·06 0·9 0·7 0·4 0·09

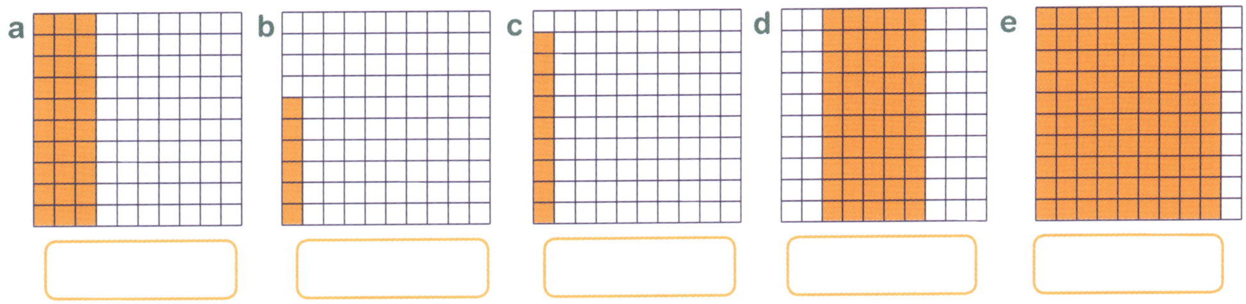

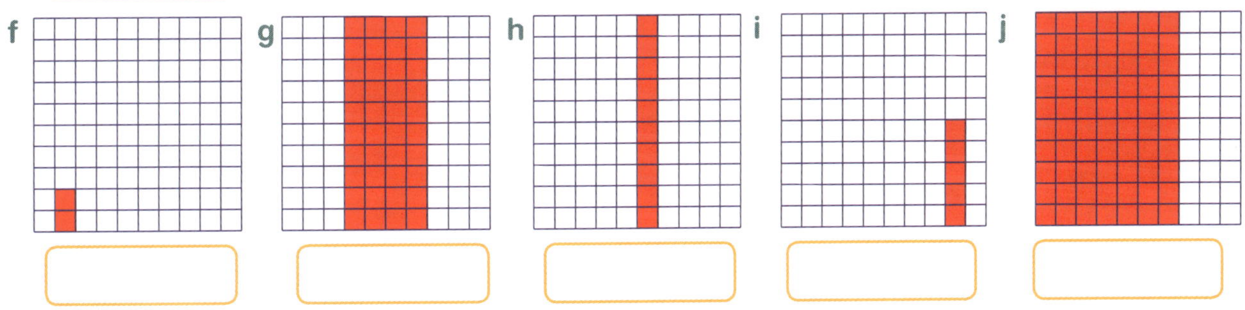

2. Colour part of each hundred square to match the given decimal.

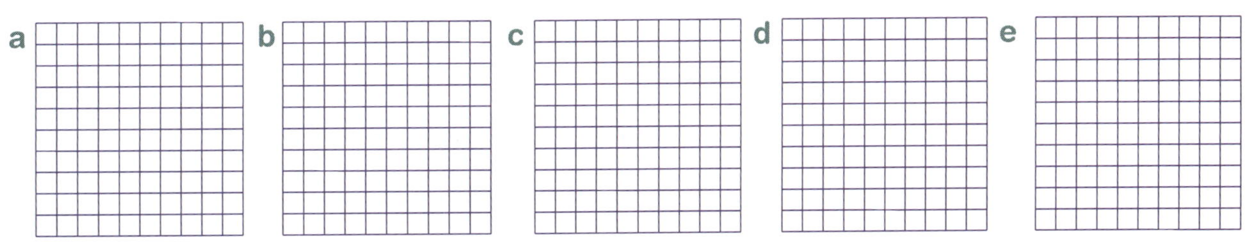

a 0·07 b 0·2 c 0·03 d 0·08 e 0·8

ACTIVITY

3. Use place-value blocks to model:

a 5 tenths b 3 hundredths c 4 tenths and 7 hundredths

d 0·3 e 0·7 f 0·02 g 0·06 h 0·4

i 0·52 j 0·89 k 0·72 l 0·67 m 0·99

0·12

1:24 Ordering Numbers to 99 999

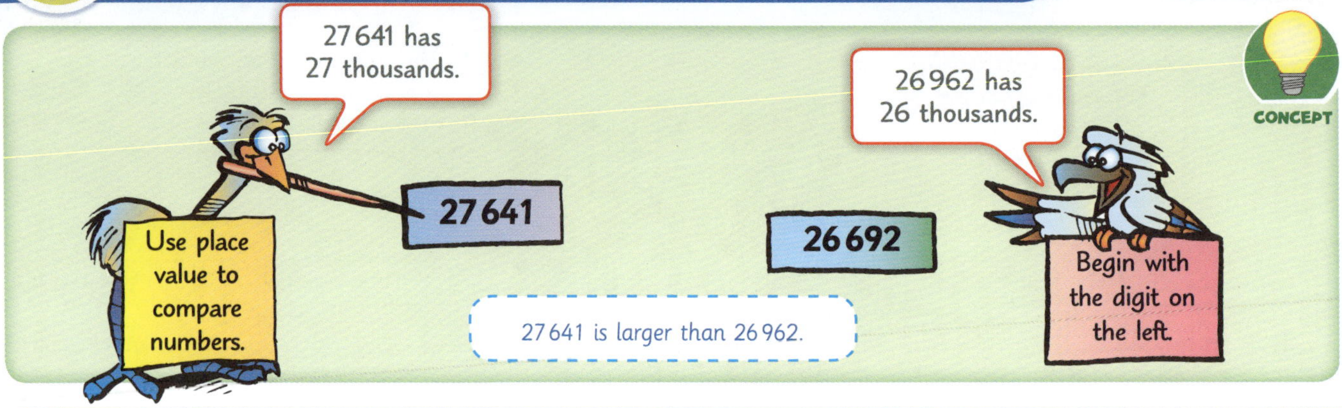

1 Circle the larger number in each pair.

a 49 768   49 713   **b** 37 281   36 281   **c** 84 971   81 302   **d** 66 438   70 923

2 Circle the smallest number in each group.

a 29 642   27 849   26 301   **b** 72 642   69 309   70 624   **c** 58 642   61 342   46 526

3 Write the numeral for:

a twenty-eight thousand seven hundred and forty-seven

b fifty thousand three hundred and seventy-eight

c eighty-two thousand four hundred and nine

d seventy-five thousand six hundred and eighty-two

e forty-four thousand nine hundred and thirty-six

4 Write the numeral for:

a 20 000 + 6 000 + 400 + 90 + 3   **b** 80 000 + 5 000 + 300 + 70 + 4

c 90 000 + 5 000 + 600 + 50 + 1   **d** 40 000 + 8 000 + 700 + 60 + 9

5 Write in words:

a 42 603

b 70 941

c 27 086

d 63 550

1:25 Reading and Writing Numbers

K is used as an abbreviation for thousand.

That means $32 000.

ACTIVITY

K comes from **kilo** which means one thousand.

- Use a dictionary to find words that use the prefix 'kilo'.
- Collect and display advertisements from the media that use abbreviations for numbers.

1 Write the numeral for:

a 40 000 + 9 000 + 700 + 60 + 9 ☐ b 70 000 + 5 000 + 200 + 30 + 8 ☐

c 80 000 + 600 + 50 + 2 ☐ d 30 000 + 9 000 + 40 + 3 ☐

2 Use expanded notation to write:

a 16 327 ☐ b 52 851 ☐

c 67 509 ☐ d 90 325 ☐

3 Write the value of each coloured digit.

a 3**4** 396 ☐ b 29 6**8**4 ☐ c 82 **5**21 ☐ d **9**6 317 ☐

e 57 4**2**3 ☐ f 4**2** 372 ☐ g **6**1 407 ☐ h 78 **6**90 ☐

4 Use < or > to show the larger number.

a 34 529 ☐ 35 014 b 62 916 ☐ 63 845 c 18 397 ☐ 17 624

d 71 310 ☐ 76 243 e 27 592 ☐ 26 991 f 49 620 ☐ 48 375

These are both true.

5 Write **true** or **false** for each.

a 79 602 < 81 310 ☐ b 48 324 > 47 213 ☐

c 23 099 > 24 302 ☐ d 67 538 < 66 435 ☐

e 81 219 < 81 921 ☐ f 56 345 > 56 435 ☐

6 Write the numeral for:

a (8 × 10 000) + (6 × 1 000) + (3 × 100) + (9 × 10) + 4 = ☐

b (5 × 10 000) + (2 × 1 000) + (7 × 100) + (3 × 10) + 1 = ☐

1:26 Place Value to Hundredths

CONCEPT

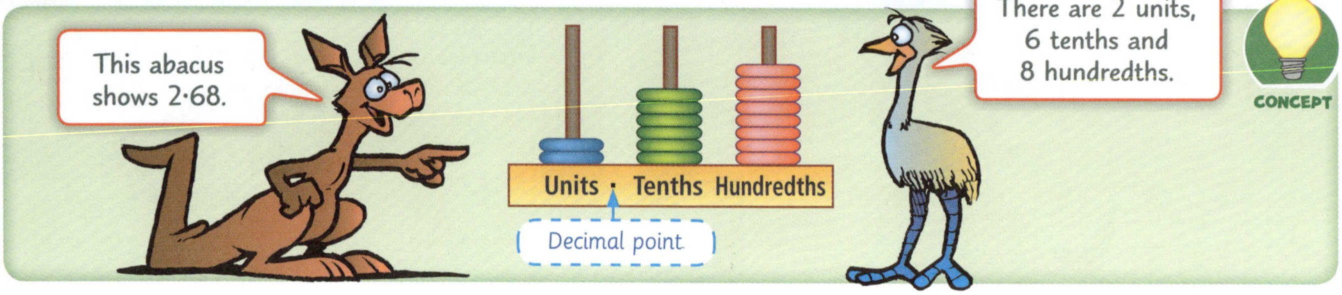

1 Write the numeral for:

a 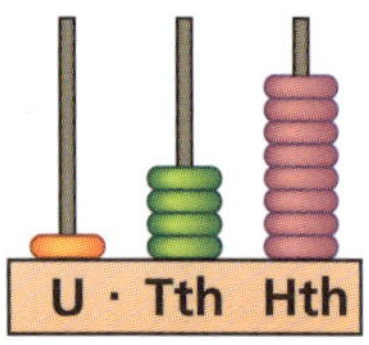

b 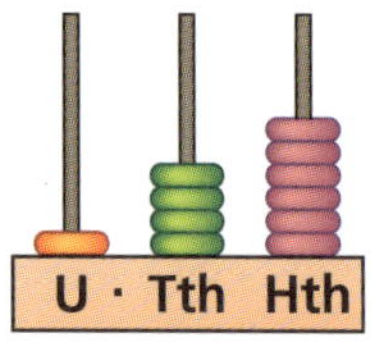

c 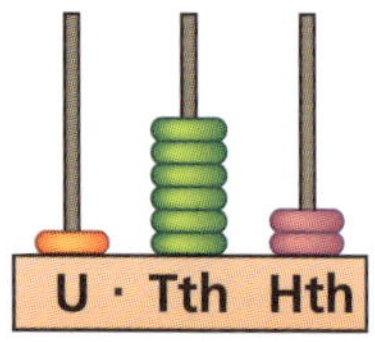

d 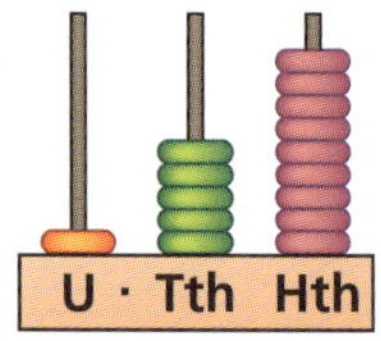

e 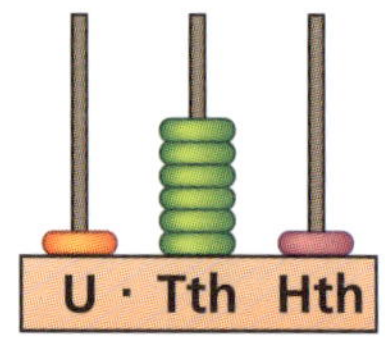

2 Write each number on the place-value chart.

a two point five nine
b six point seven three
c nine point four one
d five point two six
e eight point four eight
f seven point one three
g six point seven nine

Units		Tenths	Hundredths
	·		
	·		
	·		
	·		
	·		
	·		
	·		

3 Write the value for each coloured digit.

a 4·3**2** b 8·**4**7 c **3**·95 d 5·3**8** e **1**·76
f 7·**5**1 g **2**·84 h 9·6**2** i 1·**69** j 6·5**1**

Digit Game

FUN SPOT

- In turn, each player receives one numeral card from a dealer and places it in one column of a place-value chart.
- Two more cards are received and placed in the same way.
- The player with the largest number wins a point.

Units		Tenths	Hundredths
6	·	4	1

© PEARSON AUSTRALIA 2018 • *AUSTRALIAN SIGNPOST MATHS 4* • ISBN 978 1 4886 2187 1

1:27 Reading and Writing Decimals

CONCEPT

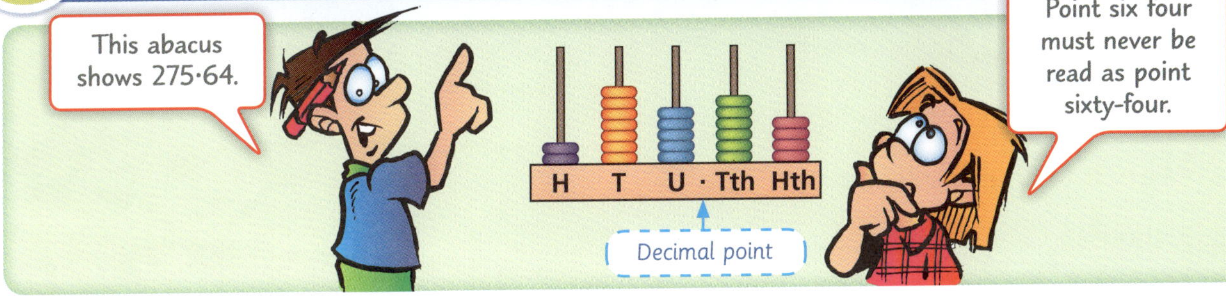

1 Write the numeral for:

a 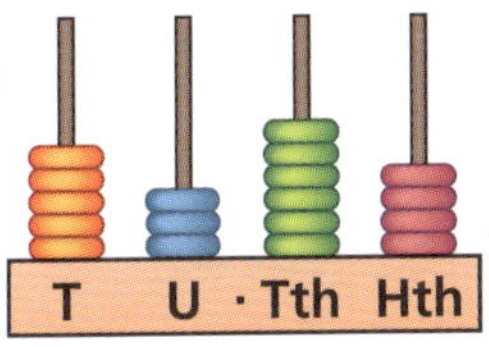

b 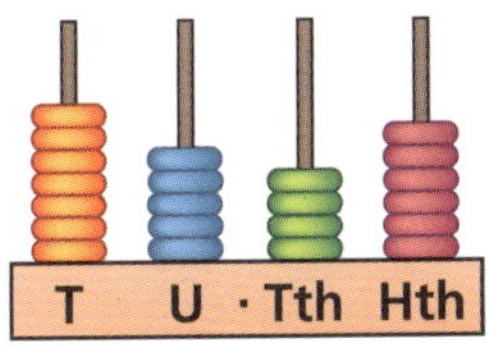

c 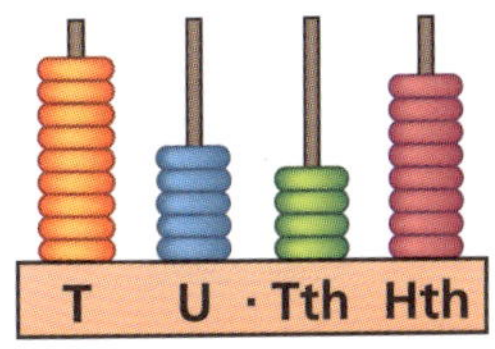

d 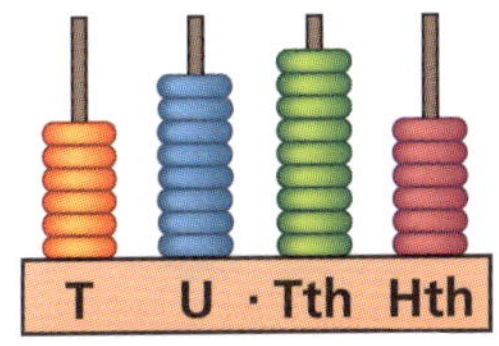

2 Use numerals to write:

a nineteen point seven five

b forty point nine one

c sixty-seven point eight two

d thirty point zero six

e fifty-two point one four

f twenty-seven point four three

3 Complete the following.

a 0·86 ☐ tenths and ☐ hundredths

b 0·09 ☐ tenths and ☐ hundredths

c 0·60 ☐ tenths and ☐ hundredths

4 Write the number represented by:

a 6 hundreds, 2 tens, 9 units, 3 tenths and 7 hundredths

b 4 hundreds, 6 tens, 3 units, 0 tenths and 9 hundredths

c 7 hundreds, 3 tens, 1 unit and 6 tenths

d 9 hundreds, 5 tens, 7 units, 4 tenths and 2 hundredths

e five and six tenths

f ten and three tenths

g two and eight tenths

h seven and nine tenths

5 Write the value for each coloured digit.

a 1**6**4·21 b 35**9**·73 c **8**47·35 d 652·**4**7

e 278·5**2** f 247·**3**8 g 5**1**6·69 h 7**5**9·57

i 316·**8**4 j 472·1**9** k 792·**6**2 l 134·8**5**

 • *AUSTRALIAN SIGNPOST MATHS 4* • ISBN 978 1 4886 2187 1

1:28 Reading and Writing Numbers

1 Write the numeral for:

a fifty-seven thousand eight hundred and four

b eighty-four thousand and forty-five

c sixty thousand one hundred and twenty-seven

d thirty-two thousand and sixty

2 Use the expanded form to write:

a 9471

b 61 308

c 29 082

d 30 657

e 83 241

f 65 402

3 Write the numeral for:

a

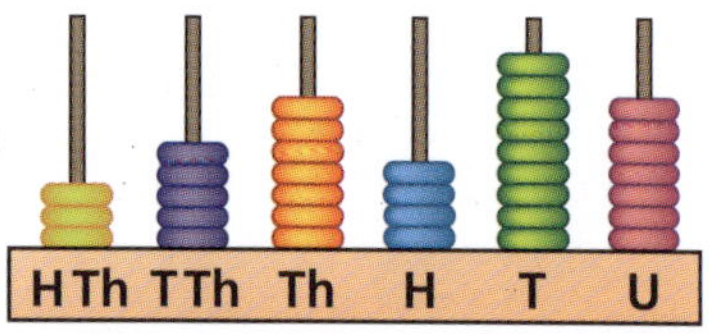

b

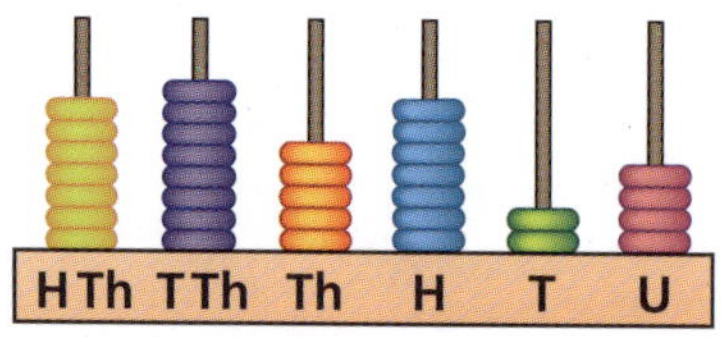

c

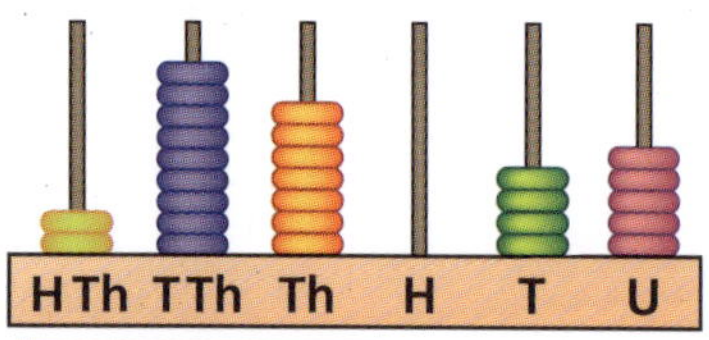

d

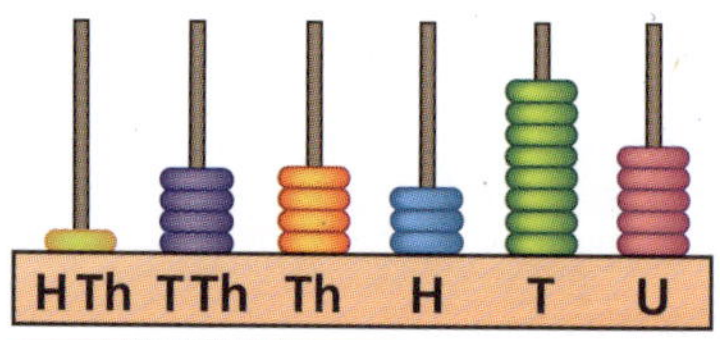

e

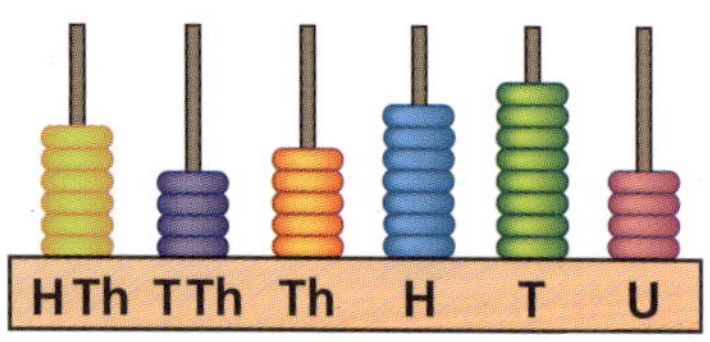

f

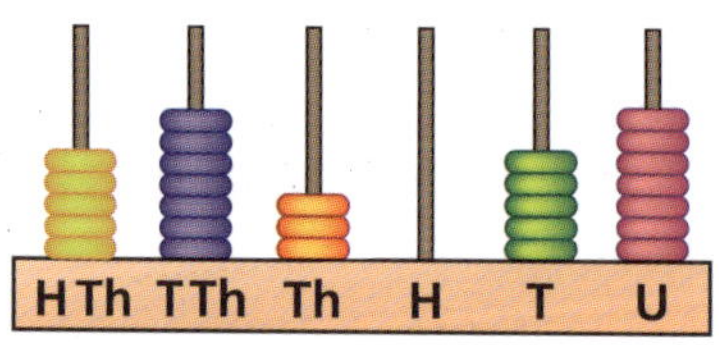

ten **ten thousands**
= 10 × 10 000
= 100 000

4 Complete:

a 1 hundred = ☐ tens

b 1 thousand = ☐ hundreds

c 1 ten thousand = ☐ thousands

d 1 hundred thousand = ☐ ten thousands

5 Use a calculator.

a 700 000 + 50 000 + 9000 + 300 + 60 + 4 =

b 500 000 + 30 000 + 2000 + 800 + 40 + 9 =

1:29 Numbers to 999 999

1 Read these numbers aloud and then write them in figures on the place-value chart.

a thirty-six thousand nine hundred and seventy-four

b five hundred and seventy-five thousand three hundred and eighty

c nine hundred and fifteen thousand four hundred and twenty-nine

	Hundred Thousands	Ten Thousands	Thousands	Hundreds	Tens	Ones
a		3	6	9	7	4
b						
c						

In part **a**, the place value of the 6 is 'thousands'. The value of the 6 is 6000.

2 Write the numeral for:

a (6 × 10 000) + (4 × 1 000) + (9 × 100) + (7 × 10) + 2

b 100 000 + 70 000 + 6 000 + 300 + 50 + 6

c 300 000 + 80 000 + 3 000 + 700 + 40 + 9

d (7 × 10 000) + (5 × 1 000) + (9 × 100) + (2 × 10) + 3

3 Write the place value for each coloured digit.

a 37 **6**19 **b** 91 **3** 683 **c** 46 81**9** **d** 19**3** 754

4 Write in words:

a 503 277

b 230 426

5 Use a calculator to find: ICT

a half of 524 288 **b** half of 964 708

c half of 729 376 **d** double 319 743

e double 437 519 **f** double 297 589

g half of 621 214 **h** double 763 213

1:30 Place Value

CONCEPT

3 8 4 1 9

= 3 8 0 0 0

\+ 4 1 9

Arrow cards help us understand the value of a number.

38 419
= 38 000 + 419

1 Complete these arrow cards.

a 2 1 5 8 6

=

\+

b 8 5 1 4 2

=

\+

2 Write the mass shown on each scale.

a

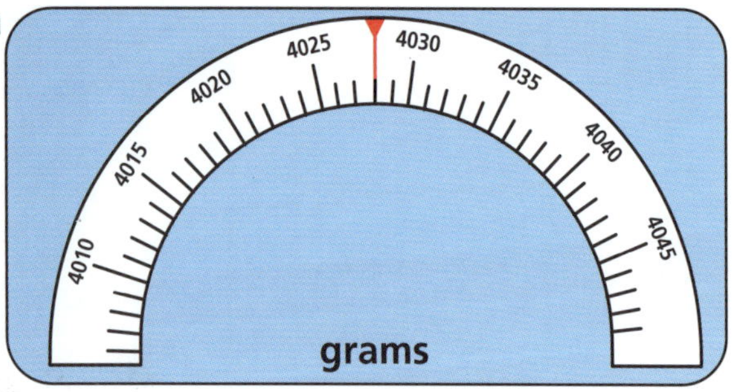

b

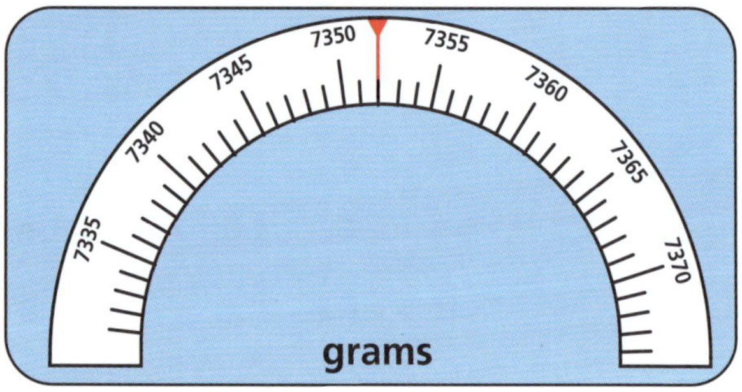

3 Each gold bar is worth $1000. What would be the value of:

a 5 bars of gold?

b 10 bars of gold?

c 20 bars of gold?

d 50 bars of gold?

e 70 bars of gold?

f 100 bars of gold?

INVESTIGATION

- Invent a system (like the one in Question 3) to represent large numbers. Be prepared to explain your system to others.

1:31 Rounding Off

1 Round off each number to the nearest thousand.

a 134 691 ☐ b 458 216 ☐ c 679 835 ☐
d 263 124 ☐ e 916 527 ☐ f 542 747 ☐
g 795 438 ☐ h 381 963 ☐ i 828 362 ☐

2 Round off each number to the nearest ten thousand.

a 252 190 ☐ b 976 824 ☐ c 534 630 ☐
d 787 531 ☐ e 143 739 ☐ f 365 498 ☐

3 Write the numeral for:

a 60 000 + 4000 + 500 + 20 + 7 ☐
b 30 000 + 9000 + 200 + 70 + 8 ☐
c 40 000 + 8000 + 900 + 80 + 1 ☐
d 90 000 + 8000 + 300 + 50 + 6 ☐
e 70 000 + 1000 + 60 + 4 ☐
f 50 000 + 700 + 60 + 1 ☐

4 Use expanded notation to write:

a 83 162 ☐ b 35 275 ☐
c 46 913 ☐ d 78 548 ☐
e 92 301 ☐ f 60 825 ☐

5 Write the numeral for:

a
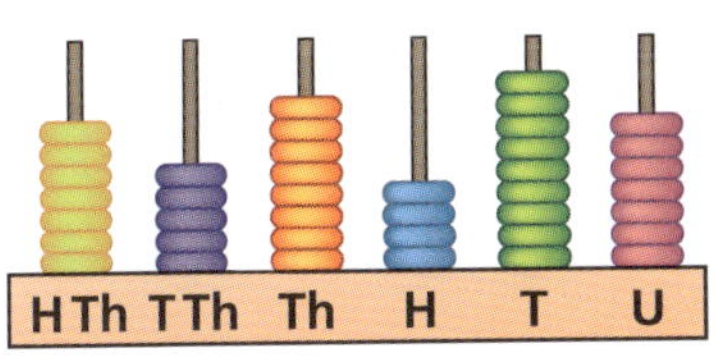

☐

b
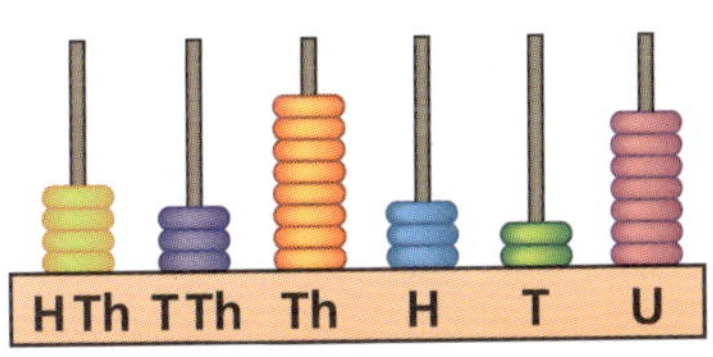

☐

c
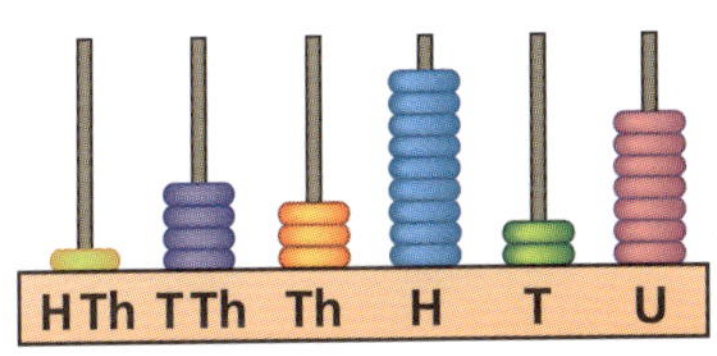

☐

6 Write the value for each coloured digit.

a 645 **2**31 ☐ b 723 4**6**9 ☐ c **4**72 975 ☐ d 1**5**7 906 ☐
e **3**79 406 ☐ f 3**9**1 973 ☐ g 24**8** 619 ☐ h **7**06 421 ☐
i 5**3**2 721 ☐ j 28**9** 578 ☐ k 932 **6**71 ☐ l **8**12 679 ☐

 • *AUSTRALIAN SIGNPOST MATHS 4* • ISBN 978 1 4886 2187 1

1:32 Fractions

CONCEPT

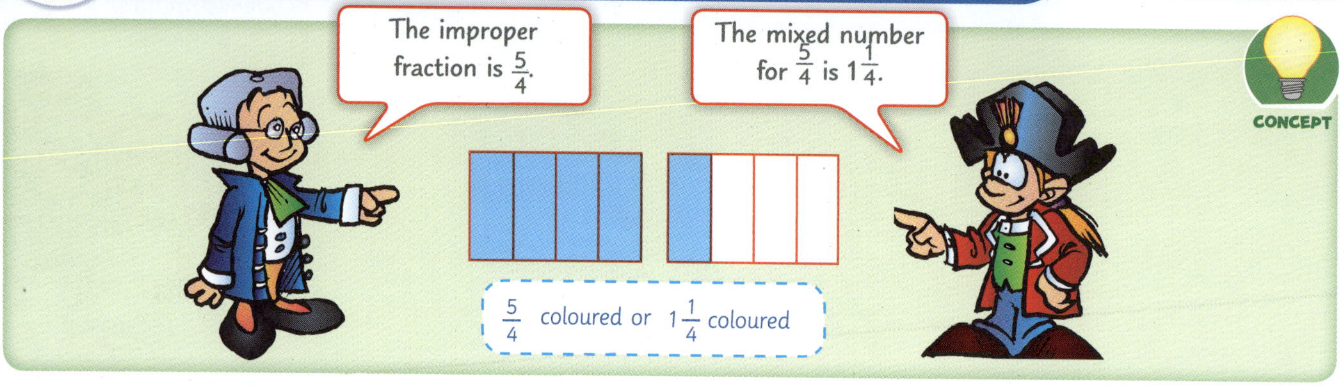

1 Write an equivalent fraction.

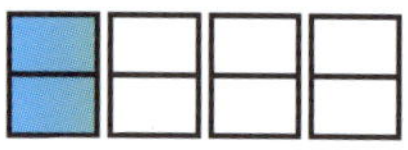

a $\frac{2}{2}$ ☐

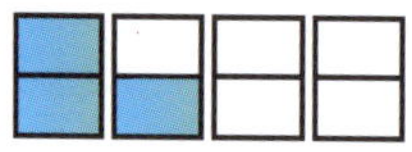

b $\frac{3}{2}$ ☐

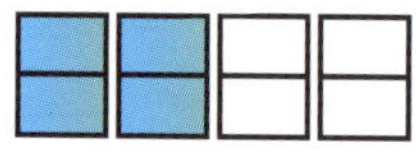

c $\frac{4}{2}$ ☐

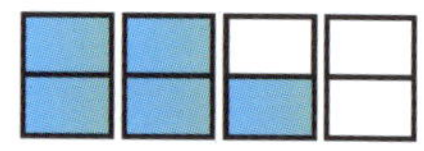

d $\frac{5}{2}$ ☐

2 Write the improper fraction and mixed number for each part.

a 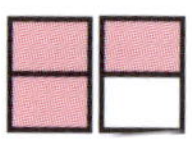

☐ or ☐

b 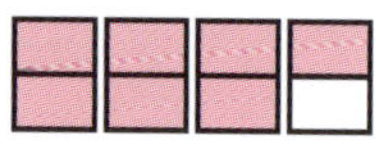

☐ or ☐

c 

☐ or ☐

d 

☐ or ☐

e 

☐ or ☐

f 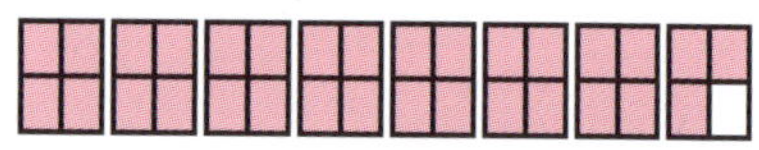

☐ or ☐

g 

☐ or ☐

3 Write the mixed number.

a $\frac{7}{5}$ ☐ b $\frac{9}{5}$ ☐ c $\frac{13}{5}$ ☐ d $\frac{11}{5}$ ☐

e $\frac{8}{5}$ ☐ f $\frac{12}{5}$ ☐ g $\frac{14}{5}$ ☐ h $\frac{17}{5}$ ☐

4 Write the improper fraction.

a $1\frac{1}{5}$ ☐ b $2\frac{1}{5}$ ☐ c $3\frac{2}{5}$ ☐ d $1\frac{4}{5}$ ☐

e $2\frac{3}{4}$ ☐ f $2\frac{5}{6}$ ☐ g $4\frac{2}{3}$ ☐ h $2\frac{5}{8}$ ☐

1:33 Fraction Patterns

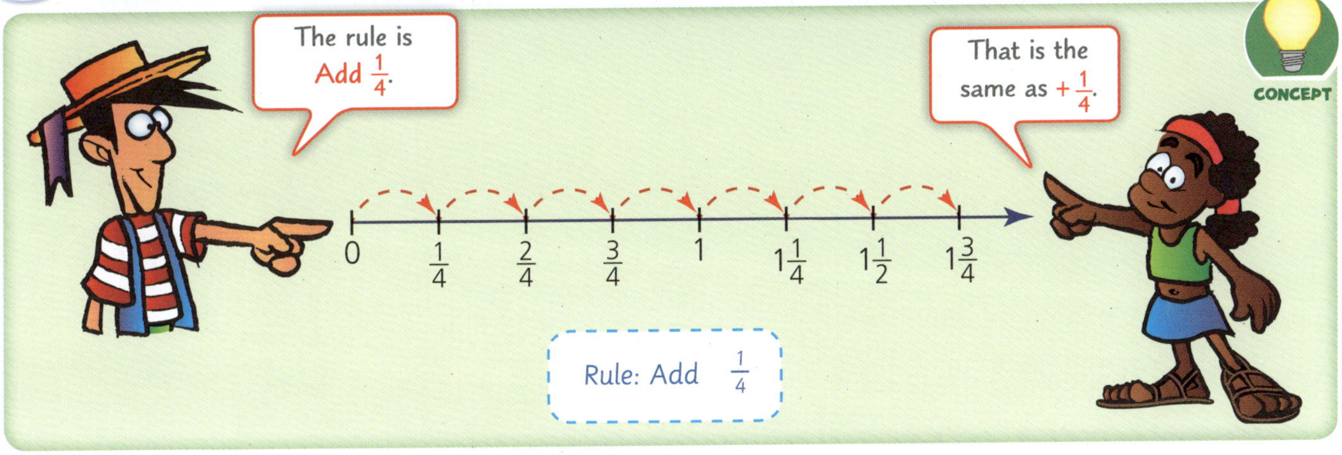

1 Complete each number line and write the rule.

a 0, $\frac{1}{2}$, ☐, ☐, 2

Rule: ☐

b 0, $\frac{1}{4}$, $\frac{1}{2}$, ☐, ☐

Rule: ☐

c 0, $\frac{1}{3}$, ☐, 1, $1\frac{1}{3}$, ☐, ☐, ☐, ☐, 3, ☐

Rule: ☐

d 0, ☐, ☐, ☐, ☐, ☐, ☐, ☐, 1, $1\frac{1}{8}$, $1\frac{2}{8}$

Rule: ☐

e 0, $\frac{1}{5}$, ☐, $\frac{3}{5}$, $\frac{4}{5}$, ☐, $1\frac{1}{5}$, ☐, ☐, ☐, ☐

Rule: ☐

f 0, ☐, ☐, ☐, ☐, ☐, ☐, ☐, $1\frac{2}{6}$, $1\frac{3}{6}$, $1\frac{4}{6}$

Rule: ☐

1:34 One Million

CONCEPT

We could use commas instead of spaces.

$999\,999 + 1 = 1\,000\,000$
$= 1{,}000{,}000$

Millions	Hund Thous	Ten Thous	Thous	Hund	Tens	Ones
	9	9	9	9	9	9

- One million is written with 6 zeros.

1 Million 0 Hund Thous 0 Ten Thous 0 Thousands 0 Hundreds 0 Tens 0 Ones

- 1 000 000 = 10 hundred thousands.

1 Read these numbers aloud and write them on the place-value chart below.

a six hundred and twenty-eight thousand four hundred and seventy-two
b nine hundred and sixty-five thousand seven hundred and twenty-three
c four hundred and eighty thousand nine hundred and thirty-five
d eight hundred and forty-nine thousand six hundred and seventy
e three hundred and sixteen thousand nine hundred and two
f one million

Practise reading these.

	Millions	Hund Thous	Ten Thous	Thous	Hund	Tens	Ones
a							
b							
c							
d							
e							
f							

2 Complete the numeral expanders for the number 935 216.

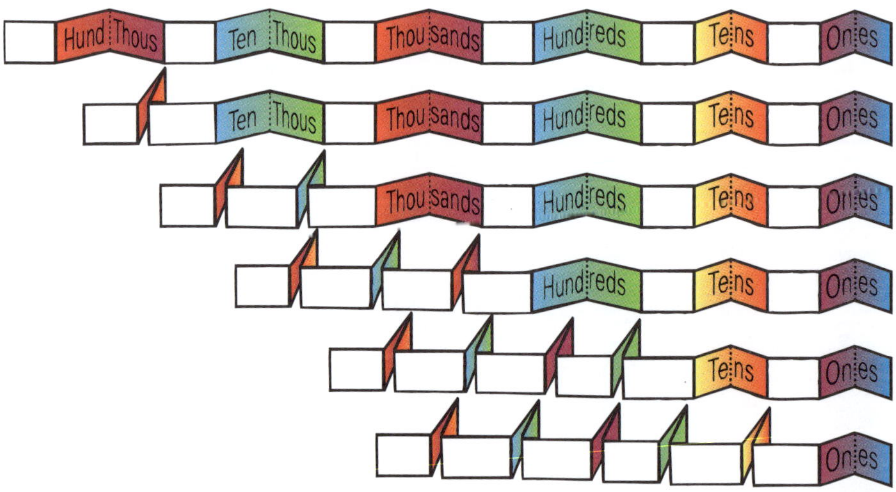

Place value helps us describe large numbers.

2:01 Number Sequences

Use skip counting to continue the sequences on this page. Add on the same number each time.

1
a 1, 2, 3, 4, ___, ___, ___, ___, ___, ___, ___
b 2, 4, 6, 8, ___, ___, ___, ___, ___, ___, ___
c 3, 6, 9, 12, ___, ___, ___, ___, ___, ___, ___
d 4, 8, 12, 16, ___, ___, ___, ___, ___, ___, ___
e 5, 10, 15, 20, ___, ___, ___, ___, ___, ___, ___
f 6, 12, 18, 24, ___, ___, ___, ___, ___, ___, ___
g 7, 14, 21, 28, ___, ___, ___, ___, ___, ___, ___
h 8, 16, 24, 32, ___, ___, ___, ___, ___, ___, ___
i 9, 18, 27, 36, ___, ___, ___, ___, ___, ___, ___
j 10, 20, 30, 40, ___, ___, ___, ___, ___, ___, ___

2
a 9, 10, 11, ___, ___
b 18, 20, 22, ___, ___
c 27, 30, 33, ___, ___
d 36, 40, 44, ___, ___
e 45, 50, 55, ___, ___
f 54, 60, 66, ___, ___
g 63, 70, 77, ___, ___
h 72, 80, 88, ___, ___
i 81, 90, 99, ___, ___
j 90, 100, 110, ___, ___

3 Consider your answers to Question 1. Explain the pattern made by the last digits of each number in part:

a b ___
b d ___
c e ___
d f ___
e h ___
f i ___
g j ___

4 Make two number sequences of your own.

a ___, ___, ___, ___
b ___, ___, ___, ___

 • *AUSTRALIAN SIGNPOST MATHS 4* • ISBN 978 1 4886 2187 1

2:02 Multiplication Tables Revision

Step 1: Have someone test you.

Step 2: For each table you don't know, make a card with the question on one side and the answer on the other.

Step 3: Carry these cards with you, testing yourself until you know them.

Use these steps to learn your 1, 2, 3, 5 and 10 times tables.

CONCEPT

1 Use skip counting to complete.

×	0	1	2	3	4	5	6	7	8	9	10
1											
2											
3											
5											
10											

2 Make up a number story for each question and then find the answers.

a 6 × 2 ☐ **b** 4 × 5 ☐

c 3 × 0 ☐ **d** 5 × 2 ☐

e 3 × 10 ☐ **f** 9 × 1 ☐

g 8 × 0 ☐ **h** 5 × 5 ☐

3 Complete these number wheels.

a

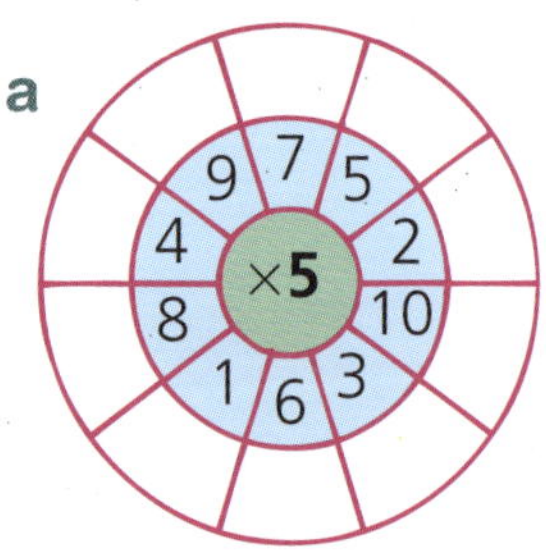

b

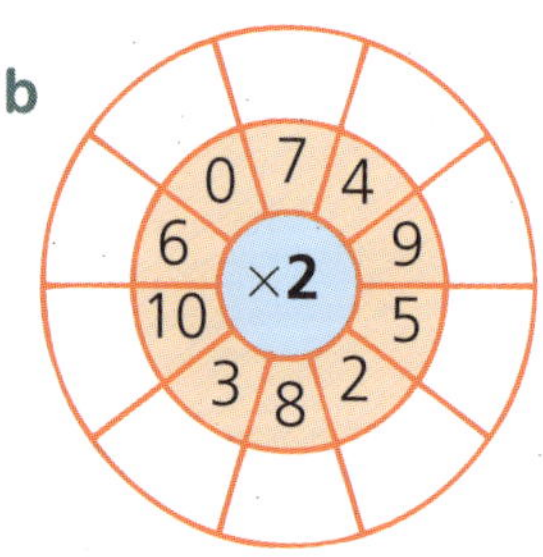

c

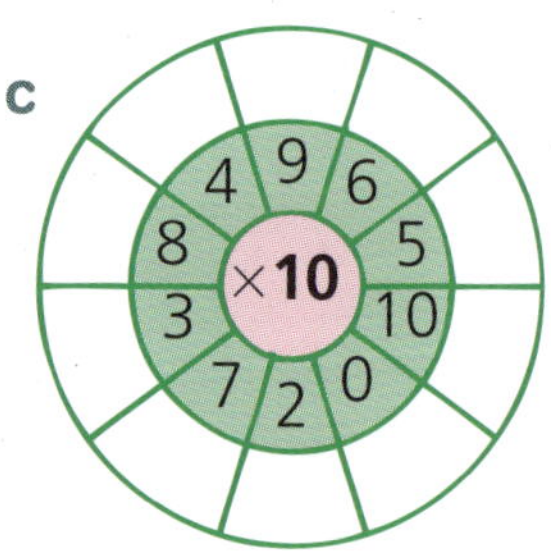

d

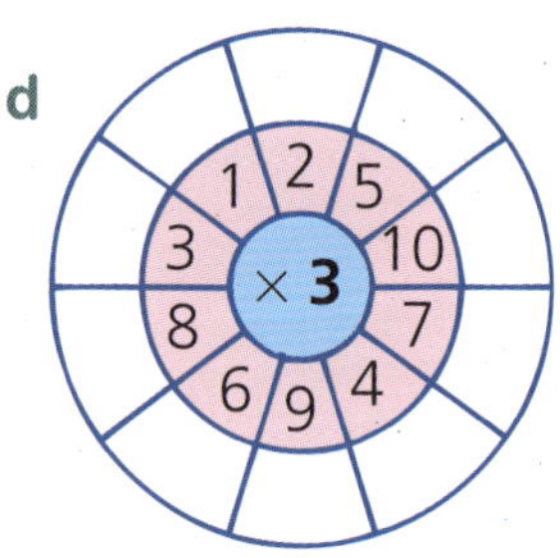

4

a $\begin{array}{r} 1 \\ \times\ 4 \\ \hline \end{array}$ **b** $\begin{array}{r} 5 \\ \times\ 6 \\ \hline \end{array}$ **c** $\begin{array}{r} 10 \\ \times\ 7 \\ \hline \end{array}$ **d** $\begin{array}{r} 2 \\ \times\ 5 \\ \hline \end{array}$ **e** $\begin{array}{r} 5 \\ \times\ 7 \\ \hline \end{array}$

f $\begin{array}{r} 2 \\ \times\ 6 \\ \hline \end{array}$ **g** $\begin{array}{r} 1 \\ \times\ 8 \\ \hline \end{array}$ **h** $\begin{array}{r} 10 \\ \times\ 6 \\ \hline \end{array}$ **i** $\begin{array}{r} 2 \\ \times\ 7 \\ \hline \end{array}$ **j** $\begin{array}{r} 0 \\ \times\ 6 \\ \hline \end{array}$

Multiplication Cards

ACTIVITY

- Cards marked 1 to 10 are placed face down in a pile.
- One card is turned at a time. The first to correctly multiply the card by 5, keeps the card. The player with the most cards wins.

 • *AUSTRALIAN SIGNPOST MATHS 4* • ISBN 978 1 4886 2187 1

2:03 Multiplication Using Pictures

1 Use the pictures to complete each number sequence.

a 5 groups of 8 hats

___ × ___ = ___

b 3 groups of 14 hats

___ × ___ = ___

c 6 groups of 9 buttons

___ × ___ = ___

d 16 groups of 3 buttons

___ × ___ = ___

e 7 groups of 6 drops

___ × ___ = ___

f 4 groups of 18 drops

___ × ___ = ___

g 2 groups of 36 drops

___ × ___ = ___

h 3 groups of 24 drops

___ × ___ = ___

2

Use the pictures to find:

a 4 groups of 5 fish

___ × ___ = ___

b 6 groups of 5 fish

___ × ___ = ___

c 3 × 5 = ___

d 3 groups of 13 ones

___ × ___ = ___

e 5 groups of 13 ones

___ × ___ = ___

f 2 × 13 = ___

g 2 groups of 6 tens

___ × ___ = ___

h 1 group of 6 tens

___ × ___ = ___

i 0 × 60 = ___

j 8 groups of 5 = ___

___ × ___ = ___

k 4 groups of 13 = ___

___ × ___ = ___

l 1 × 13 = ___

2:04 Number Facts, × 2, × 4

1 Complete these tables using the models.

a $1 \times 2 =$ ☐

b $2 \times 2 =$ ☐

c $3 \times 2 =$ ☐

d $4 \times 2 =$ ☐

e $5 \times 2 =$ ☐

f $6 \times 2 =$ ☐

g $7 \times 2 =$ ☐

h $8 \times 2 =$ ☐

i $9 \times 2 =$ ☐

j $10 \times 2 =$ ☐

k $1 \times 4 =$ ☐

l $2 \times 4 =$ ☐

m $3 \times 4 =$ ☐

n $4 \times 4 =$ ☐

o $5 \times 4 =$ ☐

2 Complete these tables using the pictures.

a

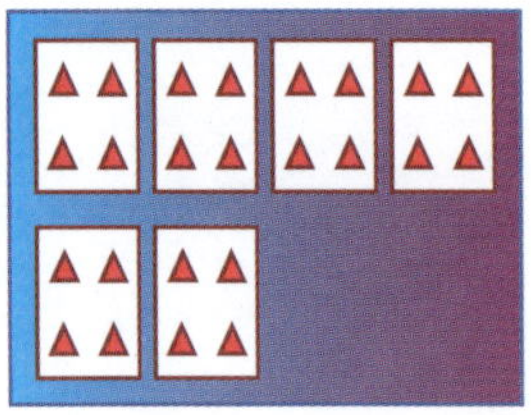

6 groups of 4 = ☐

b

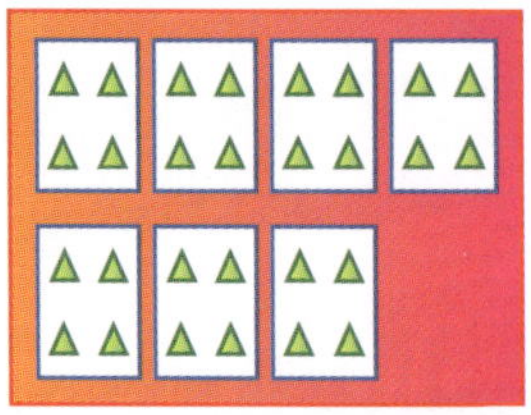

7 groups of 4 = ☐

c

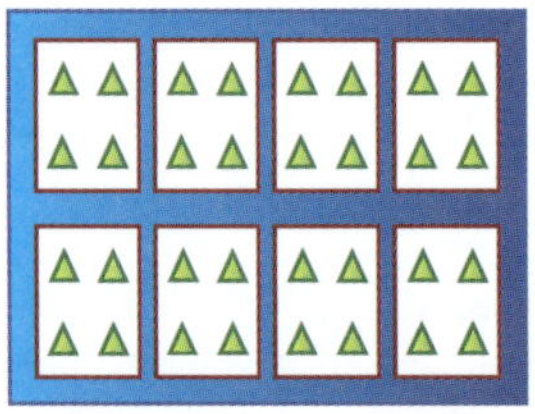

8 groups of 4 = ☐

d

9 groups of 4 = ☐

e

10 groups of 4 = ☐

3 Try to complete these from memory.

a

×	3	10	2	1	5	4	7	9	8	6
2										

b

×	3	10	2	1	5	4	7	9	8	6
4										

c

d

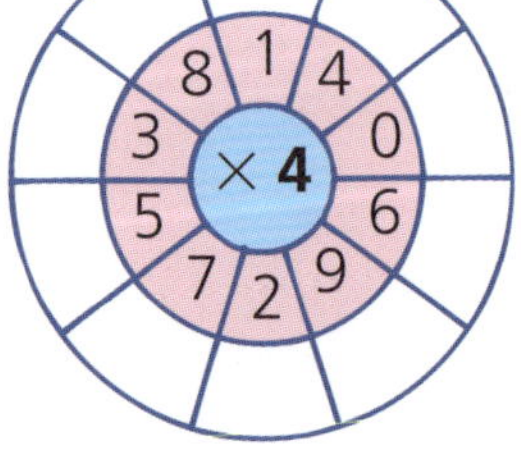

 • *AUSTRALIAN SIGNPOST MATHS 4* • ISBN 978 1 4886 2187 1

2:05 Number Facts, × 2, × 4

CONCEPT

The × 4 tables are double the × 2 tables.

5 × 2 = 10

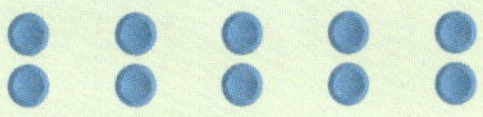

5 × 4 = 20

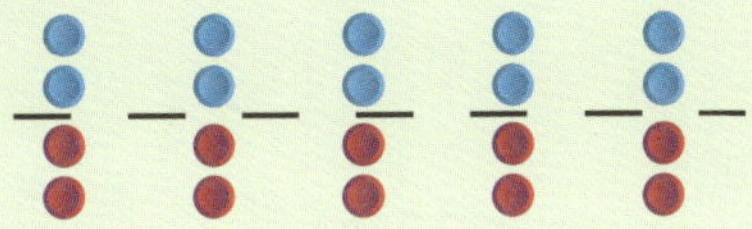

5 × 4 is double 5 × 2.

1 The top line of pictures shows the two times tables. Double this to find the four times tables.

a 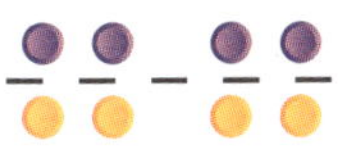

2 × 2 = ☐

2 × 4 = ☐

b 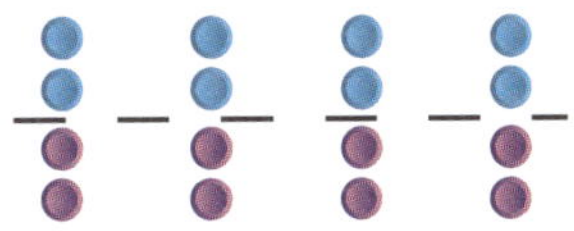

4 × 2 = ☐

4 × 4 = ☐

c 

3 × 2 = ☐

3 × 4 = ☐

d 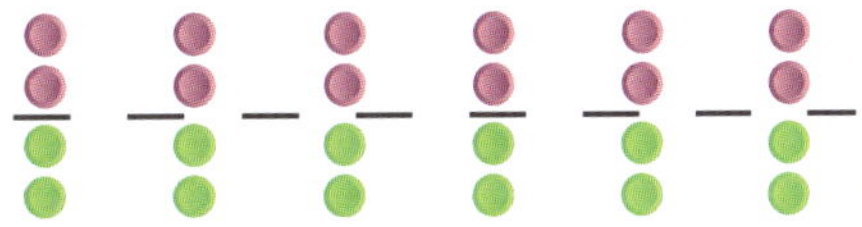

6 × 2 = ☐

6 × 4 = ☐

2 Use your answers for × 2 to find the answers for × 4.

a 1 × 2 = ☐
1 × 4 = ☐

b 7 × 2 = ☐
7 × 4 = ☐

c 10 × 2 = ☐
10 × 4 = ☐

d 8 × 2 = ☐
8 × 4 = ☐

3 Complete the × 2 tables first. Use your answers for the × 2 tables to answer the × 4 tables.

×	6	4	2	8	1	5	0	10	3	7	9
2											
4											

FUN SPOT

Circle Tables Challenge

- Students sit in a circle. One student stands behind someone in the circle.
- The student standing challenges the student sitting in front of them to multiply the number on a dice by 4. The dice is then thrown.
- The first one to answer correctly continues around the circle, standing behind the next person in the circle to challenge them. Dotted dice can be used to help students. Use your two times tables to help you.

2:06 Addition to 99, No Trading

CONCEPT

I Show	I Think	I Write
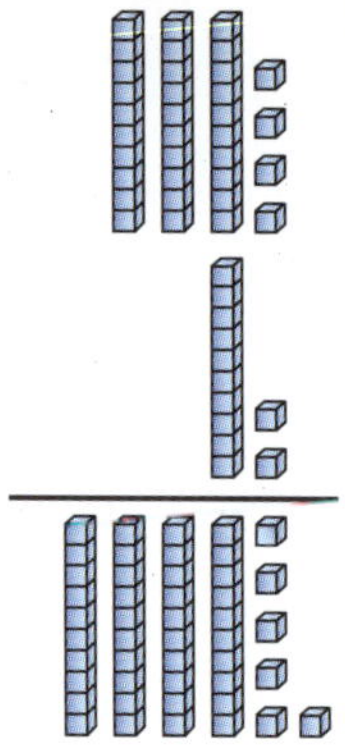	3 tens and 4 ones + 1 ten and 2 ones = 4 tens and 6 ones	Tens | Ones 3 | 4 + 1 | 2 4 | 6

1 Use the split strategy or place-value blocks to answer these.

a

Tens	Ones
9	2
+	6

b

Tens	Ones
4	2
+ 5	1

c

Tens	Ones
3	5
+ 6	2

d

Tens	Ones
1	5
+ 2	3

e

Tens	Ones
8	1
+ 1	2

f

Tens	Ones
1	4
+ 2	4

g

Tens	Ones
1	7
+ 3	0

h

Tens	Ones
5	0
+ 2	7

i

Tens	Ones
$5	0
+ $1	9

j

Tens	Ones
$2	3
+ $1	5

k

Tens	Ones
$4	4
+ $4	4

l

Tens	Ones
$2	7
+ $5	1

2

a 7 tens and 2 ones
\+ 2 tens and 3 ones

b 6 tens and 0 ones
\+ 2 tens and 9 ones

c 3 tens and 2 ones
\+ 5 tens and 6 ones

d 3 tens and 4 ones
\+ 1 ten and 2 ones

e 5 tens and 3 ones
\+ 2 tens and 5 ones

f 2 tens and 1 one
\+ 4 tens and 8 ones

3 **a** I paid $23 for a shirt and $56 for pants. How much did I spend? ☐

b There are 14 boys and 13 girls in our class. How many in our class? ☐

c 14 horses and 32 cows are on our farm. How many animals altogether? ☐

d I saved $41. Alana saved $37. How much did we save? ☐

 ISBN 978 1 4886 2187 1

2:07 Addition, No Trading

CONCEPT

I Show	I Think	I Write
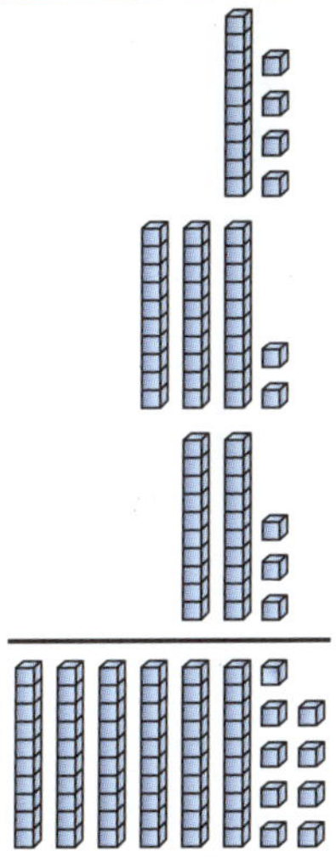	1 ten and 4 ones + 3 tens and 2 ones + 2 tens and 3 ones = 6 tens and 9 ones	Tens | Ones 1 | 4 3 | 2 + 2 | 3 **6** | **9**

1 Model the question, write it as an algorithm, then find the answer.

a 23 candles
11 candles
54 candles
How many altogether?

Tens	Ones

b 30 lollies
25 lollies
14 lollies
How many altogether?

Tens	Ones

2 Use the split strategy or place-value blocks to answer these.

a

Tens	Ones
3	1
1	6
+ 2	2

b

Tens	Ones
2	3
4	2
+ 1	0

c

Tens	Ones
4	1
2	2
+ 1	3

d

Tens	Ones
2	4
1	4
+ 3	1

e

Tens	Ones
3	0
1	4
+ 3	2

f

Tens	Ones
4	2
1	5
+ 2	1

g

Tens	Ones
2	5
3	1
+ 1	3

h

Tens	Ones
3	2
4	1
+ 1	5

i

Tens	Ones
$5	1
$1	4
+ $2	2

j

Tens	Ones
$3	3
$4	0
+ $1	5

k

Tens	Ones
$2	0
$3	6
+ $2	1

l

Tens	Ones
$4	3
$3	1
+ $2	5

2:08 Addition to 99 with Trading

Thirty-eight ducks and twenty-six chickens were in the yard. How many birds were there altogether?

Find:
How many birds?

Number sentence:
38 + 26 = ☐

Answer:
64 birds were in the yard.

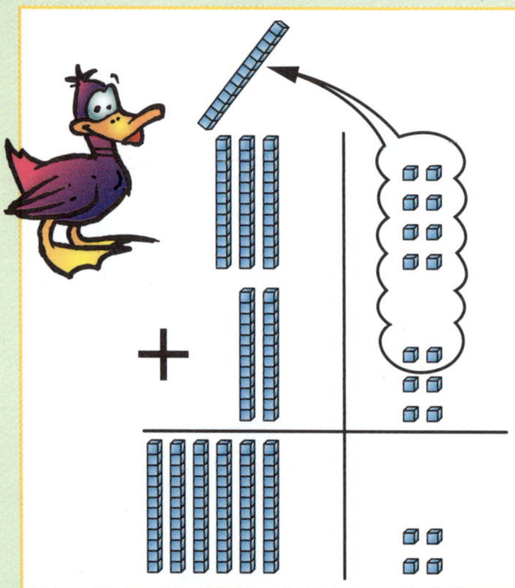

	Tens	Ones
	1	
	3	8
+	2	6
	6	4

CONCEPT

Show the trading on these algorithms.

1

a

Tens	Ones
2	2
+ 1	9

b

Tens	Ones
3	3
+ 5	7

c

Tens	Ones
1	5
+ 5	8

d

Tens	Ones
2	6
+ 5	8

e

Tens	Ones
1	6
+ 2	7

f

Tens	Ones
4	4
+ 3	8

g

Tens	Ones
6	2
+ 2	8

h

Tens	Ones
7	6
+ 1	9

i

Tens	Ones
	3
+ 8	9

j

Tens	Ones
8	1
+	9

k

Tens	Ones
2	5
+ 2	6

l

Tens	Ones
5	7
+ 2	7

2

a

Tens	Ones
$6	7
+ $	4

b

Tens	Ones
$	8
+ $5	4

c

Tens	Ones
$2	8
+ $3	8

d

Tens	Ones
$4	5
+ $1	7

 ISBN 978 1 4886 2187 1

2:09 Addition to 99 with Trading

1

a

	Tens	Ones
	4	8
+	4	6

b

	Tens	Ones
	7	4
+	1	6

c

	Tens	Ones
	2	2
+	7	5

d

	Tens	Ones
	1	5
+	5	9

e

	Tens	Ones
	8	6
+	1	3

f

	Tens	Ones
	4	9
+	3	7

g

	Tens	Ones
	5	7
+	2	8

h

	Tens	Ones
	2	6
+	6	6

i

	Tens	Ones
	$6	4
+	$2	9

j

	Tens	Ones
	$1	8
+	$5	3

k

	Tens	Ones
	$3	8
+	$3	8

l

	Tens	Ones
	$4	9
+	$2	2

Example

How many rulers were in the classroom if 27 were on the desks and 35 were still in the cupboard?

Find:

How many rulers?

Number sentence:

27 + 35 = ☐

Answer:

There were 62 rulers in the classroom.

Working:

	Tens	Ones
	1	
	2	7
+	3	5
	6	2

CONCEPT

2 Answer these questions, setting them out as shown above.

a A group of emus has 13 females and 17 males. How many emus are there altogether?

Answer: ☐ emus

Tens	Ones

b There were 37 kiwis in the zoo. 46 more were hatched. How many kiwis were in the zoo?

Answer: ☐ kiwis

Tens	Ones

c A woman needed 20 metres of pink ribbon and 34 metres of blue ribbon. How much ribbon did she need?

Answer: ☐ metres

Tens	Ones

d 28 finches and 12 doves were kept in a large aviary. How many birds were there altogether?

Answer: ☐ birds

Tens	Ones

2:10 Addition with 2-Digit Numbers

CONCEPT

23 + 24 + 17

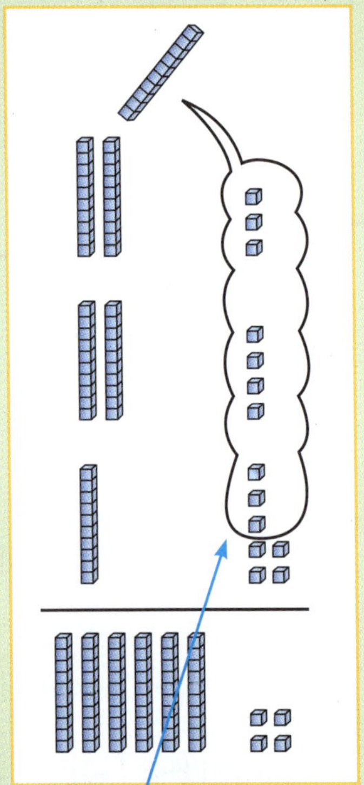

Trade 10 ones for one ten.

Record trading like this.

Tens	Ones
1	
2	3
2	4
+ 1	7
6	4

Check your answers with a calculator.

1 a

Tens	Ones
1	3
2	4
+ 2	5

b

Tens	Ones
2	6
3	3
+ 1	5

c

Tens	Ones
3	4
2	7
+ 3	0

d

Tens	Ones
2	3
1	4
+ 3	6

e

Tens	Ones
4	4
3	4
+ 1	8

f

Tens	Ones
3	5
2	7
+ 1	2

2 a

Tens	Ones
3	9
2	4
+ 1	3

b

Tens	Ones
2	8
3	6
+ 1	5

c

Tens	Ones
2	7
3	7
+ 2	8

d

Tens	Ones
3	7
2	9
+ 1	6

e

Tens	Ones
$1	4
$1	8
$2	6
+ $3	3

f

Tens	Ones
$1	2
$2	5
$1	7
+ $2	6

3 Adding 3 odd numbers gives an ______ number.

INVESTIGATION

2:11 Addition Involving Hundreds

CONCEPT

I Show

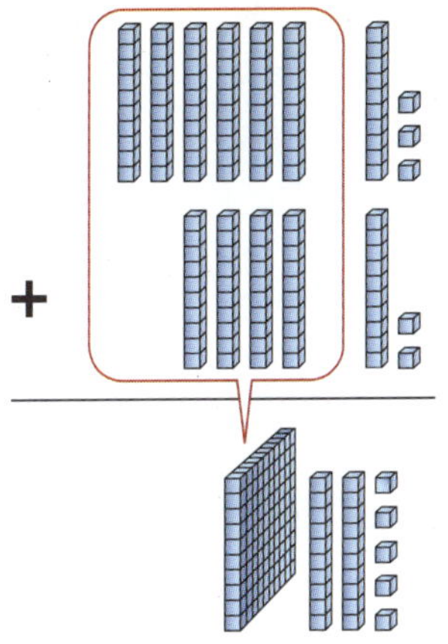

I Write

	Hund	Tens	Ones
	1		
		7	3
+		5	2
	1	2	5

1 Model these with place-value blocks, then fill in the cards.

a 61 books
88 books
☐ + ☐ = ☐
How many books altogether? ☐

b 94 stickers
53 stickers
☐ + ☐ = ☐
How many stickers altogether? ☐

2

a

	Hund	Tens	Ones
		5	8
+		6	0

b

	Hund	Tens	Ones
		9	2
+		4	6

c

	Hund	Tens	Ones
		8	5
+		7	2

d

	Hund	Tens	Ones
		6	3
+		4	3

e

	Hund	Tens	Ones
		7	4
+		8	5

f

	Hund	Tens	Ones
		3	7
+		9	1

g

	Hund	Tens	Ones
		9	4
+		6	1

h

	Hund	Tens	Ones
		6	3
+		6	4

i

	Hund	Tens	Ones
		7	2
+		5	7

2:12 Number Facts, × 8

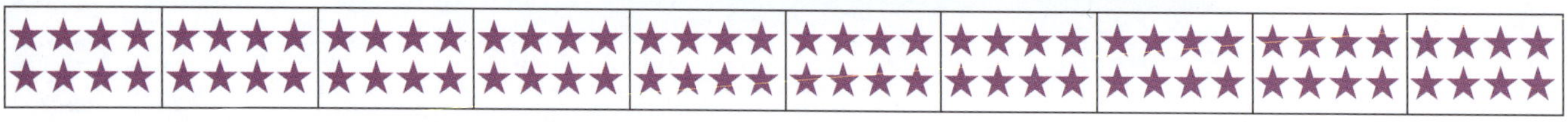

1 Count the groups of 8 to fill in the boxes.

a 1 × 8 = ☐ **b** 2 × 8 = ☐ **c** 3 × 8 = ☐
d 4 × 8 = ☐ **e** 5 × 8 = ☐ **f** 6 × 8 = ☐
g 7 × 8 = ☐ **h** 8 × 8 = ☐ **i** 9 × 8 = ☐

Doubling the answers to '×4' gives the answers to '×8'.

j 10 × 4 = ☐ so 10 × 8 = ☐ **k** 1 × 4 = ☐ so 1 × 8 = ☐
l 2 × 4 = ☐ so 2 × 8 = ☐ **m** 3 × 4 = ☐ so 3 × 8 = ☐
n 6 × 4 = ☐ so 6 × 8 = ☐ **o** 5 × 4 = ☐ so 5 × 8 = ☐

2 Complete:

a 3 groups of 8 = ☐ **b** 6 rows of 8 = ☐ **c** 5 eights = ☐
d 4 groups of 8 = ☐ **e** 7 lots of 8 = ☐ **f** 8 eights = ☐
g 1 group of 8 = ☐ **h** 9 rows of 8 = ☐ **i** 2 × 8 = ☐
j 0 groups of 8 = ☐ **k** 10 rows of 8 = ☐ **l** 11 × 8 = ☐

3 Try to complete these tables from memory.

a

×	4	0	2	8	3	7	5	9	1	6
2										

b

×	5	1	9	8	2	10	6	4	0	7
3										

c

×	4	0	2	8	3	7	5	9	1	6
4										

d

×	4	2	8	10	5	9	6	1	7	3
5										

e

×	4	0	2	8	3	7	5	9	1	6
8										

f

×	4	2	8	10	5	9	6	1	7	3
10										

The Constant Function on a Calculator

ICT

- Press the keys [8] [×] [×] [1] [=] [2] [=] [3] [=] [4] [=] [5] [=] …
  Did the eight times table come into the display?
- Press [5] [×] [×] [1] [=] [2] [=] and so on, to get the five times table.
- Use the same method to show other tables on your calculator.

 • *AUSTRALIAN SIGNPOST MATHS 4* • ISBN 978 1 4886 2187 1

2:13 Number Facts, × 2, × 4, × 8

INVESTIGATION

Hundred Chart

1	2	3	4	5	6	7	8	9	10
11	12	13	14	15	16	17	18	19	20
21	22	23	24	25	26	27	28	29	30
31	32	33	34	35	36	37	38	39	40
41	42	43	44	45	46	47	48	49	50
51	52	53	54	55	56	57	58	59	60
61	62	63	64	65	66	67	68	69	70
71	72	73	74	75	76	77	78	79	80
81	82	83	84	85	86	87	88	89	90
91	92	93	94	95	96	97	98	99	100

1. a Circle the table of twos.
   b Cross the table of fours.
   c Colour the table of eights.
   d × 4 is ______ × 2 table.
   e × 8 is ______ × 4.
   f Are all the multiples of 8 coloured? ______

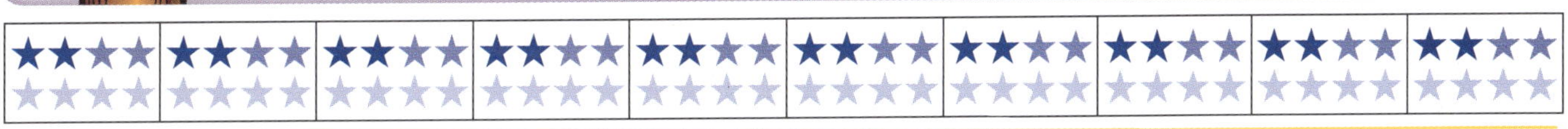

2. Use the groups of 2, 4 and 8 above to find your answers. Use the doubling strategy to help you.

a 6 × 2 = ______ b 2 × 2 = ______ c 3 × 2 = ______ d 5 × 2 = ______
6 × 4 = ______ 2 × 4 = ______ 3 × 4 = ______ 5 × 4 = ______
6 × 8 = ______ 2 × 8 = ______ 3 × 8 = ______ 5 × 8 = ______

3. Complete the × 2 tables first. Use your answers for the × 2 tables to answer the × 4 and × 8 tables.

×	10	3	2	7	0	6	1	5	4	9	8
2											
4											
8											

← × 4 is double × 2.
← × 8 is double × 4.

× 4 and × 8 tables are easy when we use doubling.

4. Race a friend to complete these. Use mental strategies to find your answer quickly.

a 4 × 8 = ______ b 1 × 2 = ______ c 9 × 2 = ______ d 10 × 2 = ______
e 7 × 8 = ______ f 0 × 8 = ______ g 7 × 4 = ______ h 2 × 8 = ______

Table Cards

FUN SPOT

- 2–4 players use cards marked 0–10 face down in a pile. A card is turned.
- The first student to multiply the card by 4 keeps the card. The person with the most cards wins.
- **Extension**: Multiply the cards by 8.

2:14 Subtraction, No Trading

CONCEPT

1 Create a number sentence and model each word problem using place-value blocks.

a 38 badges, 7 sold.
How many are left?

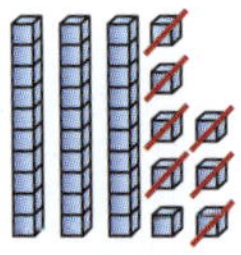

[=]

b 36 girls, 4 boys.
How many more girls than boys?

[=]

c 47 birds, 35 fly away.
How many remain?

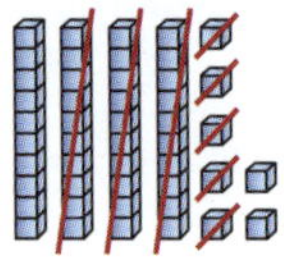

[=]

d 74 needed, 61 collected.
How many more to collect?

[=]

e 56 books, 34 covered.
How many more to cover?

[=]

f 35 lollies, 22 eaten.
How many are left?

[=]

2

a

Tens	Ones
5	7
– 3	4

b

Tens	Ones
5	6
– 2	1

c

Tens	Ones
4	7
– 1	5

d

Tens	Ones
7	8
– 4	1

e

Tens	Ones
7	5
– 7	4

f

Tens	Ones
6	9
– 4	3

g

Tens	Ones
4	4
– 1	1

h

Tens	Ones
6	3
– 5	3

i

Tens	Ones
$8	4
– $8	0

j

Tens	Ones
$7	3
– $4	3

k

Tens	Ones
$9	9
– $3	2

l

Tens	Ones
$6	8
– $2	4

- Use a calculator to check your answers.

2:15 Subtraction with Trading

Trading Down

1 34 cakes, 18 eaten. How many left?

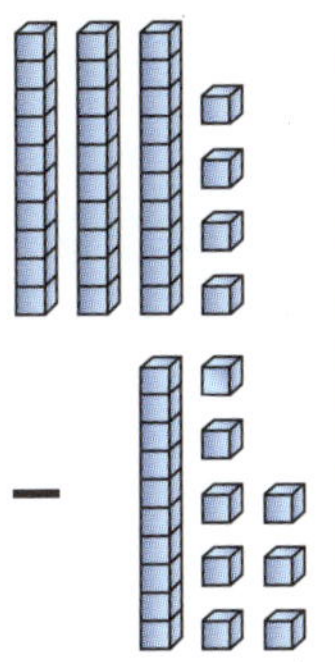

Tens	Ones
3	4
− 1	8

2 I can't take 8 ones from 4 ones. Trade 1 ten for 10 ones.

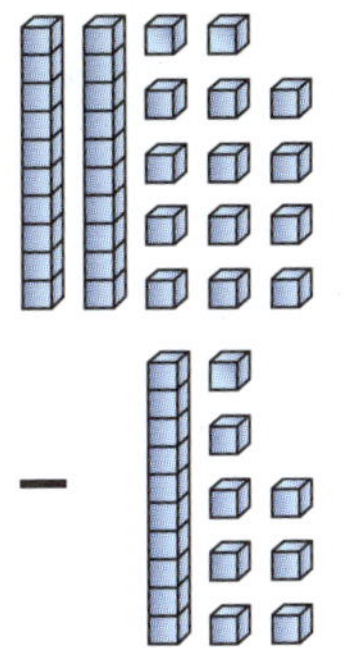

Tens	Ones
2	14
~~3~~	~~4~~
− 1	8

3 14 ones take away 8 ones leaves 6 ones. 2 tens take away 1 ten leaves 1 ten.

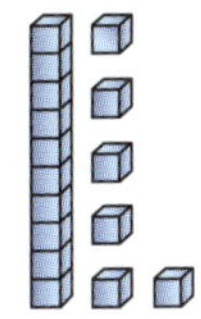

Answer
16 cakes were left uneaten.

Tens	Ones
2	14
~~3~~	~~4~~
− 1	8
1	6

Check by adding the answer to the bottom row.

1 You may use place-value blocks to do these. Estimate your answer first.

a

Tens	Ones
4	3
− 1	7

b

Tens	Ones
6	5
− 2	9

c

Tens	Ones
9	1
− 5	4

d

Tens	Ones
7	2
− 2	8

e

Tens	Ones
4	6
− 2	7

f

Tens	Ones
5	6
− 3	8

g

Tens	Ones
9	1
− 1	8

h

Tens	Ones
8	6
− 3	9

i

Tens	Ones
7	2
− 3	4

j

Tens	Ones
6	1
− 4	3

2 Explore ways to find the missing numbers.

a 76 − ☐ = 24

b 60 − ☐ = 18

c 83 − ☐ = 26

d 75 − ☐ = 38

e 90 − ☐ = 19

f ☐ − 35 = 46 **g** ☐ − 57 = 42 **h** ☐ − 28 = 43

2:16 Subtraction with Trading to 99

CONCEPT

Tens	Ones
5	10
~~6~~	~~0~~
−	8
5	2

Tens	Ones
5	10
~~6~~	~~0~~
− 3	8
2	2

1 Show trading down to 1 ten.

a Tens Ones: 5 | 0

b Tens Ones: 7 | 0

c Tens Ones: 4 | 0

d Tens Ones: 9 | 0

2 Use the method above or place-value blocks to answer these. Use your calculator to check each.

a Tens Ones: 6 0 − 4

b Tens Ones: 8 0 − 7

c Tens Ones: 5 0 − 6

d Tens Ones: 9 0 − 8

e Tens Ones: \$7 0 − \$5 6

f Tens Ones: \$4 0 − \$2 7

g Tens Ones: \$8 0 − \$3 5

h Tens Ones: \$9 0 − \$4 2

3 Explore ways to find the missing numbers then fill in the boxes.

ICT

a 40 − ☐ = 21

b 70 − ☐ = 27

c 60 − ☐ = 34

d 90 − ☐ = 49

e 80 − ☐ = 42

f ☐ − 23 = 27

g ☐ − 45 = 35

h ☐ − 17 = 43

i ☐ − 38 = 12

j ☐ − 61 = 29

2:17 Subtraction with Trading to 99

CONCEPT

1

a

Tens	Ones
3	1
− 2	5

b

Tens	Ones
2	3
− 1	6

c

Tens	Ones
4	6
− 3	8

d

Tens	Ones
3	4
− 2	7

e

Tens	Ones
5	2
− 3	4

f

Tens	Ones
7	5
− 5	9

g

Tens	Ones
6	7
− 4	8

h

Tens	Ones
5	2
− 3	7

i

Tens	Ones
8	4
− 6	8

j

Tens	Ones
6	8
− 4	9

k

Tens	Ones
9	1
− 5	3

l

Tens	Ones
7	3
− 2	9

m

Tens	Ones
3	7
−	9

n

Tens	Ones
4	4
− 1	5

o

Tens	Ones
3	8
− 1	9

p

Tens	Ones
6	2
−	7

q

Tens	Ones
$5	5
− $2	9

r

Tens	Ones
$7	2
− $5	4

s

Tens	Ones
$8	0
− $4	3

ICT

Use a calculator to check your answers.

2:18 Subtraction with Trading to 99

1

a

Tens	Ones
6	0
–	8

b

Tens	Ones
5	0
–	4

c

Tens	Ones
9	0
–	7

d

Tens	Ones
4	0
– 1	6

e

Tens	Ones
3	0
– 2	1

f

Tens	Ones
7	0
– 3	5

g

Tens	Ones
8	0
– 6	2

h

Tens	Ones
2	0
– 1	3

i

Tens	Ones
6	0
– 4	4

j

Tens	Ones
9	0
– 5	9

2

a

Tens	Ones
8	7
–	8

b

Tens	Ones
3	2
–	4

c

Tens	Ones
6	5
–	7

d

Tens	Ones
5	1
–	6

e

8	1
– 2	6

f

3	7
– 1	8

g

5	2
– 2	7

h

\$4	6
– \$2	9

i

\$6	5
– \$4	8

j

\$7	1
– \$3	6

3 Use the jump strategy to answer these questions.

a 52 – 34 = ☐

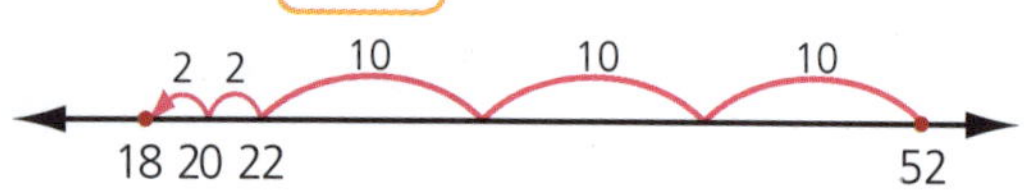

b 73 – 25 = ☐

73

c 81 – 45 = ☐

d 62 – 27 = ☐

e 44 – 16 = ☐

2:19 Addition Problems to 99

Problem Solving
How many pencils does Molly have if she has 24 in her bag and 47 in her desk?

Find: How many pencils?

Number sentence: 24 + 47 = ☐

Answer: Molly has 71 pencils.

Working

Tens	Ones
1	
2	4
+ 4	7
7	1

CONCEPT

1 Answer these questions, setting them out as shown above.

a Luke's test had 27 mistakes. Naomi's had 22 mistakes. How many mistakes do they have together? ☐ mistakes

b Brianna had 56 pet ants. Jordan caught 8 more and gave them to her. How many did she have then? ☐ ants

c Wen, an ancient Chinese king, began the first zoo 3 000 years ago. He received 56 animals from the north and 27 from the south. How many animals did he receive altogether? ☐ animals

d At night, an owl can see about 100 times better than a human. In one week an owl caught 53 mice. In the next week it caught 38. How many mice did it catch altogether? ☐ mice

e At a waterhole, Michelle photographed 31 magpie geese, 12 Burdekin ducks and 8 pied herons. How many birds did she photograph altogether? ☐ birds

f A family of 18 bandicoots lived near 13 possums and 6 native cats. How many animals were there altogether? ☐ animals

Tens Ones

Tens Ones

2 a Alan saw three varieties of finch in one paddock. There were 35 zebra finches, 15 double-bar finches and 27 spice finches. How many were there altogether? ☐ finches

b On Phillip Island, 37 penguins came ashore before 6 pm. In the next hour 8 more arrived. How many had arrived by 7 pm? ☐ penguins

c Consecutive numbers follow one after the other. Find the sum of the consecutive numbers 28, 29 and 30. ☐ is the sum

d In a Test cricket series, Eric batted three times. His scores were 44, 28 and 19. What was his total score? ☐ runs

Tens Ones

 • *AUSTRALIAN SIGNPOST MATHS 4* • ISBN 978 1 4886 2187 1

2:20 Number Facts, × 3, × 6

The **× 6** tables are double the **× 3** tables.

5 × 3 = 15 5 × 6 = 30

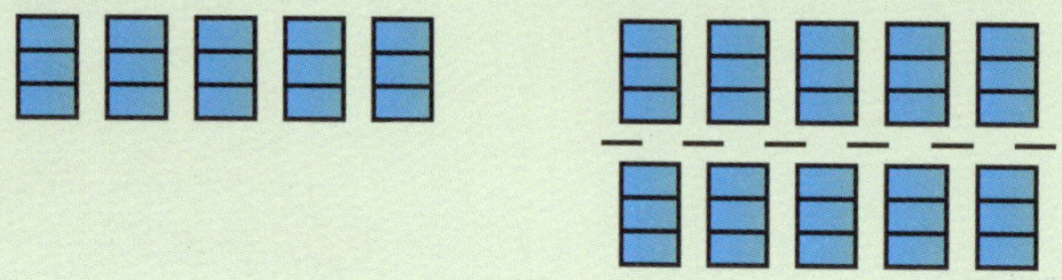

CONCEPT

1 The top line of pictures shows the × 3 tables. Double this to find the × 6 tables.

a 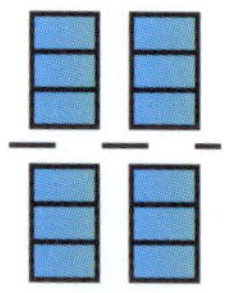 2 × 3 = ☐ 2 × 6 = ☐

b 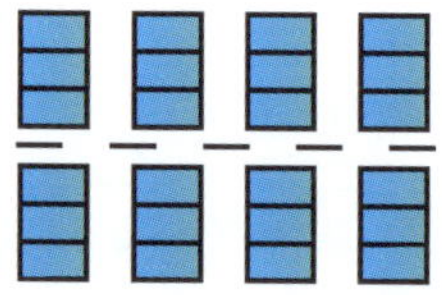 4 × 3 = ☐ 4 × 6 = ☐

c 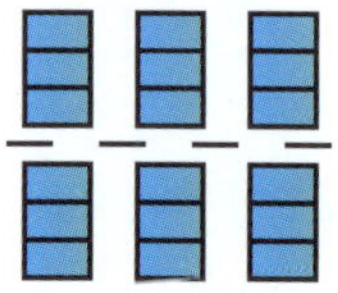 3 × 3 = ☐ 3 × 6 = ☐

d 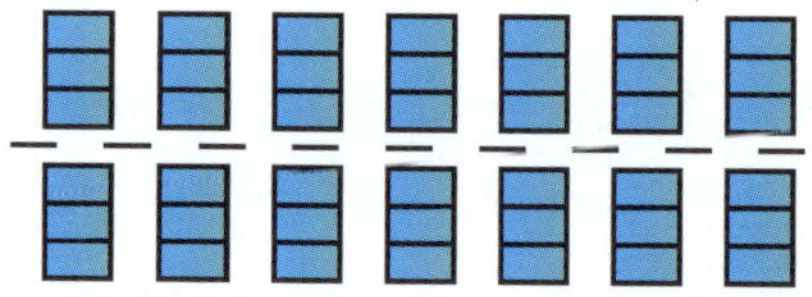 7 × 3 = ☐ 7 × 6 = ☐

2 Use your answers for × 3 to find the answers for × 6.

a 1 × 3 = ☐ 1 × 6 = ☐

b 10 × 3 = ☐ 10 × 6 = ☐

c 8 × 3 = ☐ 8 × 6 = ☐

d 11 × 3 = ☐ 11 × 6 = ☐

3 Complete the × 3 tables first. Use your answers for the × 3 tables to answer the × 6 tables.

×	1	2	3	4	5	6	7	8	9	10
3										
6										

The answers in the × 6 row are multiples of 6.

Table Cards

FUN SPOT

- 2–4 players use cards marked 0–10, placed face down in a pile.
- A card is turned. The first student to multiply the card by 3 keeps the card. The person with the most cards wins.
- **Extension**: Multiply the cards by 6.

 ISBN 978 1 4886 2187 1

2:21 Number Facts, × 3, × 6

1 Count on to complete this pattern of 3. These are multiples of 3.

☐, ☐, ☐, ☐, ☐, ☐, ☐, ☐, ☐, ☐

2 Complete these tables, using the pattern above.

a $3 \times 3 =$ ☐ **b** $5 \times 3 =$ ☐ **c** $2 \times 3 =$ ☐ **d** $1 \times 3 =$ ☐ **e** $4 \times 3 =$ ☐

f $6 \times 3 =$ ☐ **g** $7 \times 3 =$ ☐ **h** $8 \times 3 =$ ☐ **i** $9 \times 3 =$ ☐ **j** $10 \times 3 =$ ☐

3 Count on to complete this pattern of 6. These are multiples of 6.

☐, ☐, ☐, ☐, ☐, ☐, ☐, ☐, ☐, ☐

4 Complete these tables, using the pattern above.

a $3 \times 6 =$ ☐ **b** $5 \times 6 =$ ☐ **c** $2 \times 6 =$ ☐ **d** $1 \times 6 =$ ☐ **e** $4 \times 6 =$ ☐

f $6 \times 6 =$ ☐ **g** $7 \times 6 =$ ☐ **h** $8 \times 6 =$ ☐ **i** $9 \times 6 =$ ☐ **j** $10 \times 6 =$ ☐

5 Use your answers above to answer these questions.

a Why is × 6 double × 3? ☐

b Does $6 \times 3 = 3 \times 6$? ☐ **c** Does $6 \times 3 = 9 \times 2$? ☐

ACTIVITY

12 blocks have been used to make these rectangular arrays. From these we can find multiplication number sentences that make 12.

$1 \times 12 = 12$ $6 \times 2 = 12$ $4 \times 3 = 12$ $3 \times 4 = 12$ $2 \times 6 = 12$ $12 \times 1 = 12$

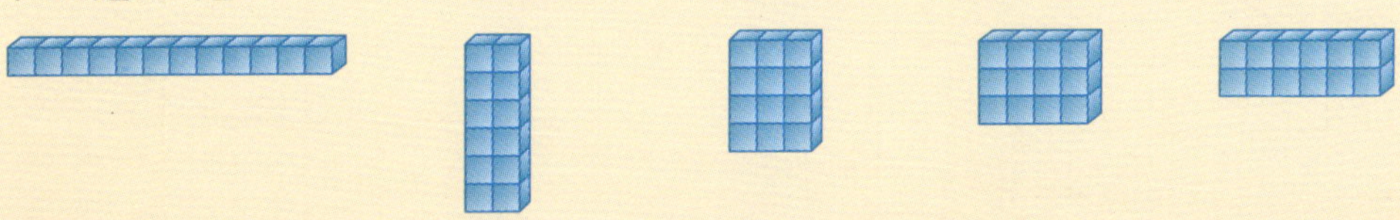

6 Use the number of blocks given to find number sentences that make:

a 24 ☐

b 15 ☐

c 30 ☐

2:22 Number Facts, × 9

1 Count by 9s to count these blocks. These are multiples of 9.

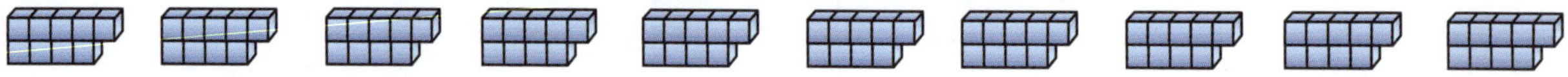

☐, ☐, ☐, ☐, ☐, ☐, ☐, ☐, ☐, ☐

2 Use the blocks above to complete these × 9 tables.

a 1 × 9 = ☐
b 2 × 9 = ☐
c 3 × 9 = ☐
d 4 × 9 = ☐
e 5 × 9 = ☐
f 6 × 9 = ☐
g 7 × 9 = ☐
h 8 × 9 = ☐
i 9 × 9 = ☐
j 10 × 9 = ☐

CONCEPT

Use your fingers to help with your × 9 tables.

For **4** × 9, put down your **4**th finger.

There are 3 fingers to the left and 6 to the right.

The answer is 36.

4 × 9 = 36

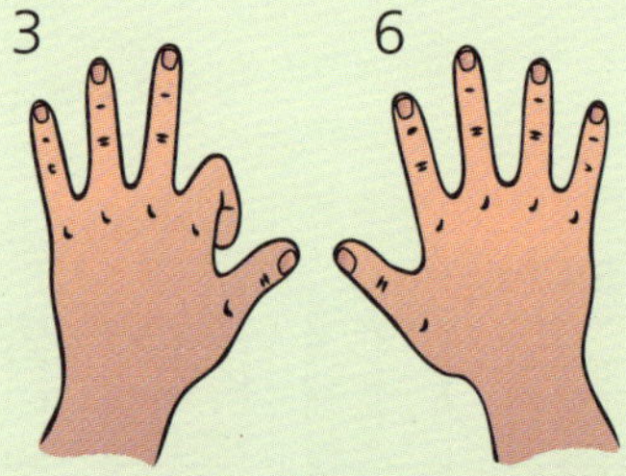

3 Use your fingers to answer these.

a 5 × 9 = ☐ **b** 9 × 9 = ☐ **c** 7 × 9 = ☐
d 8 × 9 = ☐ **e** 4 × 9 = ☐ **f** 1 × 9 = ☐
g 3 × 9 = ☐ **h** 6 × 9 = ☐ **i** 2 × 9 = ☐

ACTIVITY

Hundred Chart

1	2	3	4	5	6	7	8	9	10
11	12	13	14	15	16	17	18	19	20
21	22	23	24	25	26	27	28	29	30
31	32	33	34	35	36	37	38	39	40
41	42	43	44	45	46	47	48	49	50
51	52	53	54	55	56	57	58	59	60
61	62	63	64	65	66	67	68	69	70
71	72	73	74	75	76	77	78	79	80
81	82	83	84	85	86	87	88	89	90
91	92	93	94	95	96	97	98	99	100

4 On this hundred chart, colour the table of nines red and the table of sixes blue.

Multiplication Grid

×	0	1	2	3	4	5	6	7	8	9	10
0											
1											
2											
3											
4											
5											
6											
7											
8											
9											
10											

5 Write in the tables that use 9. Now write in the tables that use 2, 4, 10, 5, 1, 0, 3, 6 and 8. Can you see a pattern in the table of nines?

 ISBN 978 1 4886 2187 1

2:23 Number Facts, × 9

CONCEPT

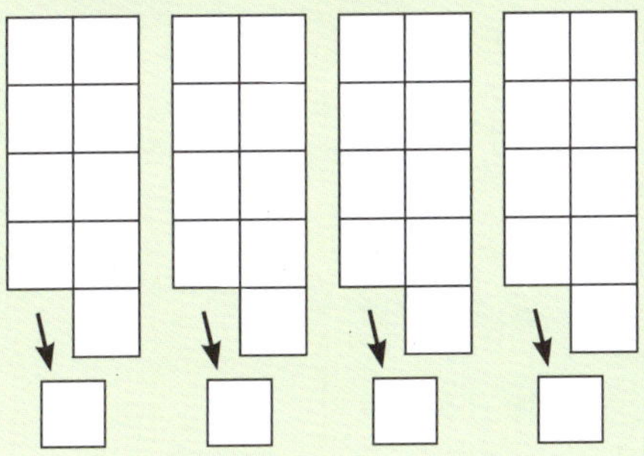

1 Fill in the boxes.

Can you see the patterns?

Add the digits in the answer.

a	1 × 9 =	10	−	1	=	☐	
b	2 × 9 =	20	−	☐	=	☐	, 1 + 8 = ☐
c	3 × 9 =	30	−	☐	=	☐	, 2 + 7 = ☐
d	4 × 9 =	40	−	☐	=	☐	, 3 + 6 = ☐
e	5 × 9 =	50	−	☐	=	☐	, 4 + 5 = ☐
f	6 × 9 =	☐	−	6	=	☐	, 5 + 4 = ☐
g	7 × 9 =	☐	−	7	=	☐	, 6 + 3 = ☐
h	8 × 9 =	☐	−	☐	=	☐	, 7 + 2 = ☐
i	9 × 9 =	☐	−	☐	=	☐	, 8 + 1 = ☐
j	10 × 9 =	☐	−	☐	=	☐	, 9 + 0 = ☐

2 Discuss patterns you can see in the nine times tables.

a What happens to the ones digits in the nines pattern? It decreases by ☐ each time.

b What happens to the tens digits in the nines pattern? It increases by ☐ each time.

Conclusion: To add 9, add ☐ then subtract ☐.

c What do the digits in each multiple of 9 add up to? ☐

d Is 135 a multiple of 9? ☐

e Is 890 a multiple of 9? ☐

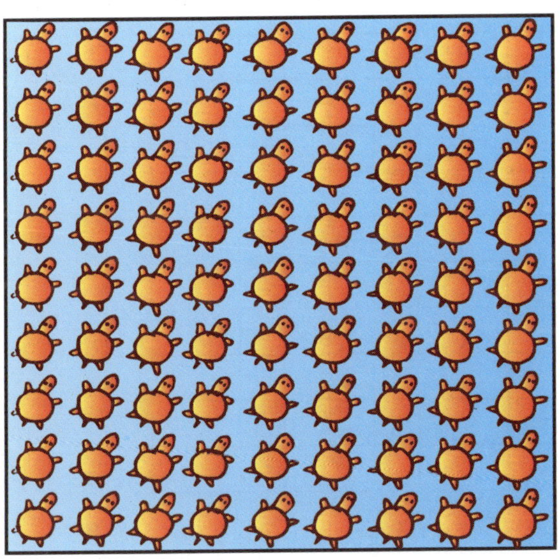

3 Use the picture to answer these questions.

a 1 row of 9 = ☐

b 2 rows of 9 = ☐

c 3 rows of 9 = ☐

d 4 rows of 9 = ☐

e 5 rows of 9 = ☐

f 6 rows of 9 = ☐

g 7 rows of 9 = ☐

h 8 rows of 9 = ☐

i 9 rows of 9 = ☐

j 10 rows of 9 = ☐

k Are 4 rows of 9 equal to 9 rows of 4? ☐

l **True** or **false**: 4 × 9 = 9 × 4? ☐

m Are 5 rows of 9 equal to 9 rows of 5? ☐

2:24 Number Facts, × 7

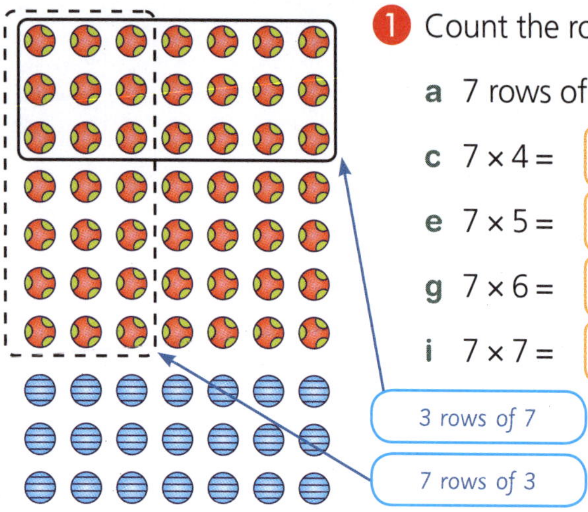

1 Count the rows and fill in the boxes.

a 7 rows of 3 = ☐ **b** 3 rows of 7 = ☐

c 7 × 4 = ☐ **d** 4 × 7 = ☐

e 7 × 5 = ☐ **f** 5 × 7 = ☐

g 7 × 6 = ☐ **h** 6 × 7 = ☐

i 7 × 7 = ☐ **j** 8 × 7 = ☐

k 9 × 7 = ☐

l 10 × 7 = ☐

Changing the order does not change the answer.

2 **a** 7 × 0 = ☐ **b** 0 × 7 = ☐ **c** 7 × 1 = ☐ **d** 1 × 7 = ☐

e 7 × 2 = ☐ **f** 2 × 7 = ☐ **g** 7 × 3 = ☐ **h** 3 × 7 = ☐

i 5 × 8 = ☐ **j** 9 × 8 = ☐ **k** 6 × 8 = ☐ **l** 8 × 8 = ☐

3

a $\begin{array}{r} 7 \\ \times\ 4 \\ \hline \end{array}$ **b** $\begin{array}{r} 7 \\ \times\ 6 \\ \hline \end{array}$ **c** $\begin{array}{r} 7 \\ \times\ 5 \\ \hline \end{array}$ **d** $\begin{array}{r} 7 \\ \times\ 10 \\ \hline \end{array}$ **e** $\begin{array}{r} 7 \\ \times\ 9 \\ \hline \end{array}$

f $\begin{array}{r} 7 \\ \times\ 8 \\ \hline \end{array}$ **g** $\begin{array}{r} 7 \\ \times\ 7 \\ \hline \end{array}$ **h** $\begin{array}{r} 8 \\ \times\ 7 \\ \hline \end{array}$ **i** $\begin{array}{r} 6 \\ \times\ 7 \\ \hline \end{array}$ **j** $\begin{array}{r} 9 \\ \times\ 7 \\ \hline \end{array}$

ACTIVITY

Hundred Chart

1	2	3	4	5	6	7	8	9	10
11	12	13	14	15	16	17	18	19	20
21	22	23	24	25	26	27	28	29	30
31	32	33	34	35	36	37	38	39	40
41	42	43	44	45	46	47	48	49	50
51	52	53	54	55	56	57	58	59	60
61	62	63	64	65	66	67	68	69	70
71	72	73	74	75	76	77	78	79	80
81	82	83	84	85	86	87	88	89	90
91	92	93	94	95	96	97	98	99	100

4 On this hundred chart, colour the table of sevens. Extend the pattern to 98.

Multiplication Grid

×	0	1	2	3	4	5	6	7	8	9	10
0											
1											
2											
3											
4											
5											
6											
7											
8											
9											
10											

5 Write in the tables that use 7. Now fill in the rest of the grid.

2:25 Multiplication Tables Review

1 Complete these multiplication webs.

a

b

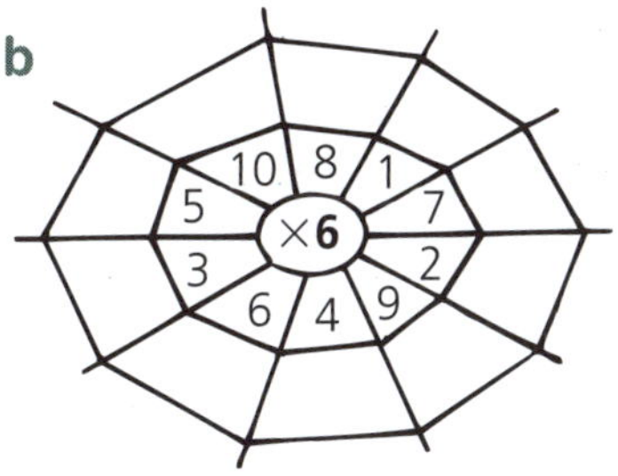

c

Write the answers to these.

2

a 4 × 7 = ☐ **b** 7 × 4 = ☐ **c** 5 × 9 = ☐

d 10 × 9 = ☐ **e** 9 × 10 = ☐ **f** 3 × 9 = ☐

g 6 × 9 = ☐ **h** 9 × 6 = ☐ **i** 4 × 9 = ☐

j 8 × 9 = ☐ **k** 7 × 9 = ☐ **l** 9 × 9 = ☐

m 9 × 3 = ☐ **n** 7 × 7 = ☐ **o** 9 × 8 = ☐

p 9 × 7 = ☐ **q** 9 × 5 = ☐ **r** 9 × 4 = ☐

For x 9 tables, do the digits always add up to 9?

3 × 9 = 27 → 2 + 7 = 9

4 × 9 = 36 → 3 + 6 = 9

3

a $\begin{array}{r} 6 \\ \times\ 2 \\ \hline \end{array}$ **b** $\begin{array}{r} 6 \\ \times\ 5 \\ \hline \end{array}$ **c** $\begin{array}{r} 6 \\ \times\ 3 \\ \hline \end{array}$

d $\begin{array}{r} 3 \\ \times\ 4 \\ \hline \end{array}$ **e** $\begin{array}{r} 3 \\ \times\ 10 \\ \hline \end{array}$ **f** $\begin{array}{r} 3 \\ \times\ 6 \\ \hline \end{array}$

g $\begin{array}{r} 3 \\ \times\ 2 \\ \hline \end{array}$ **h** $\begin{array}{r} 3 \\ \times\ 8 \\ \hline \end{array}$

CONCEPT

The answer to a multiplication question is called the **product**.

4

a The product of 7 and 6 = ☐ **b** The product of 0 and 6 = ☐

c The product of 7 and 3 = ☐ **d** The product of 5 and 9 = ☐

e The product of 6 and 6 = ☐ **f** The product of 8 and 6 = ☐

5

a

×	3	7	4	5	2	10	6	9	8
4									
8									

b

×	3	7	4	5	2	10	6	9	8
6									
7									

2:26 Addition to 999

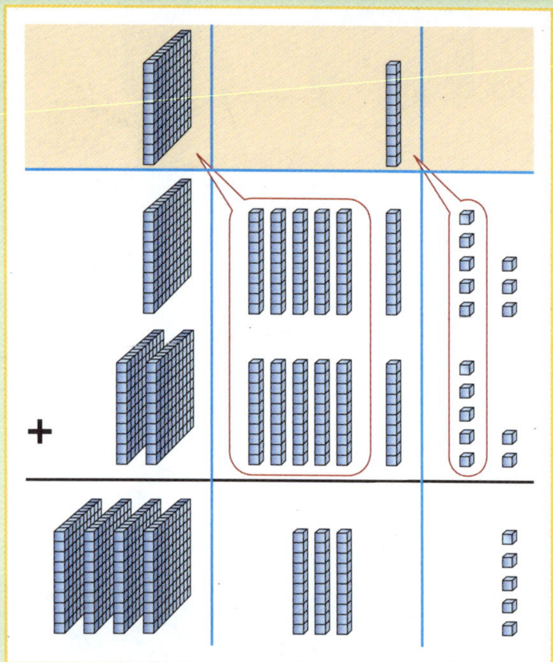

168 sheep and 267 cattle are on sale.
How many animals are there altogether?

	Hund	Tens	Ones
	1	1	
	1	6	8
+	2	6	7
	4	3	5

There are 435 animals altogether.
We can check the answer by rounding to the nearest 100 (or 10).
200 + 300 = 500
435 is reasonably close to 500.

CONCEPT

1

a

	Hund	Tens	Ones
		6	1
+		9	3

b

	Hund	Tens	Ones
		7	2
+		4	5

c

	Hund	Tens	Ones
		3	0
+		7	8

d

	Hund	Tens	Ones
	2	5	8
+		2	5

e

	Hund	Tens	Ones
		7	6
+	7	1	4

f

	Hund	Tens	Ones
	3	6	9
+		1	9

g

	Hund	Tens	Ones
	4	3	6
+		8	0

h

	Hund	Tens	Ones
	4	2	5
+		8	5

i

	Hund	Tens	Ones
		8	7
+	3	4	1

j

	Hund	Tens	Ones
	1	8	8
+	1	4	3

k

	Hund	Tens	Ones
	3	6	5
+	1	4	9

l

	Hund	Tens	Ones
	4	5	9
+	2	7	4

2 Check your answers by rounding each number to the nearest 100 (or 10) then adding.

 • *AUSTRALIAN SIGNPOST MATHS 4* • ISBN 978 1 4886 2187 1

2:27 Addition to 999

CONCEPT

We can round off to check our answers.

We round down if the digit is 4 or less. We round up if the digit is 5 or more.

When both numbers are rounded down, the answer is an **underestimate**.

319 + 443 = 862

300 + 400 = 700

The estimate is less than the actual answer.

When both numbers are rounded up, the answer is an **overestimate**.

482 + 257 = 739

500 + 300 = 800

The estimate is greater than the actual answer.

1

a

	H	T	U
	2	8	3
+		4	5

b

	H	T	U
	5	3	1
+		7	0

c

	H	T	U
		8	2
+	7	3	3

d

	H	T	U
	6	5	4
+		5	9

e

	H	T	U
		8	7
+	4	7	4

f

	H	T	U
	3	6	4
+		9	2

g

	H	T	U
	1	3	8
+	2	0	4

h

	H	T	U
	3	1	6
+	1	4	7

i

	H	T	U
	3	4	6
+	2	2	4

j

	H	T	U
	1	8	4
+	1	0	9

2

a

	H	T	U
	2	4	8
+	1	1	7

b

	H	T	U
		7	4
+	5	3	1

c

	H	T	U
	4	1	9
+	2	0	9

d

	H	T	U
	1	6	8
+	6	1	1

e

	H	T	U
	2	4	8
+	1	6	6

f

	H	T	U
	2	2	7
+	5	9	2

2:28 Writing the Addition Algorithm

CONCEPT

Hund	Tens	Ones
1	1	
5	8	3
+ 1	7	9
7	6	2

Hund	Tens	Ones
1	1	
5	8	3
+ 1	7	9
7	6	2

H	T	U
1	1	
5	8	3
+ 1	7	9
7	6	2

1	1	
5	8	3
+ 1	7	9
7	6	2

762 pieces of fruit were sold altogether.

1

a 250 + 84
b 534 + 128
c 154 + 207
d 92 + 340
e 245 + 370

f 365 + 166
g 209 + 583
h 345 + 312
i 507 + 220
j 365 + 397

k 188 + 133
l 365 + 366
m 509 + 296
n 381 + 259
o 777 + 185

2 Estimate each answer and use the estimate to check your answers.

a Luke watched 287 minutes of television on Saturday and 145 minutes on Sunday. How much did he watch altogether? Give your answer in minutes. ☐ minutes

b Church attendances in the two services on Sunday were 156 and 218. How many people attended altogether? ☐ people

c This year farmer McDonald sold 180 bags of corn. Last year he sold 17 bags more than this. How many bags were sold in the two years? ☐ bags

d Rex and Lyn bought the house next door and added 346 square metres to their original block of 575 square metres. How large is the block now? ☐ m^2

3 Use estimation to check your answers to Question 1.

2:29 Finding Missing Numbers

Write the missing number in each sentence.

1 a ☐ + 8 = 10 b ☐ + 3 = 7 c ☐ + 8 = 15

d 9 + ☐ = 12 e 8 + ☐ = 17 f 10 + ☐ = 10

g ☐ + 11 = 15 h ☐ + 9 = 14 i ☐ + 12 = 12

j 14 + ☐ = 20 k 18 + ☐ = 23 l 17 + ☐ = 19

m ☐ + 9 = 20 n ☐ + 15 = 25 o ☐ + 11 = 40

2 a 8 – ☐ = 1 b 9 – ☐ = 0 c 15 – ☐ = 10 d 10 – ☐ = 10

e ☐ – 3 = 7 f ☐ – 10 = 10 g ☐ – 7 = 8 h ☐ – 5 = 4

i 11 – ☐ = 4 j 11 – ☐ = 7 k 14 – ☐ = 6 l 14 – ☐ = 8

m ☐ – 8 = 4 n ☐ – 3 = 9 o ☐ – 1 = 19 p ☐ – 11 = 20

3 a ☐ × 6 = 18 b ☐ × 5 = 35 c ☐ × 4 = 24 d ☐ × 3 = 27

e 4 × ☐ = 20 f 7 × ☐ = 21 g 9 × ☐ = 90 h 8 × ☐ = 40

i ☐ × 10 = 80 j ☐ × 6 = 36 k ☐ × 4 = 28 l ☐ × 5 = 35

m 7 × ☐ = 56 n 8 × ☐ = 72 o 5 × ☐ = 45 p 9 × ☐ = 63

4 a ☐ ÷ 2 = 8 b ☐ ÷ 4 = 3 c ☐ ÷ 5 = 8

d 24 ÷ ☐ = 3 e 24 ÷ ☐ = 4 f 24 ÷ ☐ = 8

g ☐ squared = 9 h ☐ squared = 64

i ☐ squared = 100 j ☐ squared = 49

5 a 6 + 7 + ☐ = 15 b 3 – 1 + ☐ = 12 c 8 + 7 + ☐ = 20

d 4 × 2 + ☐ = 10 e 6 × 5 – ☐ = 26 f 21 ÷ 3 + ☐ = 13

g 2 × ☐ + 1 = 9 h 3 × ☐ + 2 = 17 i 2 × ☐ + 1 = 7

6 a 5 × (4 + 6) = (5 × ☐) + (5 × 6) b 3 × (9 + ☐) = (3 × 9) + (3 × 2)

c 7 × (3 + 4) = (7 × 3) + (7 × ☐) d 4 × (☐ + 2) = (4 × 8) + (4 × 2)

2:30 Subtraction without Trading to 999

CONCEPT

438 students.
216 were in uniform.
How many were out of uniform?

I Show

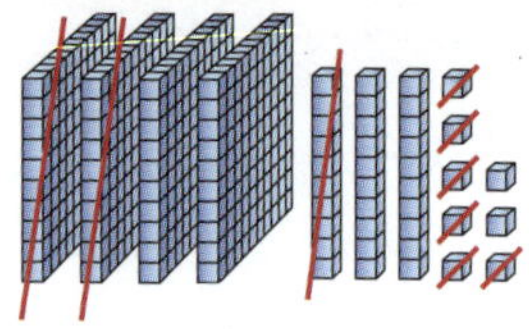

Take away 6 ones.
Take away 1 ten.
Take away 2 hundreds.

I Write

	Hund	Tens	Ones
	4	3	8
–	2	1	6
	2	2	2

222 students out of uniform

1 **a** 4 hundreds 7 tens 6 ones
– 2 hundreds 3 tens 1 one

b 8 hundreds 9 tens 7 ones
– 2 hundreds 5 tens 2 ones

c

	Hund	Tens	Ones
	8	5	2
–	7	2	1

d

	Hund	Tens	Ones
	9	6	7
–	2	3	4

e

	Hund	Tens	Ones
	9	9	8
–	4	6	5

f

	Hund	Tens	Ones
	3	4	2
–		1	1

g

	Hund	Tens	Ones
	6	9	8
–		2	4

h

	Hund	Tens	Ones
	9	4	5
–		1	1

i

	Hund	Tens	Ones
	6	8	6
–			4

j

	Hund	Tens	Ones
	7	4	7
–			4

k

	Hund	Tens	Ones
	3	6	9
–			7

l

	Hund	Tens	Ones
	5	5	9
–	4	3	0

m

	Hund	Tens	Ones
	1	7	3
–		5	0

n

	Hund	Tens	Ones
	6	4	6
–	2	0	3

o

	Hund	Tens	Ones
	7	7	8
–	6	0	0

p

	Hund	Tens	Ones
	5	2	9
–	1	0	0

q

	Hund	Tens	Ones
	1	8	3
–	1	1	3

r

	Hund	Tens	Ones
	5	8	7
–	3	8	2

2 Estimate to check each answer.

- Make up a problem to match each of the sums in parts **c**, **d**, **e** and **f**.

FUN SPOT

 • *AUSTRALIAN SIGNPOST MATHS 4* • ISBN 978 1 4886 2187 1

NUMBER & ALGEBRA

2:31 Subtraction with Trading to 999

CONCEPT

28 of my 384 stamps are Chinese. How many are not Chinese?

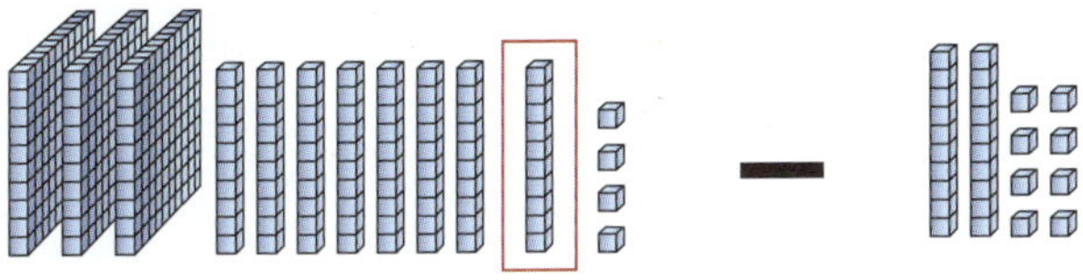

We can't take 8 ones from 4 ones, so trade 1 ten for 10 ones.

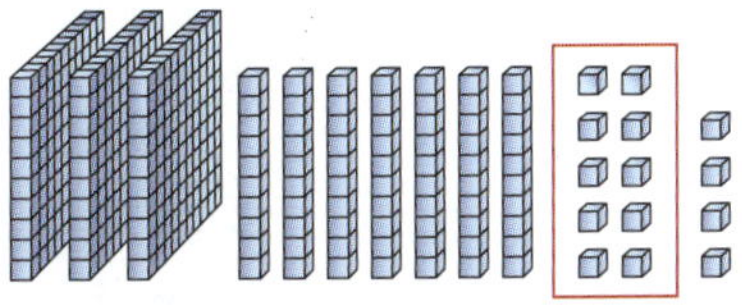

From this we take 2 tens and 8 ones.

	Hund	Tens	Ones
		7	14
	3	~~8~~	~~4~~
–		2	8
	3	5	6

1

a

	H	T	U
	5	8	1
–		5	3

b

	H	T	U
	3	4	3
–		2	8

c

	H	T	U
	2	9	6
–		6	7

d

	H	T	U
	4	6	0
–		3	9

e

	H	T	U
	2	7	3
–			9

f

	H	T	U
	1	6	6
–		5	8

g

	H	T	U
	8	4	4
–			8

h

	H	T	U
	9	7	3
–		6	5

i

	H	T	U
	7	5	0
–		3	9

j

	H	T	U
	3	4	1
–		2	6

k

	H	T	U
	2	8	0
–		3	2

l

	H	T	U
	6	4	7
–		1	9

m 835 – 108

n 891 – 836

o 520 – 415

p 684 – 76

q 214 – 208

r 784 – 50

s 391 – 375

Use estimation to check your answers.

2:32 Subtraction with Trading to 999

CONCEPT

I had \$524 and spent \$245. How much do I have left?

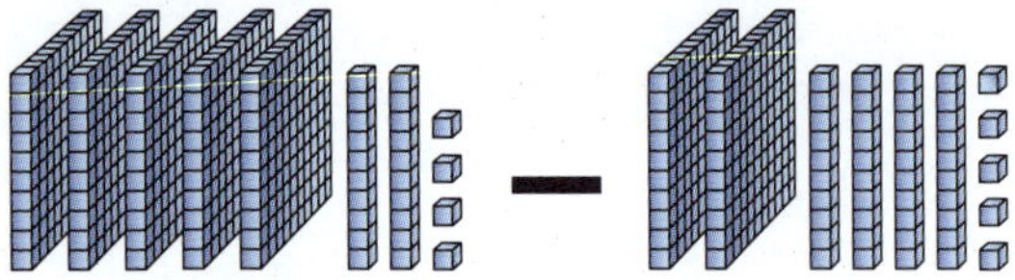

I trade 1 ten for 10 ones, and 1 hundred for 10 tens.

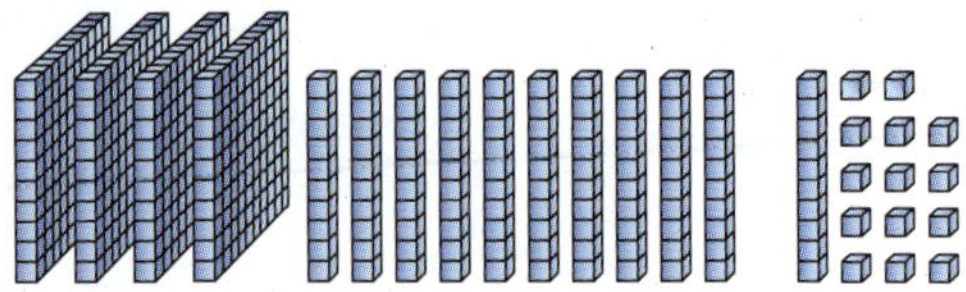

From this we take 2 hundreds, 4 tens and 5 ones.

Hund	Tens	Ones
	11	
4	~~1~~	14
~~5~~	~~2~~	~~4~~
− 2	4	5
2	7	9

Use rounding to check your answers.

1

a

H	T	U
3	1	4
−	5	6

b

H	T	U
2	4	6
−	6	7

c

H	T	U
1	3	2
−	9	8

d

H	T	U
7	3	5
−	8	9

e

H	T	U
4	3	0
− 1	5	6

f

H	T	U
3	4	1
− 2	7	8

g

H	T	U
6	2	4
− 1	3	5

h

H	T	U
9	2	2
− 3	3	3

i

H	T	U
7	1	5
− 5	7	7

j

H	T	U
4	7	0
− 2	9	3

k

H	T	U
8	4	2
− 1	8	3

l

H	T	U
9	1	1
− 3	4	4

m

6	1	4
− 5	8	5

n

5	2	8
− 4	2	9

o

7	5	5
− 6	6	8

p

4	1	5
− 3	8	7

q

7	0	6
− 2	4	9

r

6	0	1
− 1	8	4

s

5	0	6
− 2	7	8

In Questions 1q–1s, start by trading 1 hundred for 10 tens.

 • *AUSTRALIAN SIGNPOST MATHS 4* • ISBN 978 1 4886 2187 1

2:33 Subtraction with Trading to 999

CONCEPT

354 of the 718 birds in the zoo are parrots. How many are not parrots?

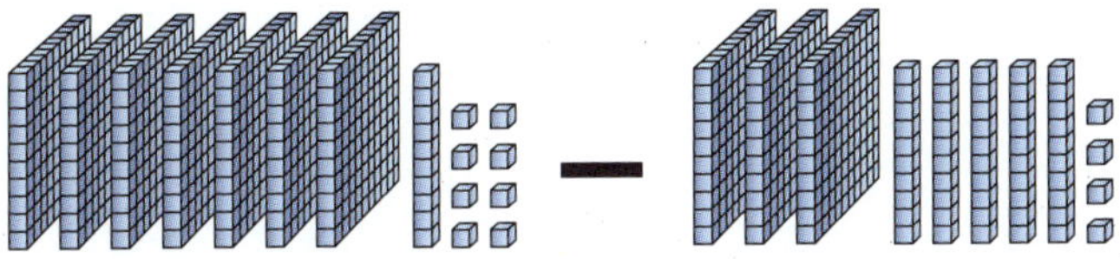

We can't take 5 tens from 1 ten, so trade 1 hundred for 10 tens.

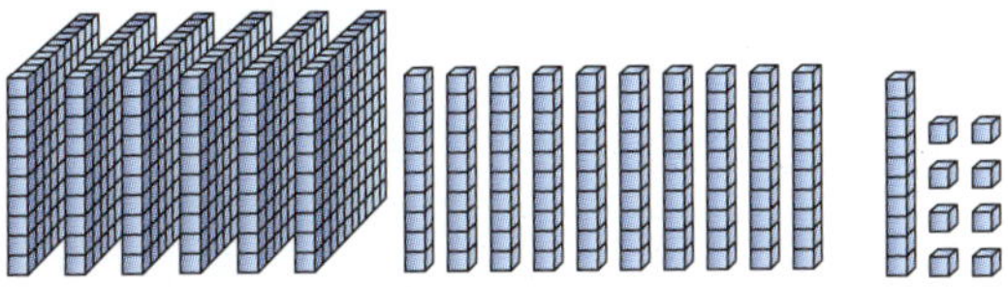

From this we take 3 hundreds, 5 tens and 4 ones.

	Hund	Tens	Ones
	6	11	
	~~7~~	~~1~~	8
−	3	5	4
	3	6	4

1

a

	H	T	U
	6	1	7
−	1	5	3

b

	H	T	U
	8	2	6
−	5	7	1

c

	H	T	U
	4	3	4
−	2	4	0

d

	H	T	U
	7	2	9
−		3	8

e

	H	T	U
	5	2	8
−	1	5	2

f

	H	T	U
	6	0	7
−	3	5	2

g

	H	T	U
	6	8	5
−		9	1

h

	H	T	U
	2	1	5
−	1	7	5

i

	H	T	U
	4	5	0
−	1	3	9

j

	H	T	U
	6	1	7
−	3	2	6

k

	H	T	U
	5	0	4
−	1	9	2

l

	H	T	U
	9	4	7
−	3	0	9

m

	7	3	3
−	1	0	8

n

	9	1	6
−	8	3	6

o

	6	0	8
−	4	1	5

p

	7	2	9
−	2	7	6

q

	4	1	4
−	2	0	8

r

	9	8	4
−	5	5	0

s

	8	6	6
−	3	7	5

Use estimation to check your answers.

  ISBN 978 1 4886 2187 1

2:34 Subtraction from Hundreds

CONCEPT

Of $600 Kim earned, he donated $155 to charity. How much money did he have left?

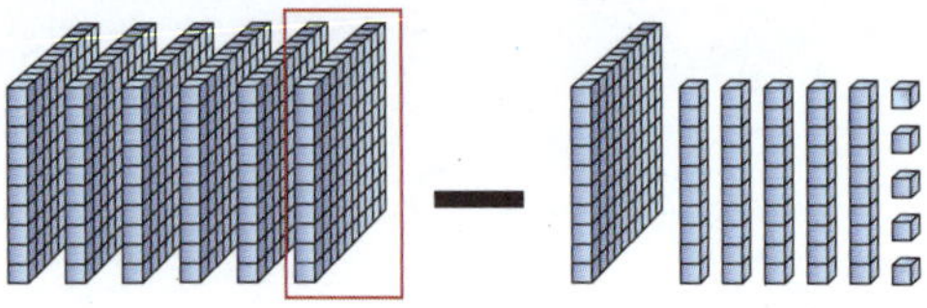

We can't take 5 ones from 0 ones, so we need to trade 1 ten for 10 ones. Since there are no tens we must first trade 1 hundred for 10 tens. Then trade 1 of the tens for 10 ones.

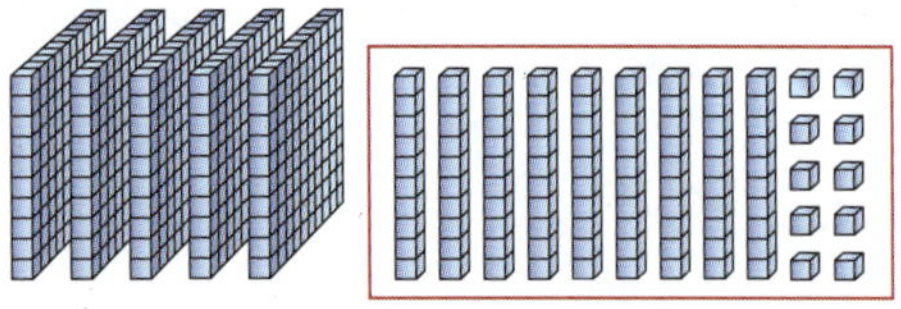

From this we take 1 hundred, 5 tens and 5 ones.

	Hund	Tens	Ones
		9	
	5	~~10~~	10
	~~6~~	~~0~~	~~0~~
−	1	5	5
	4	4	5

1

a
	H	T	U
	1	0	0
−		2	7

b
	H	T	U
	2	0	0
−		3	8

c
	H	T	U
	9	0	0
−		6	1

d
	H	T	U
	6	0	0
−		5	5

e
$$\begin{array}{r} 700 \\ -\,302 \\ \hline \end{array}$$

f
$$\begin{array}{r} 900 \\ -\,793 \\ \hline \end{array}$$

g
$$\begin{array}{r} 400 \\ -\,106 \\ \hline \end{array}$$

h
$$\begin{array}{r} 300 \\ -\,253 \\ \hline \end{array}$$

2

a
$$\begin{array}{r} 600 \\ -\,534 \\ \hline \end{array}$$

b
$$\begin{array}{r} 500 \\ -\,399 \\ \hline \end{array}$$

c
$$\begin{array}{r} 800 \\ -\,674 \\ \hline \end{array}$$

d
$$\begin{array}{r} 700 \\ -\,668 \\ \hline \end{array}$$

3 Estimate each answer and use the estimates to check your answers.

a Chloe needed 600 points to qualify for a state team. She earned only 487. How far short of her target was she? ☐ points

b Blackburn Post Office was sent 900 stamp albums to sell. If 481 were sold in the first month, how many were left? ☐ albums

c In our class library we have 400 books. I have read 147. How many have I not read? ☐ books

2:35 Subtraction from Hundreds Strategy

An easy way to subtract from 700 is to write 700 as 699 + 1.

$$
\begin{aligned}
& 700 - 372 \\
&= 699 + 1 - 372 \\
&= 699 - 372 + 1 \\
&= 327 + 1 \\
&= 328
\end{aligned}
$$

OR

$$
\begin{array}{r} 700 \\ -\ 137 \\ \hline 563 \\ \hline \end{array}
\quad \longrightarrow \quad
\begin{array}{r} 699 + 1 \\ -\ 137 \qquad \\ \hline 562 + 1 \\ \hline = 563 \qquad \end{array}
$$

CONCEPT

1

a (H T U)
$$\begin{array}{r} 9\ \ 9\ \ +1 \\ \not{1}\ \not{0}\ \not{0} \qquad \\ -\ \ 3\ \ 7 \qquad \\ \hline +1 \\ \hline = \boxed{} \end{array}$$

b (H T U)
$$\begin{array}{r} 2\ \ 9\ \ 9\ \ +1 \\ \not{3}\ \not{0}\ \not{0} \qquad \\ -\ \ 4\ \ 8 \qquad \\ \hline +1 \\ \hline = \boxed{} \end{array}$$

c (H T U)
$$\begin{array}{r} 8\ \ 9\ \ 9\ \ +1 \\ \not{9}\ \not{0}\ \not{0} \qquad \\ -\ \ 7\ \ 1 \qquad \\ \hline +1 \\ \hline = \boxed{} \end{array}$$

d
$$\begin{array}{r} +1 \\ \not{7}\ \not{0}\ \not{0} \qquad \\ -\ 4\ \ 6\ \ 9 \qquad \\ \hline +1 \\ \hline = \boxed{} \end{array}$$

e
$$\begin{array}{r} +1 \\ \not{5}\ \not{0}\ \not{0} \qquad \\ -\ 2\ \ 8\ \ 5 \qquad \\ \hline +1 \\ \hline = \boxed{} \end{array}$$

f
$$\begin{array}{r} +1 \\ \not{2}\ \not{0}\ \not{0} \qquad \\ -\ 1\ \ 3\ \ 7 \qquad \\ \hline +1 \\ \hline = \boxed{} \end{array}$$

g
$$\begin{array}{r} +1 \\ 4\ \ 0\ \ 0 \qquad \\ -\ 3\ \ 2\ \ 1 \qquad \\ \hline +1 \\ \hline = \boxed{} \end{array}$$

h
$$\begin{array}{r} +1 \\ 6\ \ 0\ \ 0 \qquad \\ -\ 2\ \ 9\ \ 3 \qquad \\ \hline +1 \\ \hline = \boxed{} \end{array}$$

i
$$\begin{array}{r} +1 \\ 8\ \ 0\ \ 0 \qquad \\ -\ 5\ \ 4\ \ 6 \qquad \\ \hline +1 \\ \hline = \boxed{} \end{array}$$

j
$$\begin{array}{r} 6\ \ 0\ \ 0 \qquad \\ -\ 2\ \ 7\ \ 1 \qquad \\ \hline + \quad \\ \hline = \boxed{} \end{array}$$

k
$$\begin{array}{r} 9\ \ 0\ \ 0 \qquad \\ -\ 3\ \ 4\ \ 9 \qquad \\ \hline + \quad \\ \hline = \boxed{} \end{array}$$

l
$$\begin{array}{r} 5\ \ 0\ \ 0 \qquad \\ -\ 1\ \ 5\ \ 6 \qquad \\ \hline + \quad \\ \hline = \boxed{} \end{array}$$

2:36 Division as Repeated Subtraction

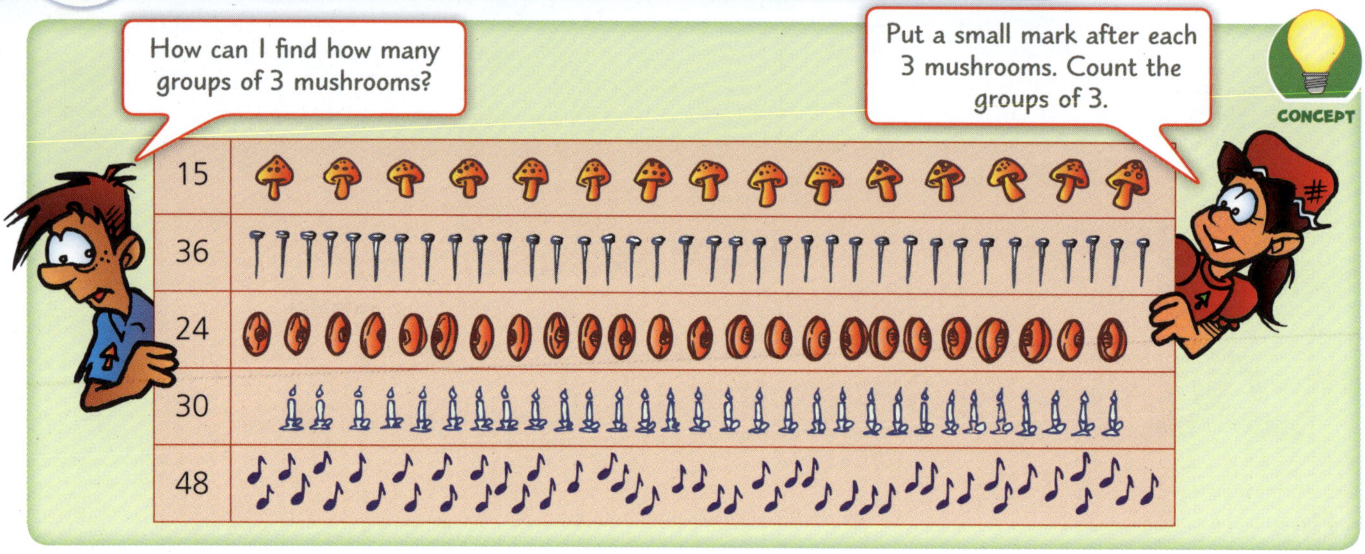

1. How many groups of 3 could be made from each of these?

a  ☐ groups

b 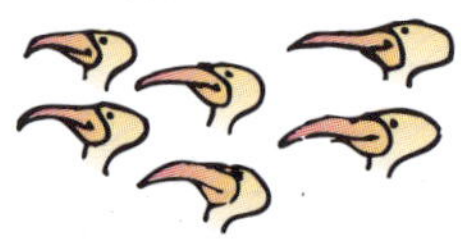 ☐ groups

c  ☐ groups

2.

a How many groups of 5 mushrooms are in 15? ☐ groups. So 15 ÷ 5 = ☐

b How many groups of 5 candles are in 30? ☐ groups. So 30 ÷ 5 = ☐

c How many groups of 3 mushrooms are in 15? ☐ groups. So 15 ÷ 3 = ☐

d How many groups of 6 tacks are in 36? ☐ groups. So 36 ÷ 6 = ☐

e How many groups of 3 candles are in 30? ☐ groups. So 30 ÷ 3 = ☐

f How many groups of 4 tacks are in 36? ☐ groups. So 36 ÷ 4 = ☐

g How many groups of 4 footballs are in 24? ☐ groups. So 24 ÷ 4 = ☐

h How many groups of 6 notes are in 48? ☐ groups. So 48 ÷ 6 = ☐

i How many groups of 9 tacks are in 36? ☐ groups. So 36 ÷ 9 = ☐

j How many groups of 8 footballs are in 24? ☐ groups. So 24 ÷ 8 = ☐

k How many groups of 8 notes are in 48? ☐ groups. So 48 ÷ 8 = ☐

3. Use counters or place-value blocks to answer these.

a 42 apples are put in groups of 6. How many groups are there? ☐, 42 ÷ 6 = ☐

b 5 cakes fill one box. How many boxes can be filled with 35 cakes? ☐, 35 ÷ 5 = ☐

c A table seats 4 people. How many tables are needed to seat 32? ☐, 32 ÷ 4 = ☐

2:37 Understanding Division

CONCEPT

You can use a number line to do division.
How many 6s in 18?

$18 \div 6 = 3$ $\quad$ $\boxed{3} \times 6 = 18$

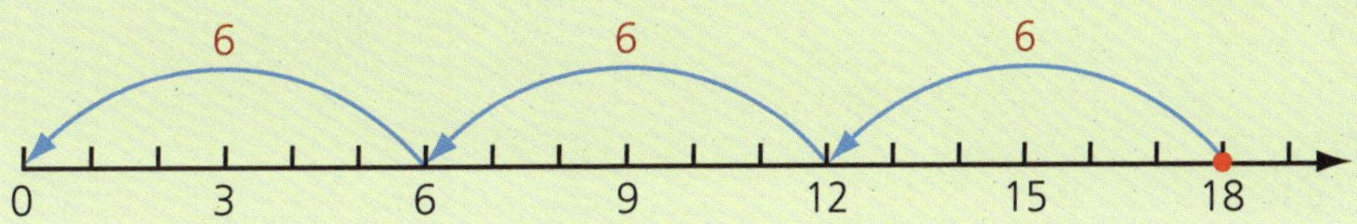

1 Use the number line above to find the answers.

a How many 3s are in 18?
$18 \div 3 =$ ☐

b How many 9s are in 18?
$18 \div 9 =$ ☐

c How many 2s are in 18?
$18 \div 2 =$ ☐

d How many 18s are in 18?
$18 \div 18 =$ ☐

2 Use this number line to find the answers.

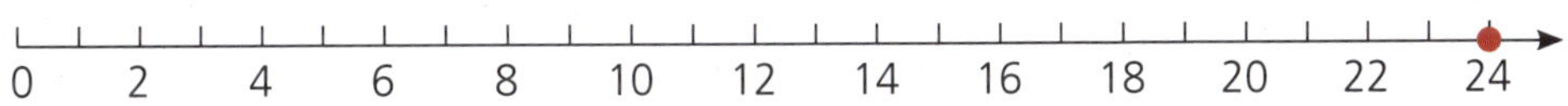

a How many 4s in 24?
$24 \div 4 =$ ☐

b How many 8s in 24?
$24 \div 8 =$ ☐

c How many 2s in 24?
$24 \div 2 =$ ☐

d How many 3s in 24?
$24 \div 3 =$ ☐

e How many 6s in 24?
$24 \div 6 =$ ☐

f How many 12s in 24?
$24 \div 12 =$ ☐

3 Use this number line to find the answers.

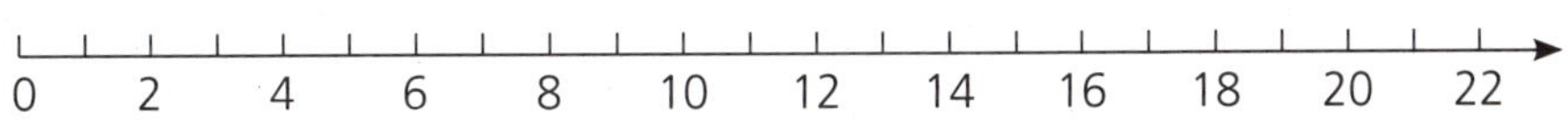

a How many 5s in 15?
$15 \div 5 =$ ☐

b How many 6s in 18?
$18 \div 6 =$ ☐

c How many 1s in 18?
$18 \div 1 =$ ☐

d How many 3s in 15?
$15 \div 3 =$ ☐

e How many 4s in 16?
$16 \div 4 =$ ☐

f How many 8s in 16?
$16 \div 8 =$ ☐

4 Write a number sentence for each problem.

a 20 stickers, 4 students
☐ = ☐
☐ each

b 16 bones, 8 dogs
☐ = ☐
☐ each

c 20 toys, 2 children
☐ = ☐
☐ each

 • • ISBN 978 1 4886 2187 1

2:38 Division Facts

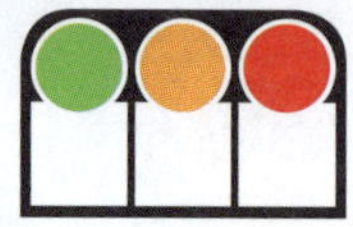

32 ÷ 8 Ask: ☐ × 8 = 32.

8 × 4 = 32

$4\overline{)32}$ with answer 8 $\qquad$ $8\overline{)32}$ with answer 4

CONCEPT

1 a 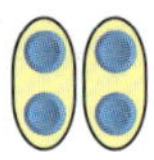
2 × 2 = ☐
4 ÷ 2 = ☐

b 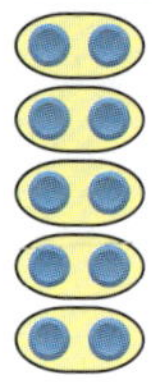
5 × 2 = ☐
10 ÷ 5 = ☐

c 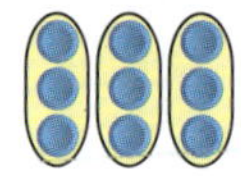
3 × 3 = ☐
9 ÷ 3 = ☐

2 a 3 × 4 = ☐
so 12 ÷ 3 = ☐

b 2 × 8 = ☐
so 16 ÷ 2 = ☐

c 7 × 4 = ☐
so 28 ÷ 7 = ☐

d 9 × 3 = ☐
so 27 ÷ 9 = ☐

e 6 × 5 = ☐
so 30 ÷ 6 = ☐

f 5 × 3 = ☐
so 15 ÷ 5 = ☐

3 Use the multiplication table to answer the division questions.

a 3 × 4 = 12
$4\overline{)12}$ $\quad$ $3\overline{)12}$

b 8 × 5 = 40
$5\overline{)40}$ $\quad$ $8\overline{)40}$

c 10 × 9 = 90
$9\overline{)90}$ $\quad$ $10\overline{)90}$

d 6 × 10 = 60
$10\overline{)60}$ $\quad$ $6\overline{)60}$

e 7 × 8 = 56
$8\overline{)56}$ $\quad$ $7\overline{)56}$

f 5 × 9 = 45
$9\overline{)45}$ $\quad$ $5\overline{)45}$

4 a 12 ÷ 6 = ☐ or ☐ × 6 = 12

b 18 ÷ 3 = ☐ or ☐ × 3 = 18

c 35 ÷ 5 = ☐ or ☐ × 5 = 35

d 30 ÷ 5 = ☐ or ☐ × 5 = 30

e 20 ÷ 2 = ☐ or ☐ × 2 = 20

f 24 ÷ 8 = ☐ or ☐ × 8 = 24

5 a We bought 80 oranges and put the same number in each of 8 buckets. How many did we put in each bucket? ☐

b We decided to give 4 toys to each girl. If we gave out 32 toys, how many girls were there? ☐

c Four boys shared 28 keys. How many did each boy get? ☐

6 Share $136 among 4 people. What is the value of one share? Show your method below. (Estimate your answer first.)

☐

Is the answer close to your estimate?

2:39 Division Facts

CONCEPT

To find 42 ÷ 6, ask:

"What times 6 gives 42?"

For 42 ÷ 6 = ☐

Ask: ☐ × 6 = 42

or

6 × ☐ = 42

If 5 × 9 = 45
then 45 ÷ 9 = 5
and 45 ÷ 5 = 9

1 Use the first number sentence to fill in the other two.

a 8 × 3 = 24
24 ÷ 3 = ☐
24 ÷ 8 = ☐

b 4 × 9 = 36
36 ÷ 4 = ☐
36 ÷ 9 = ☐

c 5 × 6 = 30
30 ÷ 6 = ☐
30 ÷ 5 = ☐

2
a ☐ × 9 = 18
b ☐ × 10 = 20
c ☐ × 3 = 21
d ☐ × 5 = 25
e ☐ × 3 = 9
f ☐ × 6 = 12
g 3 × ☐ = 6
h 2 × ☐ = 10
i 7 × ☐ = 14

3
a 40 ÷ 5 = ☐
b 63 ÷ 9 = ☐
c 48 ÷ 8 = ☐
d 32 ÷ 4 = ☐
e 3 ÷ 3 = ☐
f 90 ÷ 10 = ☐
g 54 ÷ 9 = ☐
h 35 ÷ 5 = ☐
i 16 ÷ 4 = ☐
j 27 ÷ 3 = ☐
k 72 ÷ 8 = ☐
l 40 ÷ 8 = ☐

4

a

18	÷ 2	=	
	÷ 3	=	
	÷ 6	=	
	÷ 9	=	

b

24	÷ 3	=	
	÷ 4	=	
	÷ 6	=	
	÷ 8	=	

c

30	÷ 3	=	
	÷ 5	=	
	÷ 6	=	
	÷ 10	=	

5 Use a calculator to check all your answers.

Find the Number Sentences

ACTIVITY

- One person draws a number line which shows several jumps, starting at zero.
- The other person writes multiplication and division number sentences to match the picture.

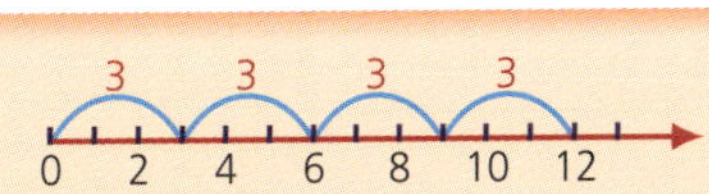

This shows:
4 × 3 = 12
12 ÷ 3 = 4
and 12 ÷ 4 = 3.

2:40 Division Facts

CONCEPT

Division is related to multiplication.

40 ÷ 8

Ask: ☐ × 8 = 40?

$5 \times 8 = 40$

$8\overline{)40}$ → 5 $5\overline{)40}$ → 8

Yes! I see.

1 Use a multiplication table to answer the division questions.

a 6 × 10 = 60 $6\overline{)60}$ $10\overline{)60}$

b 5 × 8 = 40 $8\overline{)40}$ $5\overline{)40}$

c 6 × 9 = 54 $9\overline{)54}$ $6\overline{)54}$

d 9 × 8 = 72 $8\overline{)72}$ $9\overline{)72}$

e 7 × 6 = 42 $6\overline{)42}$ $7\overline{)42}$

f 4 × 9 = 36 $9\overline{)36}$ $4\overline{)36}$

2

a ☐ × 2 = 12   **b** 6 × ☐ = 18   **c** ☐ × 4 = 20   **d** 5 × ☐ = 10

e ☐ × 3 = 18   **f** 7 × ☐ = 56   **g** ☐ × 9 = 9   **h** 6 × ☐ = 36

i ☐ × 5 = 35   **j** 4 × ☐ = 32   **k** ☐ × 4 = 16   **l** 3 × ☐ = 27

3

a 16 ÷ 2 = ☐   **b** 25 ÷ 5 = ☐   **c** 30 ÷ 6 = ☐   **d** 24 ÷ 6 = ☐

e 30 ÷ 3 = ☐   **f** 24 ÷ 4 = ☐   **g** 35 ÷ 7 = ☐   **h** 90 ÷ 9 = ☐

i 24 ÷ 8 = ☐   **j** 48 ÷ 6 = ☐   **k** 64 ÷ 8 = ☐   **l** 36 ÷ 6 = ☐

m 27 ÷ 3 = ☐   **n** 50 ÷ 5 = ☐   **o** 63 ÷ 9 = ☐   **p** 30 ÷ 5 = ☐

q 45 ÷ 5 = ☐   **r** 72 ÷ 9 = ☐   **s** 54 ÷ 6 = ☐   **t** 90 ÷ 10 = ☐

u 81 ÷ 9 = ☐   **v** 48 ÷ 8 = ☐   **w** 16 ÷ 8 = ☐   **x** 30 ÷ 5 = ☐

4 Use a calculator to check all your answers.

ACTIVITY

×	1	2	3	4	5	6
1						
2						
3						
4						
5						
6						

Dice Multiplication

- Take turns to throw two dice.
- Multiply the numbers showing.
- Write the answer in the table (each player uses a different colour).
- The first to get 3 answers in a line wins.

Note: 3 × 1 and 1 × 3 are made by 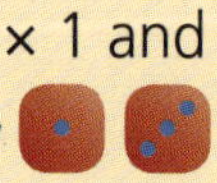 .

3 × 4 makes 12.

2:41 Division Using the Multiplication Grid

CONCEPT

×	0	1	2	3	4	5	6	7	8	9	10
0	0	0	0	0	0	0	0	0	0	0	0
1	0	1	2	3	4	5	6	7	8	9	10
2	0	2	4	6	8	10	12	14	16	18	20
3	0	3	6	9	12	15	18	21	24	27	30
4	0	4	8	12	16	20	24	28	32	36	40
5	0	5	10	15	20	25	30	35	40	45	50
6	0	6	12	18	24	30	36	42	48	54	60
7	0	7	14	21	28	35	42	49	56	63	70
8	0	8	16	24	32	40	48	56	64	72	80
9	0	9	18	27	36	45	54	63	72	81	90
10	0	10	20	30	40	50	60	70	80	90	100

5) 35 means 35 divided by 5.

5) 35 → 7 ← Answer

Use the multiplication grid above to answer these division questions.

1

a 15 ÷ 3 = ☐ b 20 ÷ 5 = ☐ c 16 ÷ 8 = ☐ d 90 ÷ 10 = ☐

e 42 ÷ 6 = ☐ f 10 ÷ 10 = ☐ g 24 ÷ 6 = ☐ h 24 ÷ 4 = ☐

i 64 ÷ 8 = ☐ j 45 ÷ 5 = ☐ k 36 ÷ 6 = ☐ l 25 ÷ 5 = ☐

m 72 ÷ 9 = ☐ n 16 ÷ 4 = ☐ o 10 ÷ 5 = ☐ p 81 ÷ 9 = ☐

q 18 ÷ 6 = ☐ r 54 ÷ 9 = ☐ s 40 ÷ 8 = ☐ t 36 ÷ 4 = ☐

u 40 ÷ 8 = ☐ v 50 ÷ 10 = ☐ w 48 ÷ 6 = ☐ x 21 ÷ 3 = ☐

2

a 2) 8 b 5) 5 c 2) 6 d 10) 30 e 6) 12

f 4) 12 g 10) 80 h 3) 9 i 9) 45 j 8) 32

k 5) 30 l 5) 15 m 5) 40 n 3) 24 o 9) 27

3 Follow this track, putting an answer in every empty place.

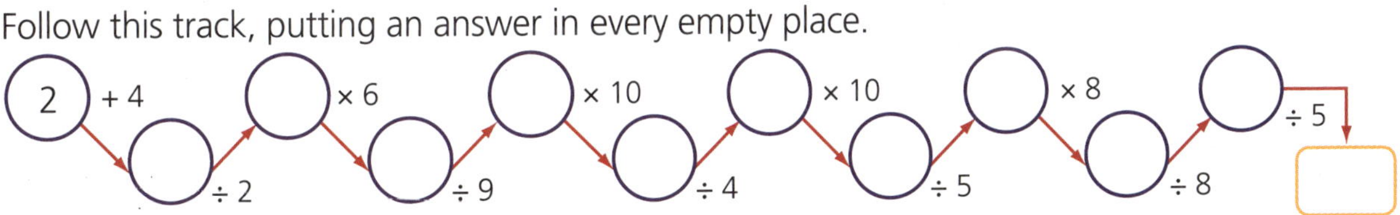

2:42 Money

- An amount of money can be made in different ways.
- We usually use the highest value notes and coins possible.

$187.50 =	$100	$50	$20	$10	$5	$2	50c	
or	$100	$20	$20	$20	$20	$5	$2	50c

The rounded boxes stand for coins.

CONCEPT

1 Make each amount using the highest value of notes and coins possible.

a $57

b $51.50

c $23

d $60.30

e $188.60

f $96.95

g $133.80

h $165.75

i $65.35

j $153.35

k $275.40

l $212.80

m $107.25

2 From a collection of coins, choose 8. Write the value of each coin and the total. Repeat the process each time.

FUN SPOT

 • *AUSTRALIAN SIGNPOST MATHS 4* • ISBN 978 1 4886 2187 1

2:43 Counting Change

CONCEPT

To give change, we count on from the price paid.

Example: What is the change from $100 if the cost is $23.90?

- Count on from $23.90.

$24… $25… $30… $50… $100

10c | $1 | $5 | $20 | $50 Change = $76.10

1 Count on and write the change given in coins and notes.

	Cost	Money Given	Coins and Notes Given as Change	Total Change
a	$6.50	Count on → $50	$7 → $8 → $10 → $30 → $50	
b	$1.20	Count on → $10	$1.30 → $1.50 → $2 → $3 → $5 → $10	
c	$8.70	Count on → $50	$8.80 → $9 → $10 → $30 → $50	
d	$3.00	Count on → $100	$5 → $10 → $30 → $50 → $100	
e	$36.40	Count on → $50		
f	$67.20	Count on → $100		

2 Find the change from $50 using only the boxes you need.

a $36.50 Change =

b $9.20 Change =

c $18.30 Change =

INVESTIGATION

3 Show three different ways of giving change from $50 if the cost is $41.95.

a

b

c

Practise giving change using play money.

2:44 Factors and Multiples

CONCEPT

The Factors of 12

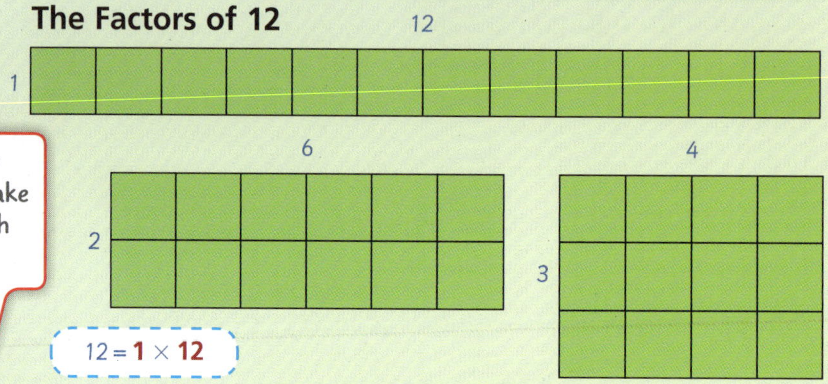

In how many ways can we make rectangles with 12 squares?

$12 = 1 \times 12$

$12 = 2 \times 6$

$12 = 3 \times 4$

These numbers are the factors of 12.
12 is a multiple of each number.

On the sides we could have 1 and 12, 2 and 6, or 3 and 4.

1 Use these rectangles to complete the number sentences below.

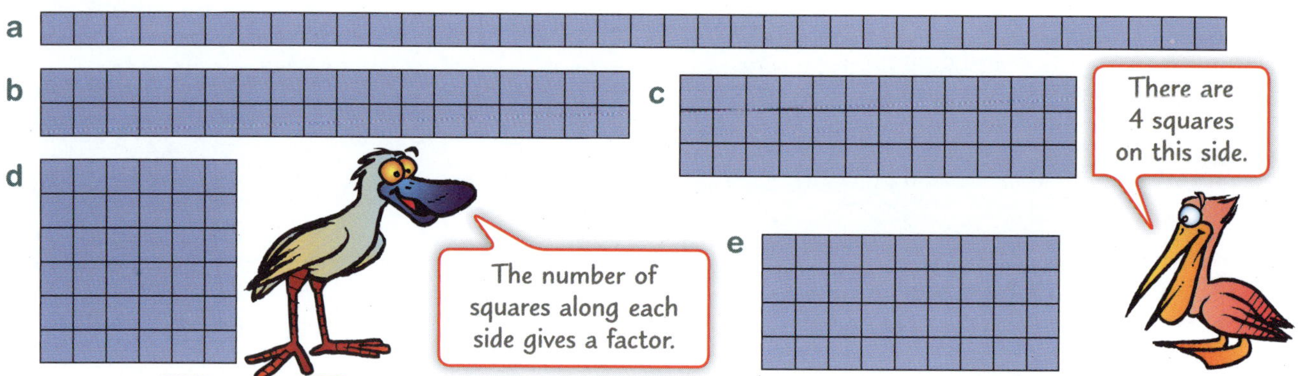

a 36 = ☐ × ☐ **b** 36 = ☐ × ☐ **c** 36 = ☐ × ☐

d 36 = ☐ × ☐ **e** 36 = ☐ × ☐

f The factors of 36 are ☐, ☐, ☐, ☐, ☐, ☐, ☐, ☐ and ☐

2 Make rectangles using place-value blocks to find all the factors of these.

a 10: ☐, ☐, ☐ and ☐ **b** 7: ☐ and ☐

c 18: ☐, ☐, ☐, ☐, ☐ and ☐ **d** 21: ☐, ☐, ☐ and ☐

3 Which of the numbers 10, 7, 18 and 21 are multiples of:

a 2? ☐ and ☐ **b** 3? ☐ and ☐ **c** 6? ☐ **d** 7? ☐ and ☐

4 If 70 = 1 × 70, 70 = 2 × 35, 70 = 5 × 14 and 70 = 7 × 10, then the factors of 70 are:

☐, ☐, ☐, ☐, ☐, ☐, ☐ and ☐

5 Write all of the factors of 24. ☐, ☐, ☐, ☐, ☐, ☐, ☐, ☐

 • *AUSTRALIAN SIGNPOST MATHS 4* • ISBN 978 1 4886 2187 1

2:45 Products and Factors

Factors come in pairs.
Any member of a product is a factor.
$10 = 1 \times 10$
$10 = 2 \times 5$

CONCEPT

1 Write a pair of factors for each number.

a 2 ☐ ☐ **b** 5 ☐ ☐

2 Write two pairs of factors for each number.

a 8 ☐ ☐ ☐ ☐ **b** 10 ☐ ☐ ☐ ☐

3 Write the product of each pair of factors.

a 5 and 2 ☐ **b** 5 and 8 ☐

c 3 and 5 ☐ **d** 7 and 5 ☐

$5 \times 8 = 8 \times 5$

4 Complete this factor table.

Product	15	15	20	20	20	40	40	40	40	35	35
Factor	1	3	1	2	4	1	2	4	5	1	5
Factor	15	5									

- Two handles are joined to make one whole.
- Two factors are multiplied to form a product.

Calculate-a-factor

ICT

- Use a calculator to divide any number by 1, 2, 3, 4 etc.

  Numbers which divide exactly are factors.

  $16 \div 1 = 16$
  $16 \div 2 = 8$
  $16 \div 3 = 5{\cdot}3333$
  $16 \div 4 = 4$
  $16 \div 5 = 3{\cdot}2$
  $16 \div 6 = 2{\cdot}6666$
  $16 \div$ ...

  Not factors

$16 = 1 \times 16$
$16 = 2 \times 8$
$16 = 4 \times 4$
1, 16, 2, 8 and 4 are factors of 16.

- Try this with numbers of your own.

 • *AUSTRALIAN SIGNPOST MATHS 4* • ISBN 978 1 4886 2187 1

2:46 Problem Solving

CONCEPT

- Problems can be solved in different ways.
- Decide which is the best way for you to perform the calculation.

Find:

Number sentence:

Answer:

1. Emily had 52 books and gave away 36. How many does she have now?

   Find:

   Number sentence:

   Answer:

2. Jeff bought 9 light bulbs at $3 each. What was the total bill?

   Find:

   Number sentence:

   Answer:

3. Sophie had 24 lollies. She shared them among 4 people. How many each?

   Find:

   Number sentence:

   Answer:

4. Carol set up 8 Christmas trees with 25 stars on each. How many stars?

   Find:

   Number sentence:

   Answer:

5. Kate bought 5 packets of 9 stickers. How many stickers altogether?

   Find:

   Number sentence:

   Answer:

6. Christie made 36 cupcakes to share between 6 people. How many each?

   Find:

   Number sentence:

   Answer:

7. Greg made 37 pizzas in May and 24 more in June. How many did he make?

   Find:

   Number sentence:

   Answer:

8. Scott shared 32 pencils among 8 children. How many each?

   Find:

   Number sentence:

   Answer:

 • *AUSTRALIAN SIGNPOST MATHS 4* • ISBN 978 1 4886 2187 1

2:47 Problem Solving

These words might help.

CONCEPT

+	*add, plus, sum, increase, total, altogether*
−	*minus, subtract, decrease, difference, less*
×	*times, product, groups of, rows of, lots of*
÷	*divide, share, how many groups, how many rows*

1. Of the 86 students at our school, 37 are girls. How many students are boys?

2. What is the sum of 23, 38 and 11?

3. 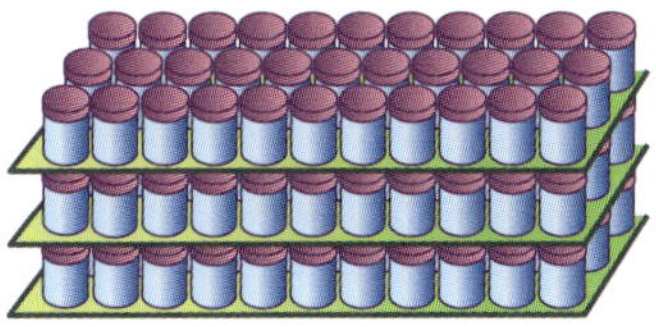
On each of these shelves we have stacked 33 jars. How many jars are there altogether?

4. What is the product of the numbers 11, 5 and 2?

Two numbers are multiplied to form a product.

5. What would be a fair share if 4 children shared 32 pens?

6. To make a wagon we must use 4 wheels. If we have 39 wheels, how many wagons can we make?

How many wheels will be left over?

7. Heather had 30 eggs but 18 were cracked. How many were not cracked?

8. Each carton of raisins contains 8 packets. How many packets of raisins would be in 7 cartons?

9. 
Which method would give me more pins?
A I share 36 pins with three other people.
B I share 30 pins with two other people. **A** or **B**?
What would be the difference in the number of pins I have?

10. 8 girls painted their fingernails and toenails red. How many nails were painted altogether?

11. We replaced half of the tyres on our four cars. How many tyres did we replace?

2:48 Working with Numbers

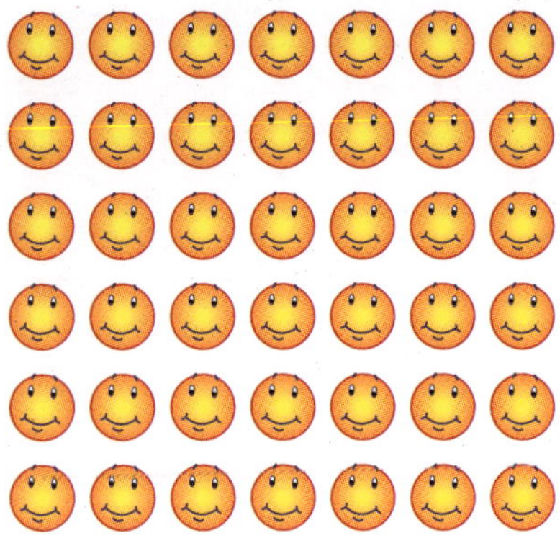

1. Estimate the number of stickers to the nearest ten. ☐
2. What two multiplication tables are shown by this array?
   ☐ × ☐ = ☐ and ☐ × ☐ = ☐
3. What two division tables are shown by this array?
   42 ÷ ☐ = ☐ and 42 ÷ ☐ = ☐
4. How many children could be given 5 stickers? ☐
5. How many more stickers are needed to make 60? ☐

6. Halve the number and keep halving for as long as you can.

   a 400 ☐ ☐ ☐ ☐ ☐ **b** 160 ☐ ☐ ☐ ☐ ☐

   c 256 ☐ ☐ ☐ ☐ ☐ ☐ ☐ ☐ ☐ ☐

   Double the number and keep doubling until you have filled the boxes.

   d 3 ☐ ☐ ☐ ☐ ☐ ☐ ☐ ☐

   e 9 ☐ ☐ ☐ ☐ ☐ ☐ ☐ ☐

INVESTIGATION

True or **false**?

2 × 72

= (2 × 70) + (2 × 2)

7. Use doubling and the fact given to answer the questions.

   a 2 × 12 = 24 4 × 12 = ☐, 8 × 12 = ☐, 16 × 12 = ☐

   b 6 × 3 = 18 12 × 3 = ☐, 24 × 3 = ☐, 48 × 3 = ☐

   c 7 × 9 = 63 14 × 9 = ☐, 28 × 9 = ☐, 56 × 9 = ☐

8. You have to design a box that holds 48 chocolates.
   There can be one or two layers, and on each layer the chocolates are arranged in equal rows.

   How many ways could you arrange the chocolates in a box? ☐

9. Show the method that you use to answer each of these questions.

   a 18 × 23 Estimate = ☐

   b 852 ÷ 4 Estimate = ☐

2:49 Number Patterns

1 Write the first four terms of each pattern.

a Start at 40 and count backwards by fives.

b Start at 137 and count forwards by tens.

c Start at 995 and count forwards by hundreds.

d Start at 1 696 and count forwards by thousands.

2 Write the tenth number in each number pattern.

a 14, 21, 28, 35, …

b 189, 179, 169, 159, …

c 40, 80, 120, 160, …

d 16, 24, 32, 40, …

e 643, 743, 843, 943, …

f 950, 900, 850, 800, …

g $1\frac{1}{4}$, $1\frac{1}{2}$, $1\frac{3}{4}$, 2, …

h $6\frac{1}{2}$, 6, $5\frac{1}{2}$, 5, …

3 Complete each number pattern.

a 20, ___, 60, 80, ___, ___

b 479, 489, ___, ___, 519, ___

c $1\frac{1}{4}$, $1\frac{1}{2}$, $1\frac{3}{4}$, ___, $2\frac{1}{4}$, ___

d 16, 24, 32, ___, 48, ___

4 Fill in the missing measurements for each bottle.

a

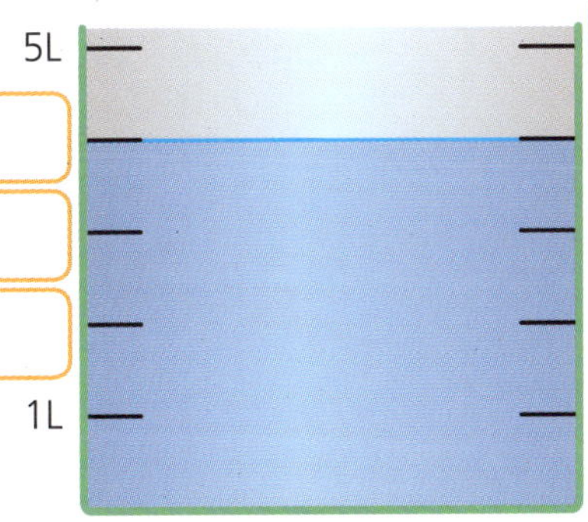

Volume of water = ___ or ___.

b

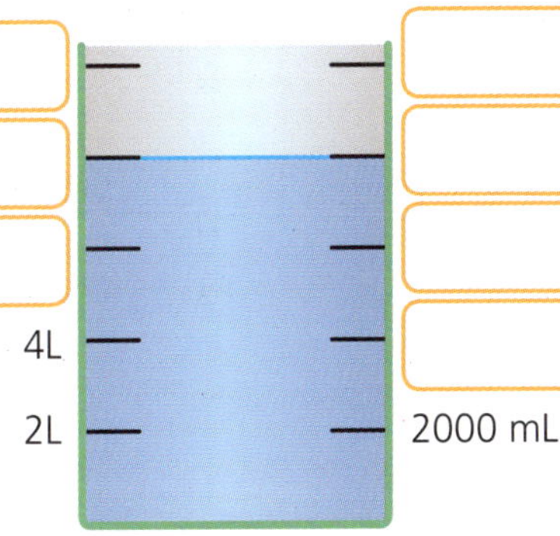

Volume of water = ___ or ___.

5 Continue each pattern.

a 1 kg, 2 kg, 3 kg, ___, ___

b 1 000 g, 2 000 g, 3 000 g, ___, ___

c 5 km, 6 km, 7 km, ___, ___

d 5 000 m, 6 000 m, 7 000 m, ___, ___

e 2 kg, 4 kg, 6 kg, ___, ___

f 2 000 g, 4 000 g, 6 000 g, ___, ___

2:50 Using Odd and Even Numbers

1 Find the rules for operating with two even numbers.

a

6 + 14		10 + 8		100 + 84	
16 + 12		32 + 16		104 + 58	
28 + 4		92 + 20		94 + 94	

even number plus even number = ______

b

44 – 42		56 – 56		36 – 8	
28 – 16		86 – 20		958 – 602	
32 – 14		198 – 8		600 – 2	

even number minus even number = ______

c

12 × 2		8 × 6		4 × 2	
8 × 4		2 × 20		6 × 6	
6 × 4		4 × 10		2 × 8	

even number times even number = ______

d

48 ÷ 2		10 ÷ 10		80 ÷ 2	
40 ÷ 4		24 ÷ 4		40 ÷ 8	
18 ÷ 6		24 ÷ 8		36 ÷ 6	

even number divided by even number = ______ or ______

Use other pairs of even numbers to check these rules.

2 Use many different examples of each case to complete this table. In each case write whether the answer would be odd or even.

Number Types	+	–	×	÷
Even and Even				odd or even
Odd and Odd				
Odd and Even				

3 Use the rules above to cross (✗) the answers that must be wrong.

6248 + 396 = 6645		8104 – 7998 = 106		366 × 47 = 17 201	
6107 ÷ 31 = 196		7319 + 2997 = 10 313		2817 – 199 = 2617	
283 × 654 = 28 353		173 × 881 = 152 416		68 019 ÷ 79 = 862	

 • *AUSTRALIAN SIGNPOST MATHS 4* • ISBN 978 1 4886 2187 1

2:51 Rounding Off Money

CONCEPT

When paying by cash, the price is rounded to the nearest 5 cents.

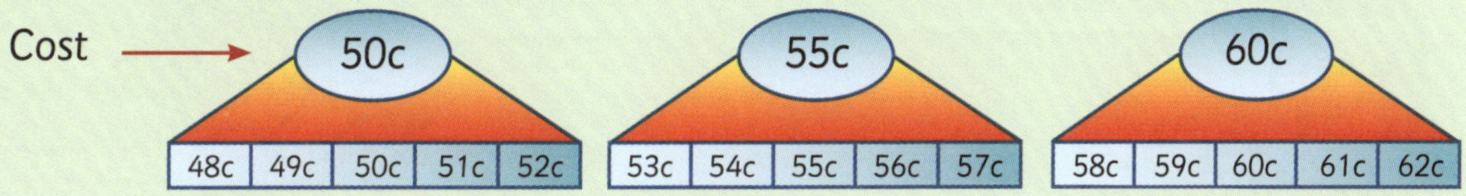

1 Round these amounts to the nearest 5 cents.

a	87c		b	$0.87		c	$1.52	
d	$6.88		e	$3.26		f	$9.45	
g	$8.90		h	$4.51		i	$7.43	
j	$5.04		k	$0.59		l	$8.74	
m	$7.13		n	$4.88		o	$0.42	
p	$8.77		q	$3.56		r	$8.04	

2 Find the total rounded cost of the items in each case.

a ice cream ($5.86), milk ($2.33), chocolate ($3.58)

b biscuits ($3.35), nuts ($12.48), raisins ($3.96)

c tea ($12.47), scissors ($2.42), calculator ($45.11)

d pens ($3.71), paper ($6.15), cereal ($7.32)

CONCEPT

To find the change, follow these steps:

Step 1 Find the total cost.

Step 2 Round to the nearest 5 cents.

Step 3 Count on, to determine the change

Step 4 Write the amount of change.

3 Complete the table.

	Cost of Items	Total	Total Rounded	Amount Given	Count on to Give Change	Change
a	$1.14, $6.35, $4.17	$11.66	$11.65	$50	5c, $11.70 30c, $12 $8, $20 $30	$38.35
b	$0.98, $7.93, $11.42			$50		
c	$12.56, $5.99, $3.04			$50		
d	$1.78, $4.29, $3.95			$20		
e	$0.75, $0.89, $1.13			$20		
f	$8.46, $12.18, $9.87			$100		
g	$37.54, $9.45, $6.69			$100		

2:52 What's the Rule?

CONCEPT

The rule here is take away 3.

14, 11, 8, ...

The next number would be 5.

The rule here is multiply by 3.

2, 6, 18 ...

The next number would be 54.

1 Write the next number in each pattern.

a 2, 4, 6, ☐ **b** 1, 5, 9, ☐ **c** 20, 18, 16, ☐

d 3, 6, 12, ☐ **e** 27, 9, 3, ☐ **f** 1, 5, 25, ☐

g 99, 98, 97, ☐ **h** 9, 16, 23, ☐ **i** 80, 40, 20, ☐

2 Write the rule for each part of Question 1.

a ☐ **b** ☐ **c** ☐

d ☐ **e** ☐ **f** ☐

g ☐ **h** ☐ **i** ☐

3 Continue each pattern by following the rule.

a Add 7. 0, ☐, ☐, ☐ **b** Subtract 5. 26, ☐, ☐, ☐

c Multiply by 4. 1, ☐, ☐, ☐ **d** Divide by 2. 88, ☐, ☐, ☐

e Add 11. 23, ☐, ☐, ☐ **f** Subtract 9. 47, ☐, ☐, ☐

g Multiply by 2. 6, ☐, ☐, ☐ **h** Divide by 3. 27, ☐, ☐, ☐

4 Write the pattern for the number of lines used in the pictures and write the rule used (like 'add 3').

a

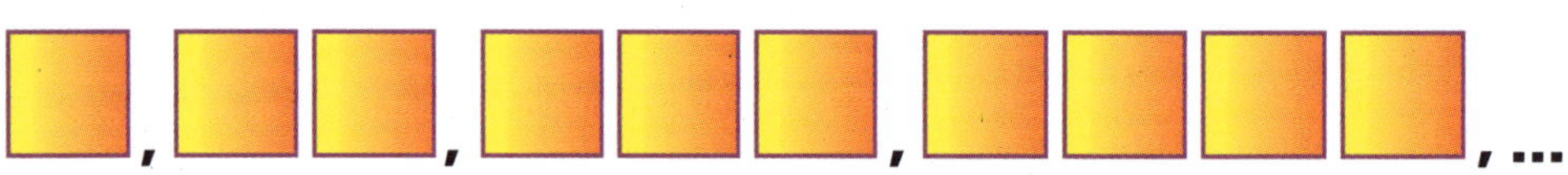

Pattern: 4, ☐, ☐, ☐, … Rule: ☐

b

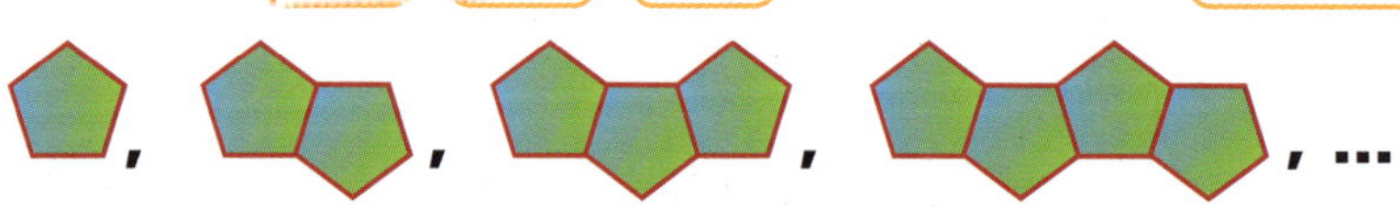

Pattern: ☐, ☐, ☐, ☐, … Rule: ☐

 • *AUSTRALIAN SIGNPOST MATHS 4* • ISBN 978 1 4886 2187 1

2:53 Multiplication Using Place-Value Blocks

CONCEPT

There are 18 stamps in each book.

How many stamps in 3 books?

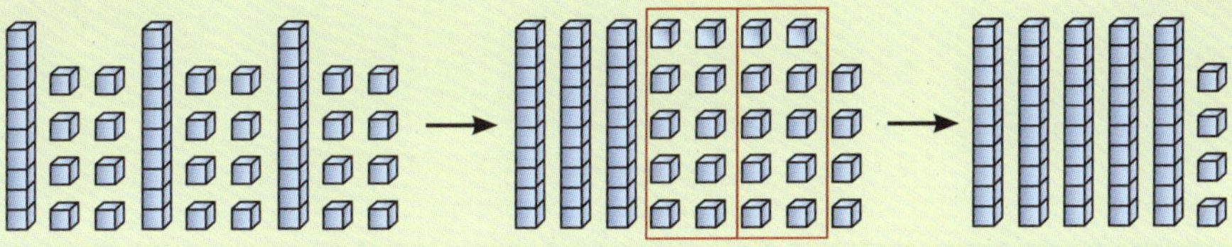

Make an estimate, then use place-value blocks to find:

1 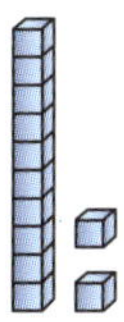

a $2 \times 12 =$ ☐ b $3 \times 12 =$ ☐ c $4 \times 12 =$ ☐

d $5 \times 12 =$ ☐ e $6 \times 12 =$ ☐ f $7 \times 12 =$ ☐

g $8 \times 12 =$ ☐ h $9 \times 12 =$ ☐ i $10 \times 12 =$ ☐

2 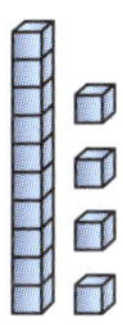

a 3 groups of 14 apples = ☐ 

b 5 rows of 14 candles = ☐ 

c 8 lots of 14 pencils = ☐ 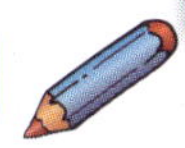

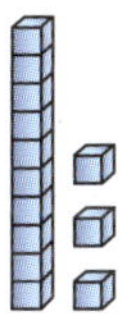

d 2 groups of 13 caps = ☐ 

e 5 rows of 13 fish = ☐ 

f 4 lots of 13 pins = ☐ 

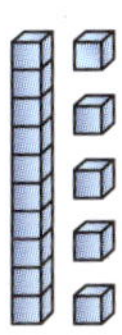

g 4 groups of 15 trains = ☐ 

h 9 rows of 15 nails = ☐ 

i 3 lots of 15 frogs = ☐ 

Ancient Egyptian Multiplication

ACTIVITY

This method involves doubling and adding.

A 1 group of 16 = **16**
B 2 groups of 16 = 32
C 4 groups of 16 = **64**
D 8 groups of 16 = **128**
E 16 groups of 16 = 256

To multiply by 13 we would add **D**, **C** and **A** (because $8 + 4 + 1 = 13$).

$13 \times 16 = \mathbf{128} + \mathbf{64} + \mathbf{16}$
$= 208$

3 Use this method and rows **A** to **E** to complete:

a $20 \times 16 =$ ☐ b $18 \times 16 =$ ☐ c $12 \times 16 =$ ☐

d $21 \times 16 =$ ☐ e $24 \times 16 =$ ☐ f $28 \times 16 =$ ☐

Use a calculator to check your answers.

 • AUSTRALIAN SIGNPOST MATHS 4 • ISBN 978 1 4886 2187 1

3:01 Analogue Time

CONCEPT

digital time: 6:50

analogue time: 10 to 7

6:50 → six fifty
6:50 → 50 minutes past 6
6:50 → 10 minutes to 7

1 Complete the labels for each time shown.

a 
:
past

b 
:
to

c 
:
to

d 
:
past

e 
:
to

f 
:
to

g 
:
past

h 
:
to

i 
:
past

j 
:
to

2 On each face, show the time given.

a 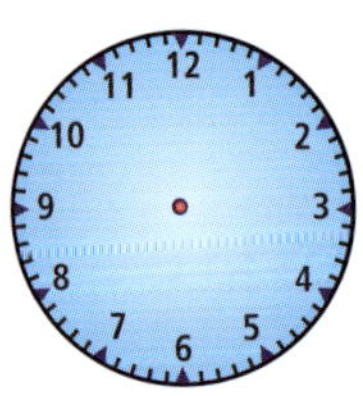
3:35

b 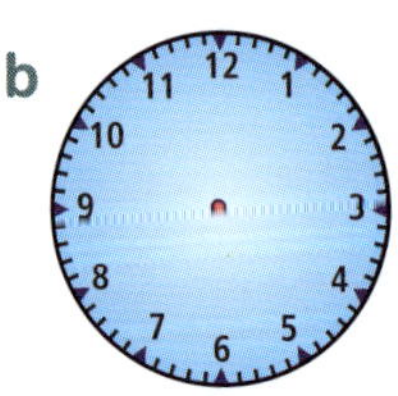
10:15

c 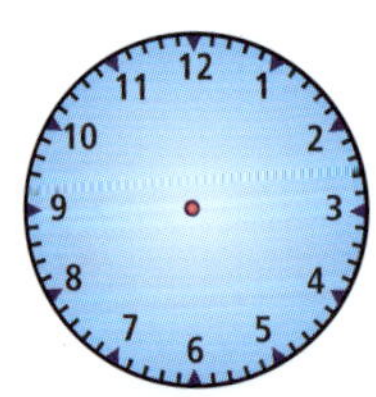
9:45

d 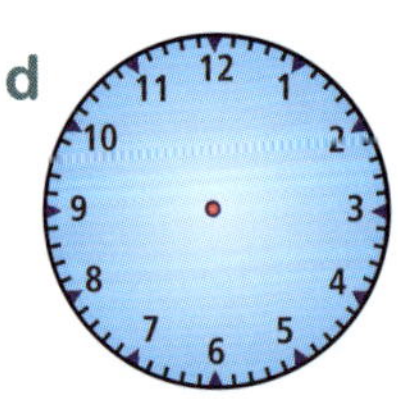
12:40

e 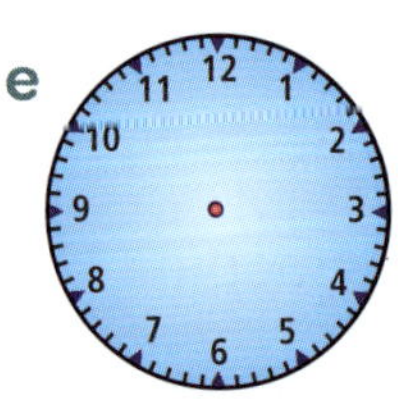
6:20

f 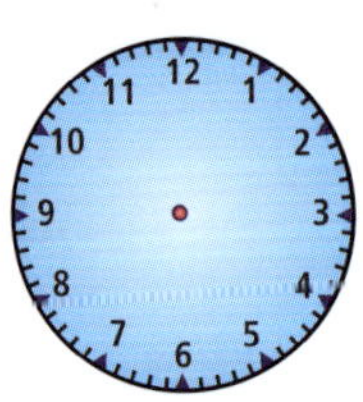
4:55

g 
8:05

h 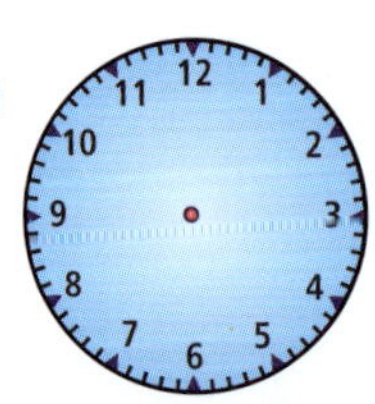
1:25

i 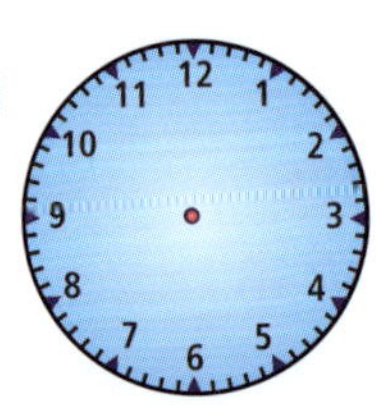
7:10

j 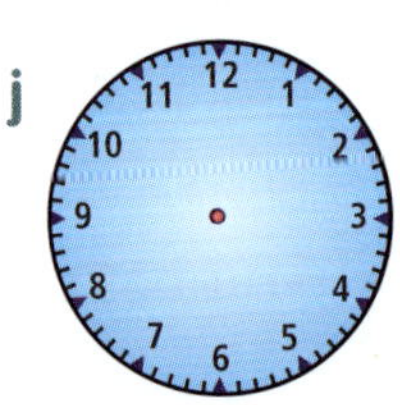
5:50

 • *AUSTRALIAN SIGNPOST MATHS 4* • ISBN 978 1 4886 2187 1

3:02 Analogue Time

CONCEPT

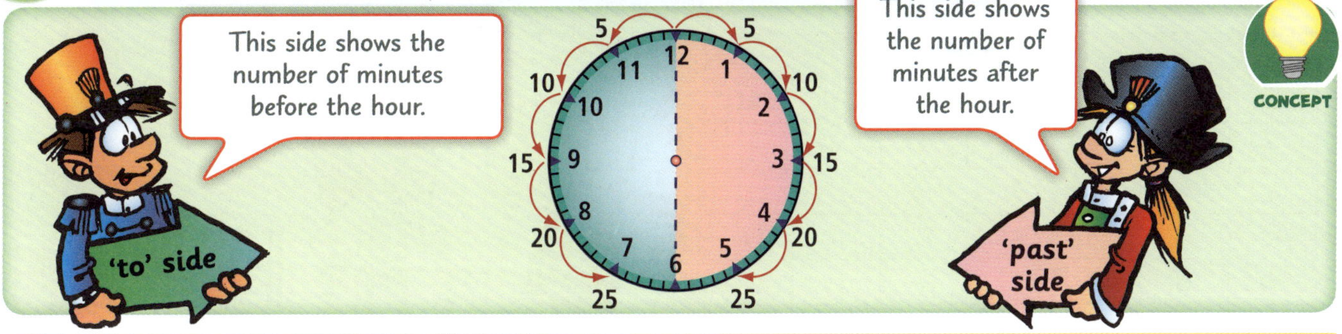

1 Complete the label for each time shown.

a  **to**

b 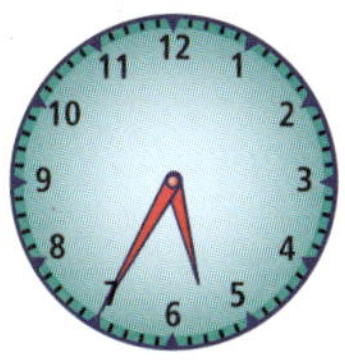 **to**

c  **past**

d 

e 

f 

g 

h 

i 

j 

2 Complete the clocks to show the given times.

a 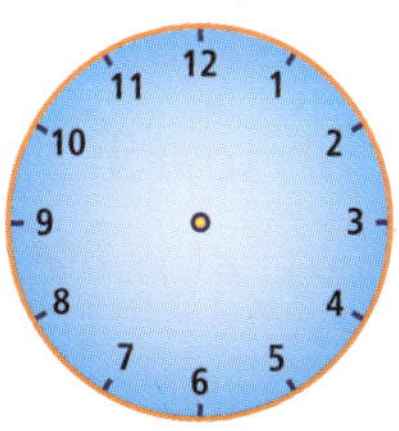 10 to 4

b 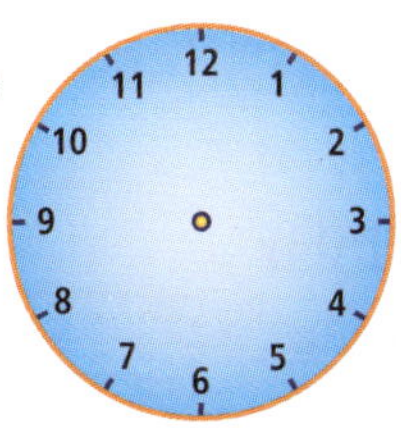 20 past 7

c  25 to 10

d 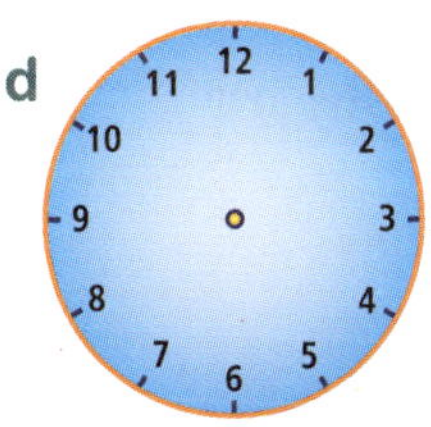 5 to 6

e  5 past 4

3 Write the time that is five minutes after:

a 5 past 8		b 10 past 7		c 10 to 12	
d 15 past 6		e 20 to 8		f 15 to 3	
g 20 past 3		h 25 to 3		i 5 to 9	

4 Write the time that is five minutes before:

a 25 past 1		b 10 past 4		c 15 to 4	
d 10 to 5		e 20 past 6		f 25 to 8	
g 15 past 10		h 20 to 11		i 5 past 2	

3:03 The Calendar

February						
Sun	Mon	Tues	Wed	Thur	Fri	Sat
		1	2	3	4	5
6	7	8	9	10	11	12
13	14	15	16	17	18	19
20	21	22	23	24	25	26
27	28	29				

Study the calendar then answer these questions.

1 a On what day of the week is the 18th?

b Write the date of the first Thursday.

c Write the date of the last Friday.

d How many Saturdays are in February?

e How many Tuesdays are there?

f On what day does February begin?

g How many days are in February on this calendar?

h How many days are in February in most years?

7 days in a week
2 weeks in a fortnight
52 weeks in a year
365 days in a year
366 days in a leap year (one year in four)
12 months in a year

2 Write the date for:

a the first Wednesday

b the first Sunday

c the last Saturday

d the second Monday

3 Write the day of:

a 10th February

b 4th February

c 19th February

d 21st February

4 Write the dates for:

a the first weekend

b the second weekend

c the third weekend

d all the Wednesdays

INVESTIGATION

Research

What is the average temperature in:

- February?
- July?

ACTIVITY

On this time line, show events that happened yesterday. **breakfast, went home, bed**

3:04 The Calendar

Thirty days has September, April, June and November
All the rest have thirty-one, except for February alone
Which has twenty-eight days clear,
and twenty-nine days each leap year.

Here is a jingle to help you remember how many days there are in each month.

A calendar shows the way a year is broken up into months, weeks and days.

January

S	M	T	W	T	F	S
		1	2	3	4	5
6	7	8	9	10	11	12
13	14	15	16	17	18	19
20	21	22	23	24	25	26
27	28	29	30	31		

February

S	M	T	W	T	F	S
					1	2
3	4	5	6	7	8	9
10	11	12	13	14	15	16
17	18	19	20	21	22	23
24	25	26	27	28		

March

S	M	T	W	T	F	S
31					1	2
3	4	5	6	7	8	9
10	11	12	13	14	15	16
17	18	19	20	21	22	23
24	25	26	27	28	29	30

April

S	M	T	W	T	F	S
	1	2	3	4	5	6
7	8	9	10	11	12	13
14	15	16	17	18	19	20
21	22	23	24	25	26	27
28	29	30				

May

S	M	T	W	T	F	S
			1	2	3	4
5	6	7	8	9	10	11
12	13	14	15	16	17	18
19	20	21	22	23	24	25
26	27	28	29	30	31	

June

S	M	T	W	T	F	S
30						1
2	3	4	5	6	7	8
9	10	11	12	13	14	15
16	17	18	19	20	21	22
23	24	25	26	27	28	29

July

S	M	T	W	T	F	S
	1	2	3	4	5	6
7	8	9	10	11	12	13
14	15	16	17	18	19	20
21	22	23	24	25	26	27
28	29	30	31			

August

S	M	T	W	T	F	S
				1	2	3
4	5	6	7	8	9	10
11	12	13	14	15	16	17
18	19	20	21	22	23	24
25	26	27	28	29	30	31

September

S	M	T	W	T	F	S
1	2	3	4	5	6	7
8	9	10	11	12	13	14
15	16	17	18	19	20	21
22	23	24	25	26	27	28
29	30					

October

S	M	T	W	T	F	S
		1	2	3	4	5
6	7	8	9	10	11	12
13	14	15	16	17	18	19
20	21	22	23	24	25	26
27	28	29	30	31		

November

S	M	T	W	T	F	S
					1	2
3	4	5	6	7	8	9
10	11	12	13	14	15	16
17	18	19	20	21	22	23
24	25	26	27	28	29	30

December

S	M	T	W	T	F	S
1	2	3	4	5	6	7
8	9	10	11	12	13	14
15	16	17	18	19	20	21
22	23	24	25	26	27	28
29	30	31				

1 Look carefully at the calendar and find which month comes before:

a February ______ **b** November ______ **c** June ______ **d** October ______
e March ______ **f** August ______ **g** May ______ **h** July ______

2 How many days are in:

a April? ______ **b** August? ______ **c** December? ______ **d** March? ______
e June? ______ **f** October? ______ **g** January? ______ **h** May? ______

3 Look carefully at the month of April. Write the day of:

a 1st April ______ **b** 30th April ______ **c** 7th April ______

4 Look carefully at the month of May. Write the date of:

a the first Sunday ______ **b** the first Wednesday ______

5 Look at the month of September. Write the day and date for:

a one week after 12th ______ **b** two days before 27th ______

6 Study the whole calendar to find the day of:

a 13th September ______ **b** 21st June ______ **c** 2nd February ______
d 31st May ______ **e** 29th August ______ **f** 3rd January ______

© PEARSON AUSTRALIA 2018 • *AUSTRALIAN SIGNPOST MATHS 4* • ISBN 978 1 4886 2187 1

3:05 Using Centimetres and Millimetres

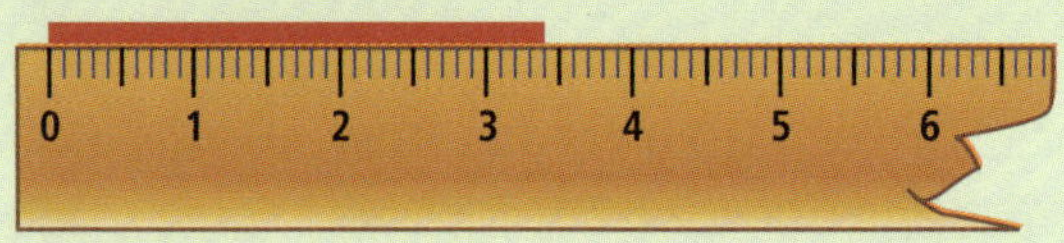

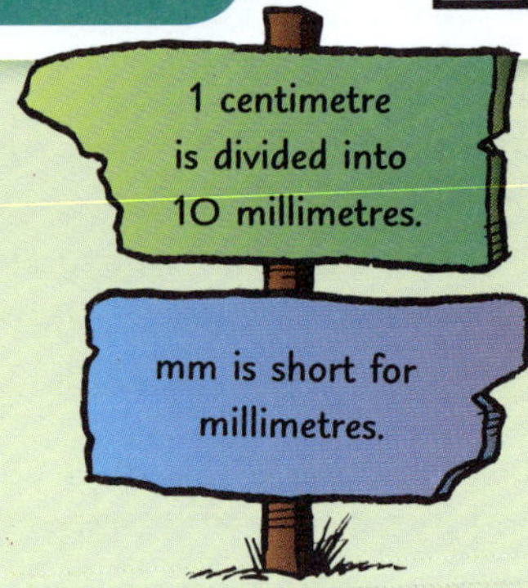

CONCEPT

1 Estimate, then measure the length of each bar.

a ______ mm b ______ mm
c ______ mm d ______ mm
e ______ mm f ______ mm

2 Write these as millimetres.

a 1 cm 9 mm ______ mm b 4 cm 7 mm ______ mm c 6 cm 1 mm ______ mm
d 3 cm 6 mm ______ mm e 5 cm 5 mm ______ mm f 7 cm 2 mm ______ mm
g 2 cm 8 mm ______ mm h 9 cm 3 mm ______ mm

3 Write these as centimetres and millimetres.

a 25 mm ______ b 68 mm ______ c 51 mm ______
d 92 mm ______ e 43 mm ______ f 87 mm ______

ACTIVITY

- Estimate, then use a ruler to measure these objects to the nearest millimetre.

Object	Estimated Length	Actual Length
length of a finger		
height of a cup		
width of an eraser		
thickness of a book		

- Use a tape measure to measure these objects to the nearest centimetre.
  - the length of a brick ______
  - the width of your desk ______
  - the height of your chair ______

 • • ISBN 978 1 4886 2187 1

3:06 Using Millimetres

1 Estimate, then measure the length of each bar.

a ☐ mm b ☐ mm

c ☐ mm d ☐ mm

e ☐ mm f ☐ mm

2 Record the length of each interval.

a ☐ mm

b ☐ mm

c ☐ mm

d ☐ mm

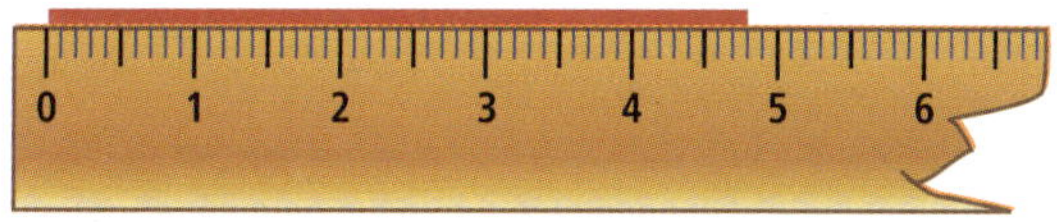

e ☐ mm

3 Write these as millimetres.

a 3 cm 2 mm ☐ b 8 cm 1 mm ☐ c 1 cm 9 mm ☐ d 5 cm 3 mm ☐

e 7 cm 4 mm ☐ f 2 cm 8 mm ☐ g 4 cm 6 mm ☐ h 9 cm 5 mm ☐

4 Write these as centimetres and millimetres.

a 38 mm ☐ b 19 mm ☐ c 47 mm ☐

d 82 mm ☐ e 51 mm ☐ f 25 mm ☐

g 64 mm ☐ h 93 mm ☐ i 76 mm ☐

5 Find the perimeter of each shape.

a  ☐

b  ☐

c 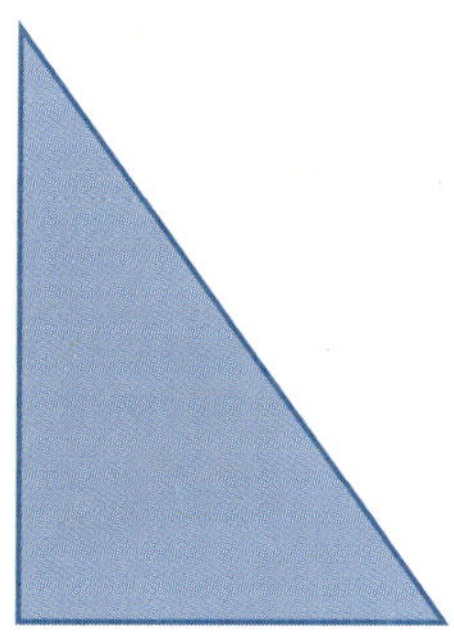 ☐

3:07 Recording Length

1 m equals 100 cm.

1 Write these as centimetres.

a 2 m 56 cm ______ b 3 m 16 cm ______

c 5 m 40 cm ______ d 7 m 81 cm ______

e 4 m 95 cm ______ f 1 m 79 cm ______

g 8 m 63 cm ______ h 6 m 28 cm ______

2 Write these as metres and centimetres.

a 119 cm ______ b 853 cm ______

c 582 cm ______ d 697 cm ______

e 935 cm ______ f 374 cm ______

3 Write these as centimetres and millimetres.

a 27 mm ______ b 36 mm ______ c 51 mm ______

d 19 mm ______ e 42 mm ______ f 85 mm ______

4 Write these as millimetres.

a 5 cm 4 mm ______ b 6 cm 9 mm ______ c 1 cm 1 mm ______

d 8 cm 7 mm ______ e 4 cm 3 mm ______ f 2 cm 6 mm ______

5 Which unit of measure (**m**, **cm** or **mm**) has been left off each measurement?

a

tree **5**

b
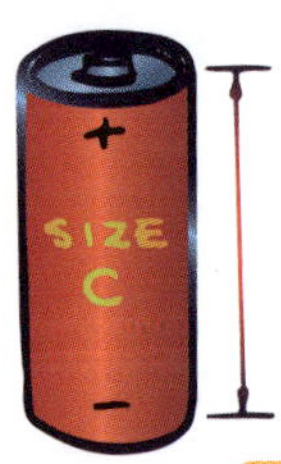

battery **5**

c

window **3**

d
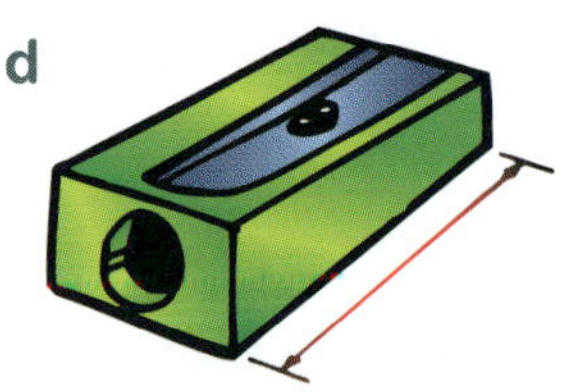
sharpener **25**

e
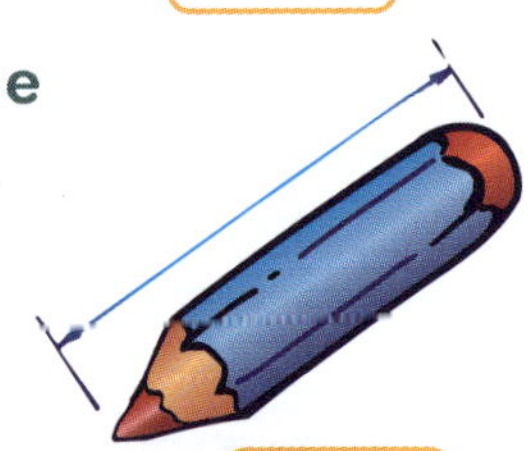
pencil **14**

f
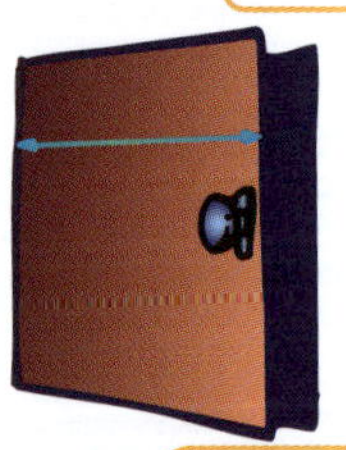
door **1**

g
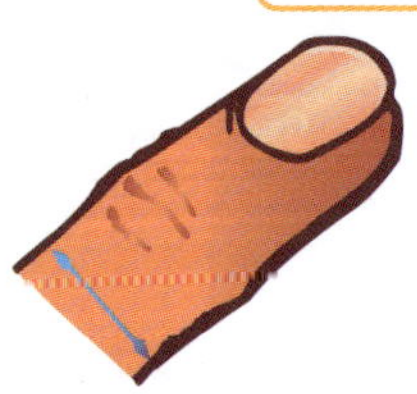
finger **12**

h

ruler **300**

3:08 Using Measurement Scales

When we use a measuring instrument, we read the scale to the nearest unit mark.

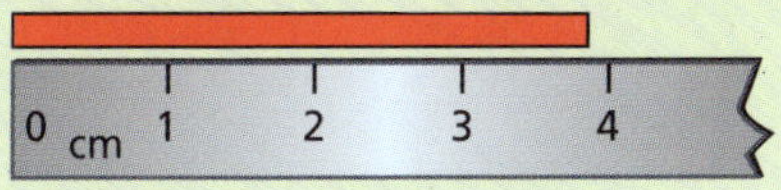

The measured length would be recorded as 4 cm.

CONCEPT

1 Record the measure of the bar using each of the rulers.

a

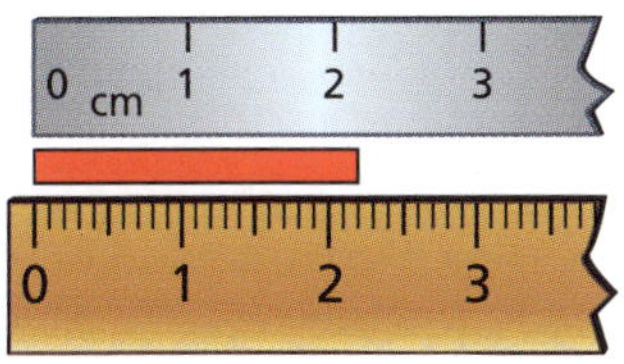

☐ cm

☐ mm

b

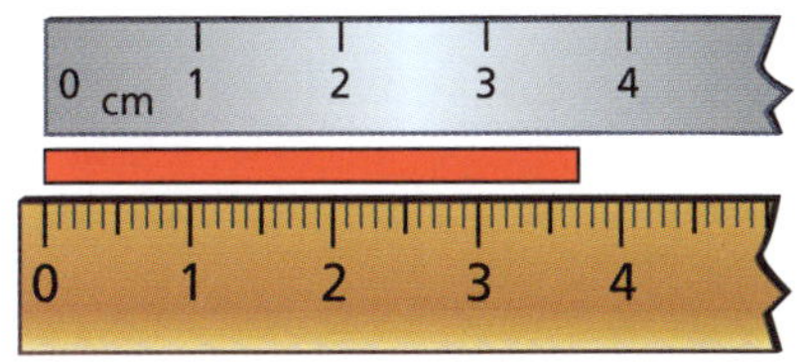

☐ cm

☐ mm

c

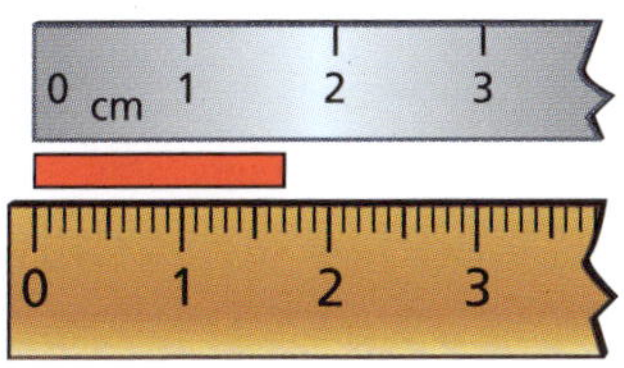

☐ cm

☐ mm

d

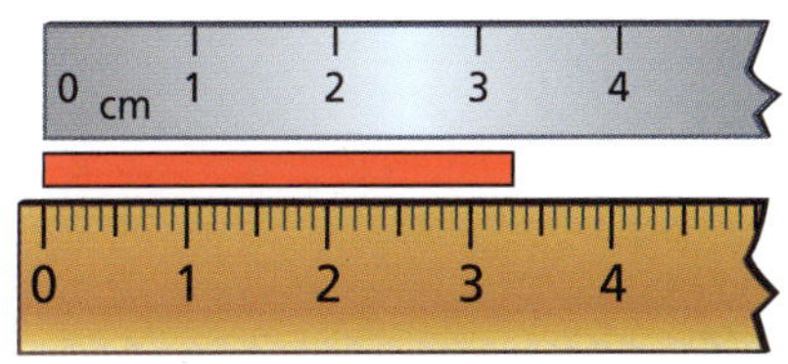

☐ cm

☐ mm

e

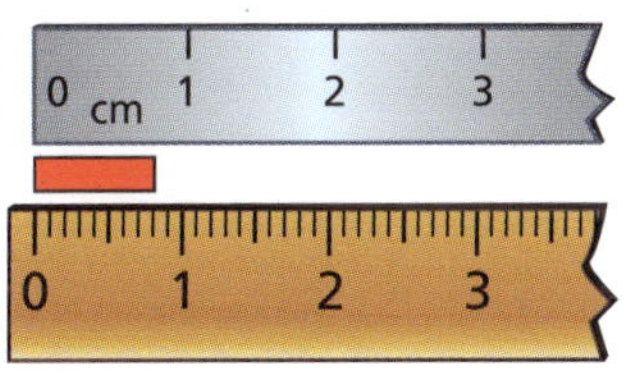

☐ cm

☐ mm

f

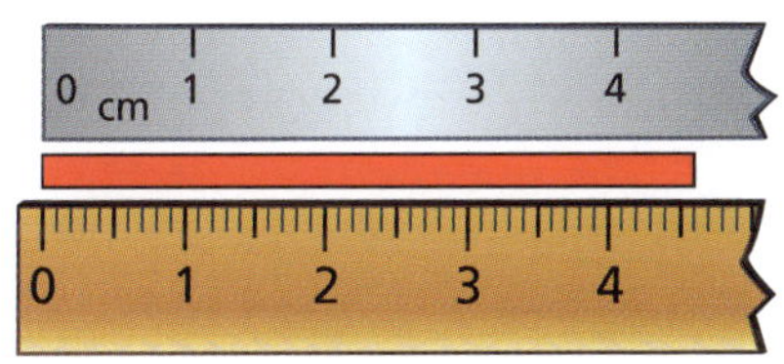

☐ cm

☐ mm

g

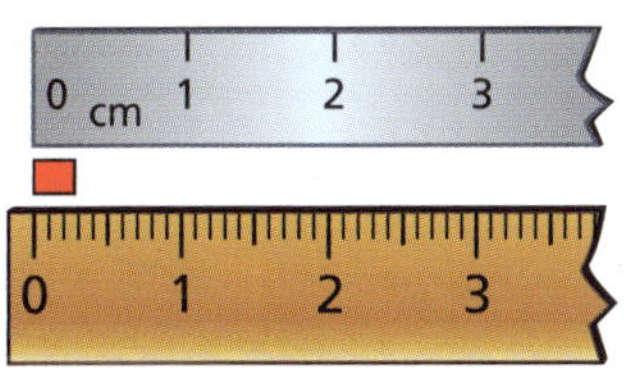

☐ cm

☐ mm

What does a measurement of 0 cm mean?

2 Record the measure shown on each scale.

a

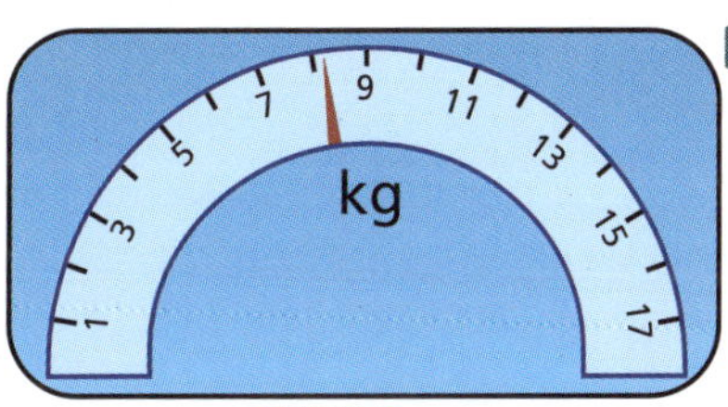

☐ kg

b

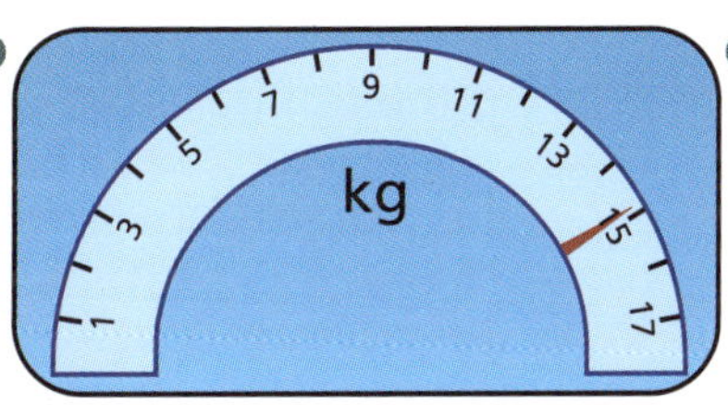

☐ kg

c

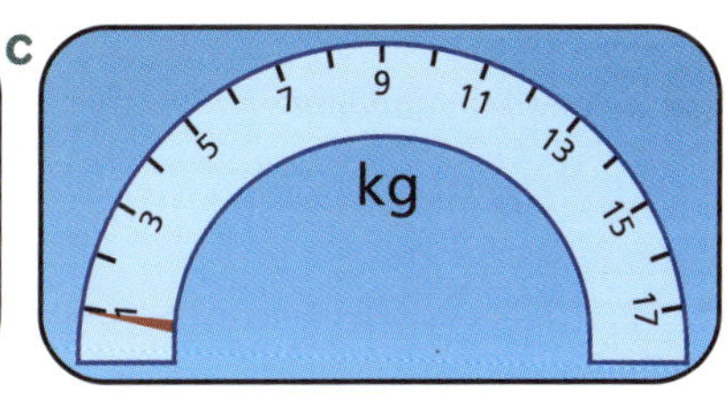

☐ kg.

d

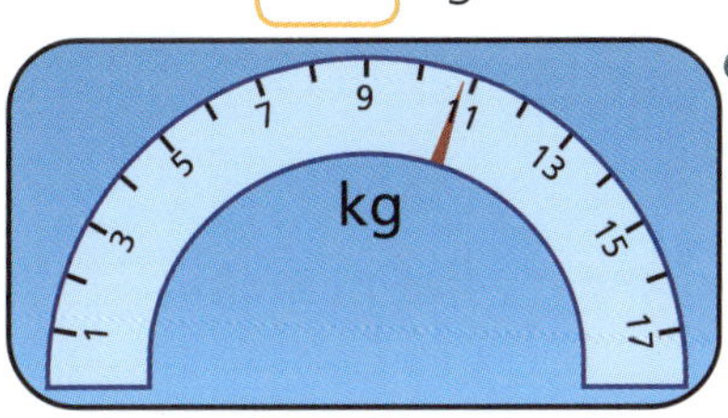

☐ kg

e

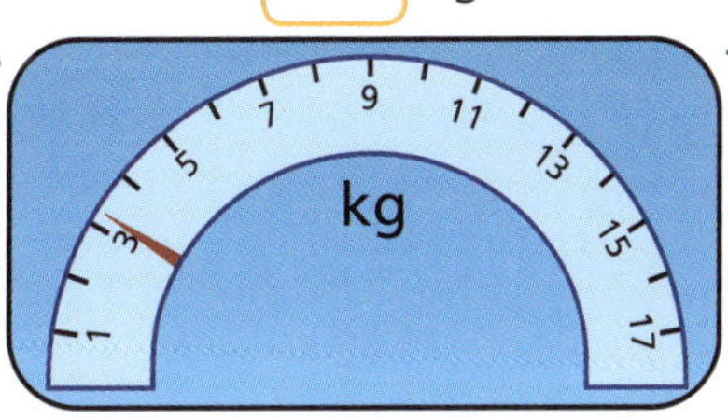

☐ kg

f

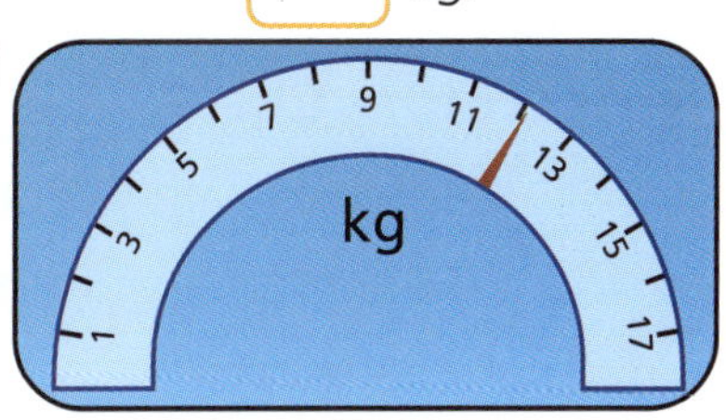

☐ kg

MEASUREMENT & GEOMETRY

3:09 Temperature

To read temperature we use the Celsius temperature scale.

0°C is the *freezing* point of water.

100°C is the *boiling* point of water.

CONCEPT

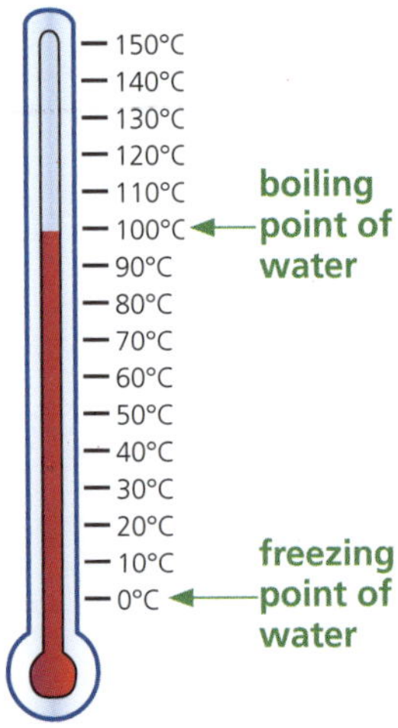

1 Draw lines to make each statement true.

The boiling point of water ●	● is cool to warm weather.
35°C to 45°C ●	● is cold weather.
25°C to 35°C ●	● is 100°C.
15°C to 25°C ●	● is 0°C.
5°C to 15°C ●	● is warm to hot weather.
The freezing point of water ●	● is very hot weather.

2 What temperature is shown on each thermometer?

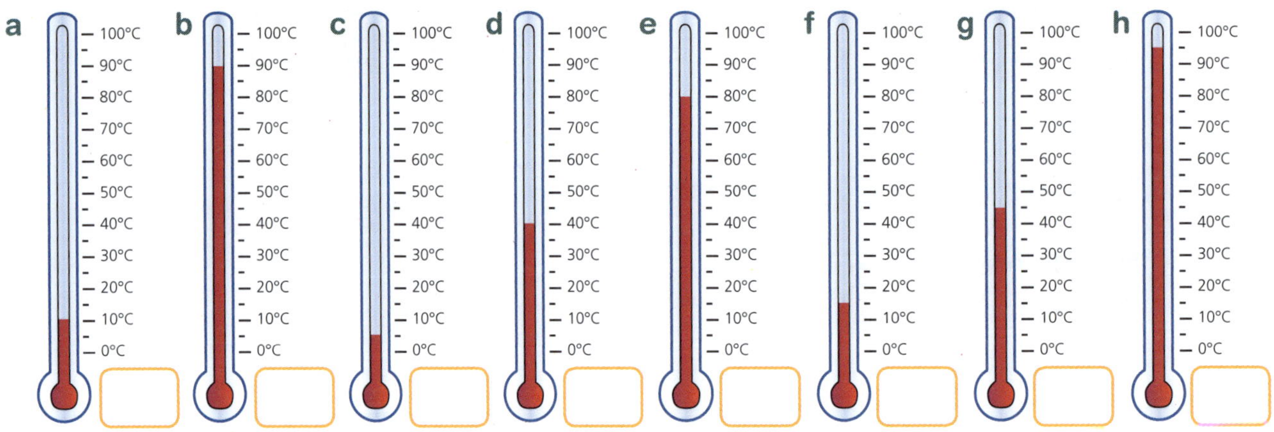

3 Each thermometer above is measuring the temperature of water.

a Which three thermometers show the water is hot?

b Which three thermometers show the water is cold?

c Which two thermometers show it is warm?

Temperature Chart

ACTIVITY

- Fill a container with warm water.
- Place a thermometer in the water and record the temperature every five minutes.
- Graph the results.
- Discuss why the temperature dropped quickly at the beginning of the experiment.

3:10 Recording Temperature

- When reading a thermometer use the nearest graduation mark.

- Read aloud these temperatures.
  - 42°C • 27°C • 19°C • 63°C • 100°C • 37°C • 84°C • 34°C
  - 16°C • 0°C • 96°C • 21°C • 39°C • 72°C • 55°C • 25°C

1 Write these measurements of temperature in short form.

a twenty-five degrees Celsius		**b** zero degrees Celsius	
c eighteen degrees Celsius		**d** fifty-eight degrees Celsius	
e eighty-six degrees Celsius		**f** thirty-seven degrees Celsius	
g forty-one degrees Celsius		**h** seventy-three degrees Celsius	

2 On each thermometer, colour the mercury to match the given temperature.

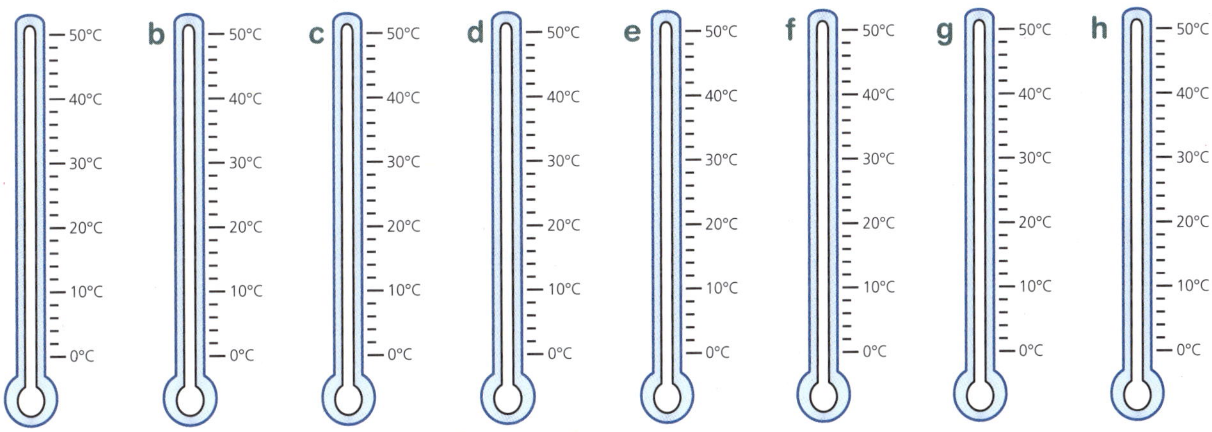

a 40°C **b** 28°C **c** 36°C **d** 14°C **e** 8°C **f** 33°C **g** 27°C **h** 15°C

3 Draw lines to match to the most suitable temperature.

a a cold glass of milk	40°C	**f** a very cold day	100°C
b a very hot day	96°C	**g** water boiling	17°C
c a warm day	70°C	**h** body temperature	67°C
d a bowl of hot soup	26°C	**i** a cup of coffee	9°C
e a pot of very hot tea	2°C	**j** a cool day	37°C

3:11 Using Millilitres

CONCEPT

ACTIVITY

- Collect containers that have labels marked in millilitres.
  List the kinds of products sold in millilitres.
  Discuss the most common shape found in the containers.
- Pour 100 mL of water from a standard measure into a large plastic bottle. Mark the level of water with a pen.
  Pour in another 100 mL of water and mark that level.
  Repeat until 1 litre has been reached.

  This is a **calibrated measure**.

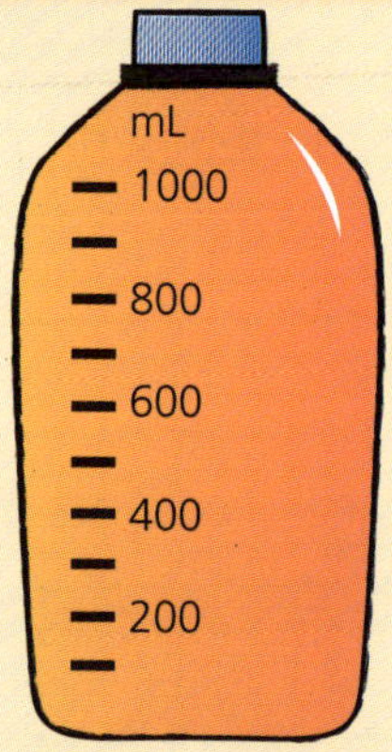

- Check the accuracy of the calibrated measure by pouring in 500 mL of water.
- Pour water from different containers into the calibrated one. Check that the capacity shown on each label is correct.

1 Which measure (**L** or **mL**) has been left off each label?

2 Would we use litres (**L**) or millilitres (**mL**) to measure:

a drink in a cup?		**b** water in a bath?	
c medicine in a teaspoon?		**d** juice in a glass?	
e water in a fish pond?		**f** petrol in our car?	
g cream in a small carton?		**h** ice cream in a large container?	

3 How many litres in:

a 1 000 mL?	**b** 4 000 mL?	**c** 2 000 mL?	**d** 6 000 mL?
e 5 000 mL?	**f** 7 000 mL?	**g** 9 000 mL?	**h** 3 000 mL?

3:12 Using Millilitres

- Use a medicine glass to measure the capacity of:
  - a teaspoon
  - a dessertspoon
  - an eggcup
  - an eye-dropper
  - a tablespoon
  - a bottle lid
- Arrange the above containers in order, from smallest to largest.
- Pour an estimated 10 mL of water into different containers.
- Use the medicine glass to check the estimates.
- Complete the table and discuss any difficulties.

Container	Estimated 10 mL	Measurement	Description of Estimate	Difference
Eggcup	10 mL	15 mL	Too big	5 mL
Tall Bottle	10 mL			
Thin Bottle	10 mL			
Short Bottle	10 mL			

1 Would we use litres or millilitres to measure:

a a bowl of soup? ☐ b a sink full of water? ☐ c a spoonful of sauce? ☐
d a cup of coffee? ☐ e a drum of oil? ☐ f a can of drink? ☐

2 How many millilitres are there in:

a 2 L? ☐ b 7 L? ☐ c 3 L? ☐ d 5 L? ☐
e 8 L? ☐ f 4 L? ☐ g $6\frac{1}{2}$ L? ☐ h $9\frac{1}{2}$ L? ☐

3 How many litres are there in:

a 1000 mL? ☐ b 7000 mL? ☐ c 5000 mL? ☐ d 3000 mL? ☐
e 6000 mL? ☐ f 4000 mL? ☐ g 2000 mL? ☐ h 8000 mL? ☐

4 What is the total capacity of each set of containers?

a
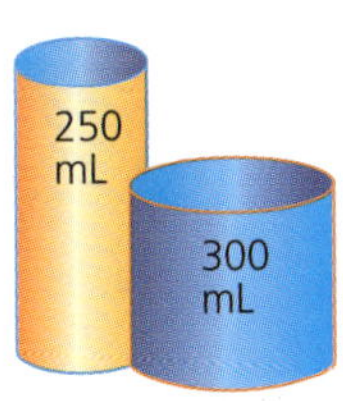

☐

b
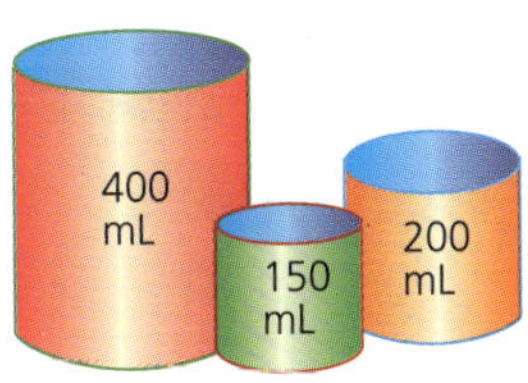

☐

c
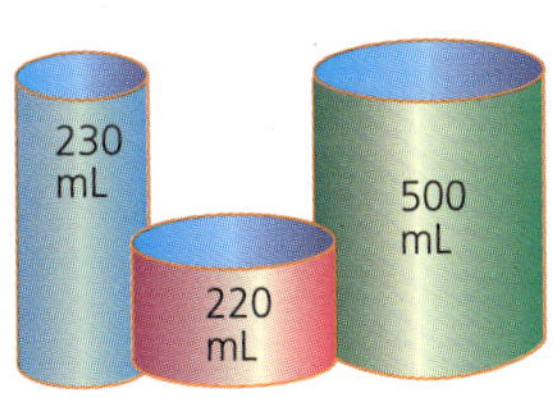

☐

d
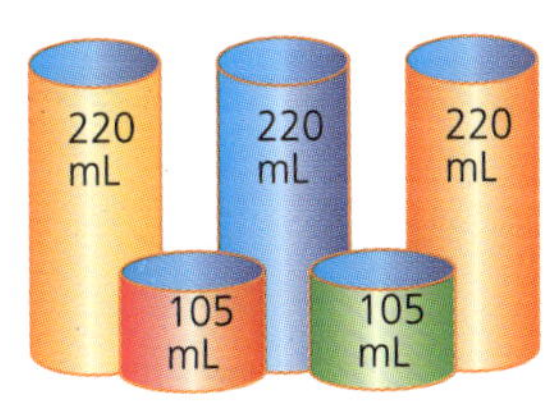

☐

 • AUSTRALIAN SIGNPOST MATHS 4 • ISBN 978 1 4886 2187 1

3:13 Area Using Informal Units

CONCEPT

The area of a shape is the amount of space within it.

To measure area we use units that are placed together so that there are no gaps or overlaps.

1. How many tiles have been used to cover each area?

a
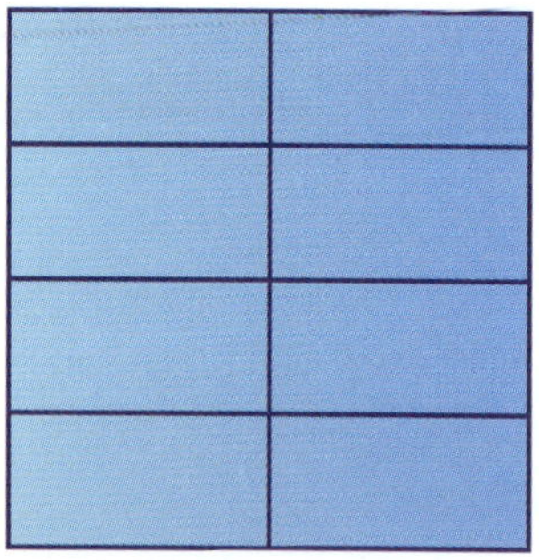

☐ rectangular tiles

b

☐ square tiles

c

☐ triangular tiles

2. Books were used to cover this desk.

   a Did the books cover the desk exactly? ☐

   b Estimate the amount of overhang. ☐

   c If we could cut the books, how many books would cover the table? ☐

   d Do you think a book would be a good unit of area? ☐

   Why or why not? ______

3. Which tile in Question 1 do you think is best for measuring area? ☐

   Why did you choose that tile? ______

Comparing Areas

INVESTIGATION

- Choose three areas of similar size.
- Choose a unit of area to compare them.
- Measure each area and list the areas in order of size, smallest to largest.

☐ ☐ ☐

 • *AUSTRALIAN SIGNPOST MATHS 4* • ISBN 978 1 4886 2187 1

3:14 The Square Centimetre

Small areas can be measured in **square centimetres**. A square centimetre is a square with each side 1 cm long. The short way to write square centimetre is **cm^2**.

CONCEPT

1 Record the area of each shape.

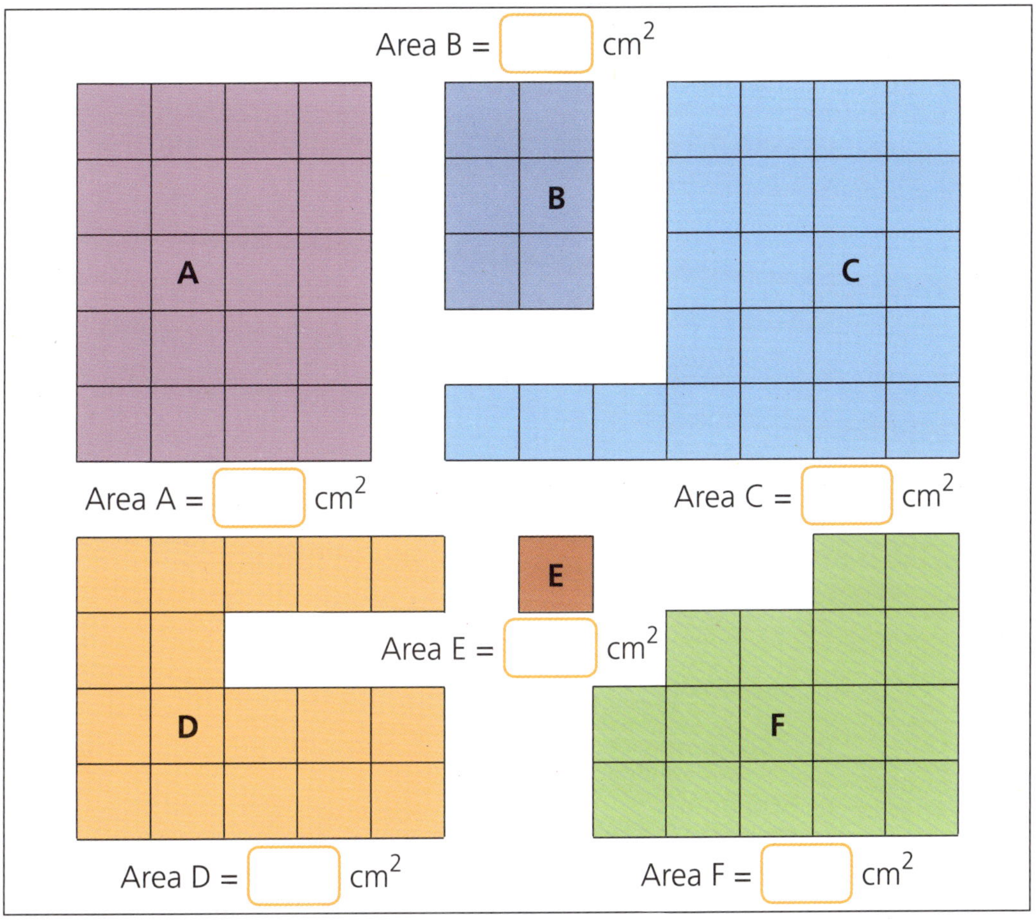

We use cm^2 to measure small areas.

2 Which shape has:

a the greatest area? ☐ **b** the least area? ☐

c an area less than **B**? ☐ **d** an area less than **A** but greater than **F**? ☐

ACTIVITY

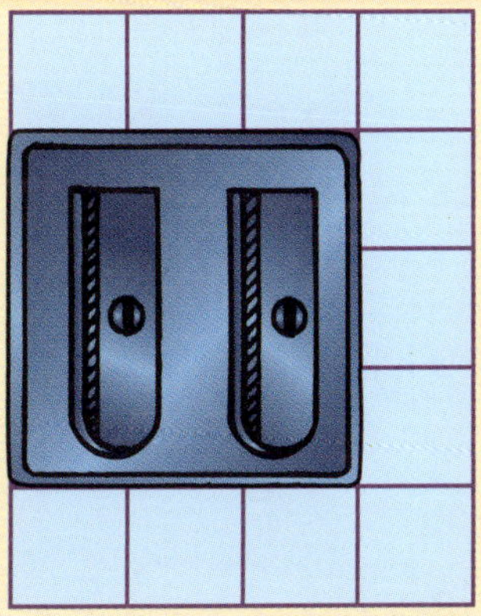

- On 1 cm grid paper, trace around some small objects. Count the number of square centimetres covered. Colour and label each area.
- Overlay some small surfaces with a 1 cm grid transparency. Count the number of square centimetres covered. Label each area measured.

sharpener 9 cm^2

Any square with more than half covered should be counted as one square.

3:15 Measuring Mass

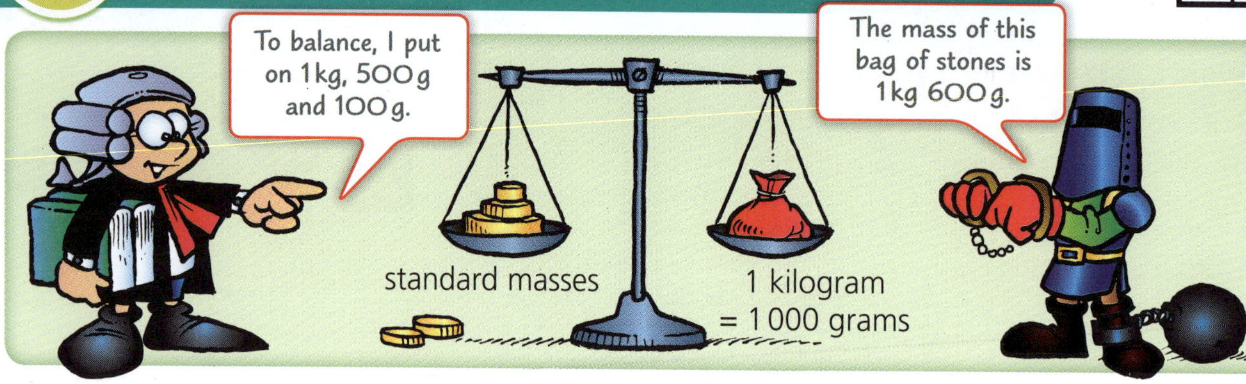

CONCEPT

ACTIVITY

- Use an equal arm balance and standard masses to measure classroom objects to the nearest 10 grams.
- Complete the table.

Object	Estimated Mass	Actual Mass	Standard Masses Used

1 Use the short form to write:

a 60 grams ☐ b 1 kilogram ☐ c 350 grams ☐

d 700 grams ☐ e 500 grams ☐ f 3 kilograms ☐

g 9 kilograms ☐ h 900 grams ☐ i 16 kilograms ☐

2 Which unit of measurement (**g** or **kg**) has been left off each item?

a 

☐

b 

☐

c 

☐

d 

☐

e 

☐

3 How many grams are there in:

a 1 kilogram? ☐ b 3 kilograms? ☐ c 5 kilograms? ☐ d 9 kilograms? ☐

4 How many kilograms are there in:

a 1000 g? ☐ b 4000 g? ☐ c 6000 g? ☐ d 8000 g? ☐

 • *AUSTRALIAN SIGNPOST MATHS 4* • ISBN 978 1 4886 2187 1

3:16 Measuring Mass

Use balance scales and standard masses to complete this table.

ACTIVITY

Mass	Objects Found
Less than 100 g	
About 500 g	
About 1 kg	
More than 1 kg	

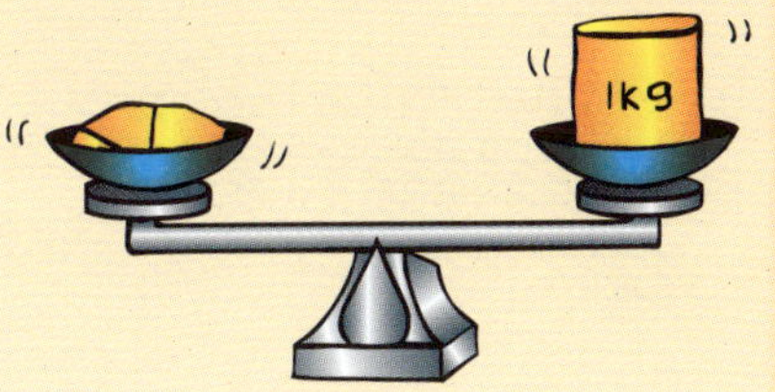

1 Use the short form to write:

a 1 kilogram
b 450 grams
c 3 kilograms
d 935 grams
e 325 grams
f 750 grams
g 1 kilogram 300 grams
h 1 kilogram 840 grams
i 1 kilogram 560 grams
j 2 kilograms 125 grams

2 How many grams are in:

a 3 kilograms?
b 5 kilograms?
c 9 kilograms?
d 7 kilograms?
e 4 kilograms?
f 6 kilograms?

3 How many kilograms are in:

a 1 000 grams?
b 8 000 grams?
c 2 000 grams?
d 4 000 grams?
e 6 000 grams?
f 5 000 grams?

4 Which unit of measure (**g** or **kg**) has been left off each item?

a
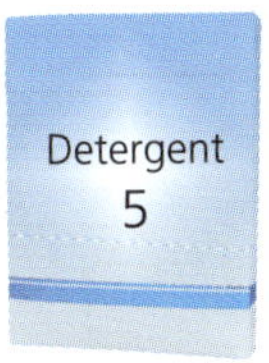

b

c

d

e

f

g

h
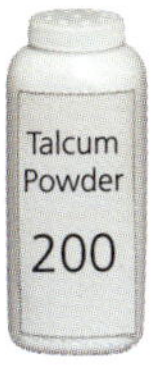

i

j

3:17 The Square Centimetre

- Use a 1 cm grid transparency to measure and compare the areas of different surfaces.
- Discuss ways to find the area when some squares are not fully covered.
- List items that have an area between 20 cm^2 and 50 cm^2.

INVESTIGATION

Area is 15 cm^2.

1 Use a 1 cm grid transparency to find the area of each shape.

a

b

c

d

e

f

2 Draw your own grid to find the area of each shape.

a

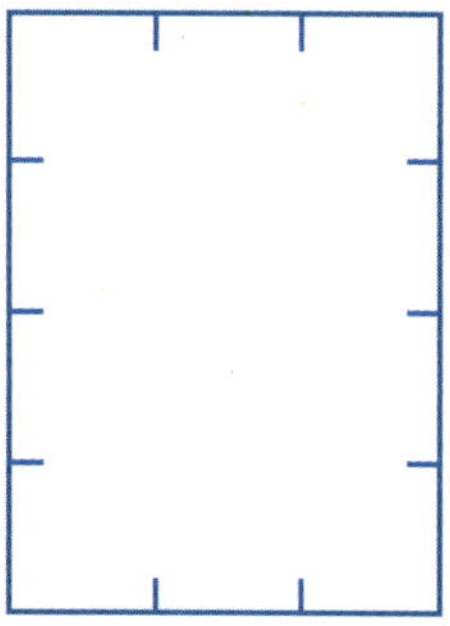

b

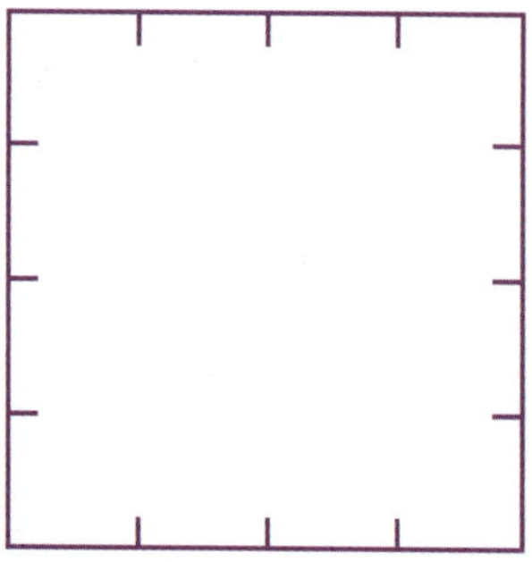

c

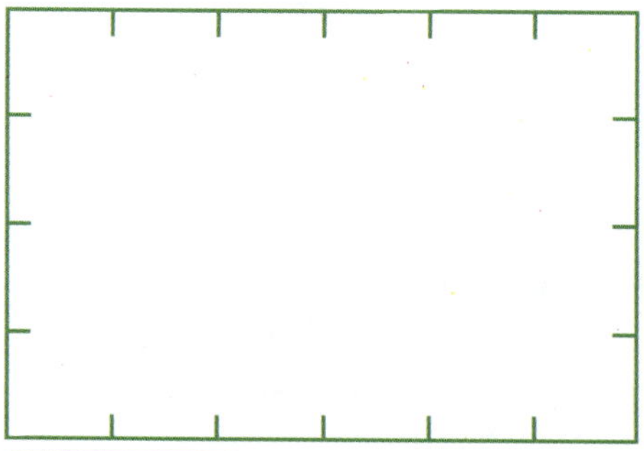

3:18 The Square Centimetre

1. Use a 1 cm grid transparency to find the area of each shape.

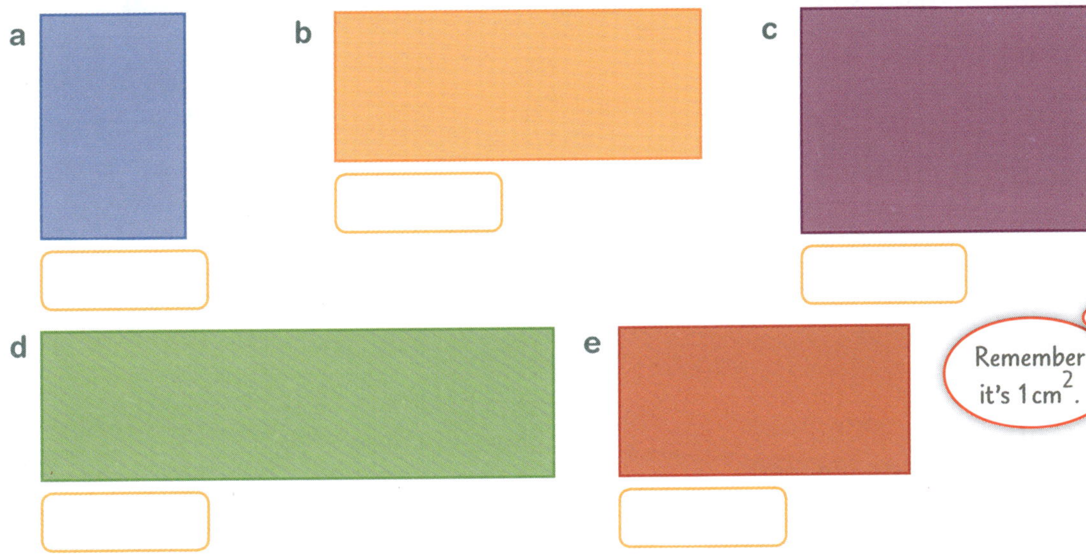

2. Draw your own grid to find the area of each shape.

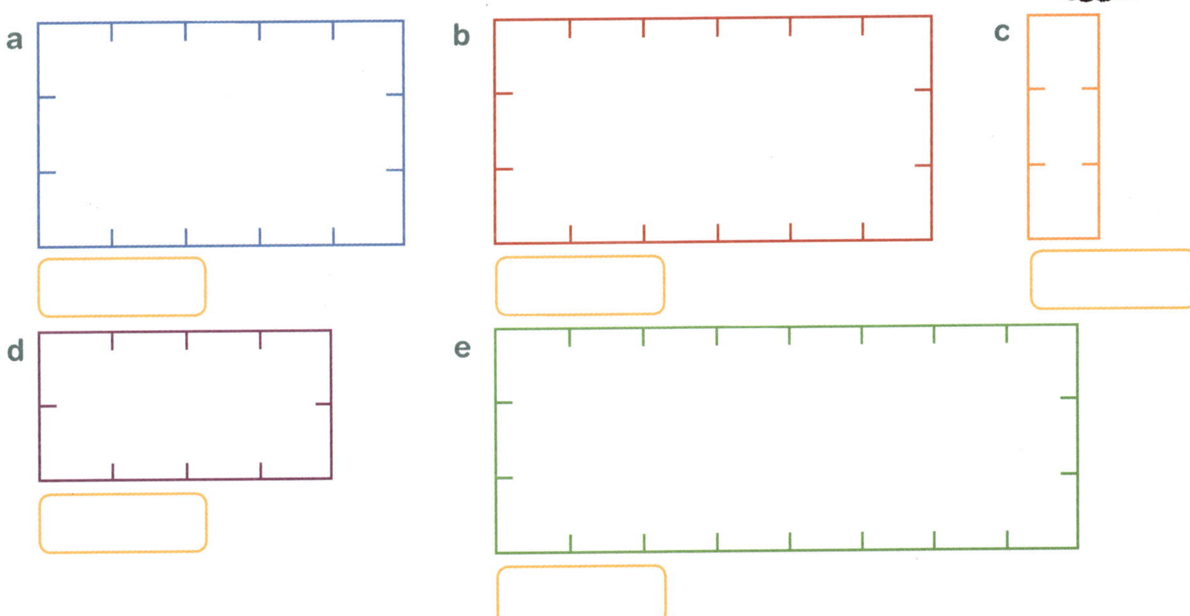

ACTIVITY

- Use a 10 cm by 10 cm grid transparency to find areas:
  - less than 100 cm^2
  - about 100 cm^2
  - more than 100 cm^2
- Record three surfaces with areas of more than 100 cm^2.

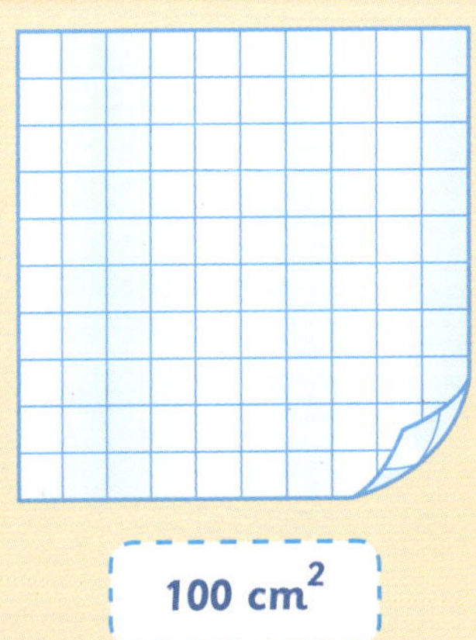

3:19 The Square Metre

Making a Square Metre

- Tape large sheets of newspaper together.
- Use a metre ruler to draw a square of side length 1 m on the paper.
- Cut out your square.

1 m

1 m

Because this square has sides of length one metre, it is a square metre.

ACTIVITY

1. List things in your classroom that are bigger than a square metre.

2. List things in your classroom that are smaller than a square metre.

3. Estimate and then check to find how many:
   a children can stand on one square metre.
   b children can sit on one square metre.
   c children can lie on one square metre.

The short way to write 1 square metre is 1 m^2.

4. Estimate and then check to find how many:
   a books are needed to cover one square metre.
   b school bags are needed to cover one square metre.
   c boxes are needed to cover one square metre.

5. Estimate and then check to find how many:
   a square metres are needed to cover a teacher's desk.
   b square metres are needed to cover a door.
   c square metres are needed to cover the floor.

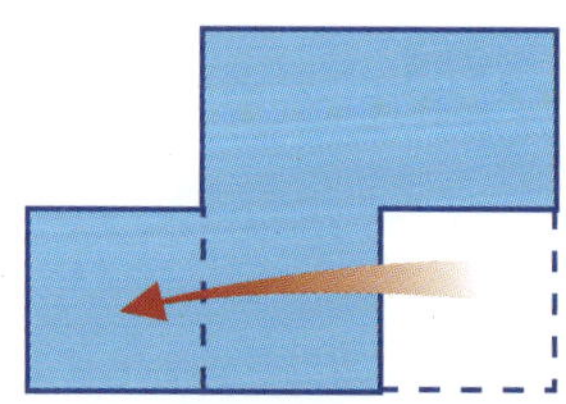

6. Cut, then join your square metre to make shapes like these.
   a Does each still cover one square metre?
   b Discuss how this could help us measure different surfaces.

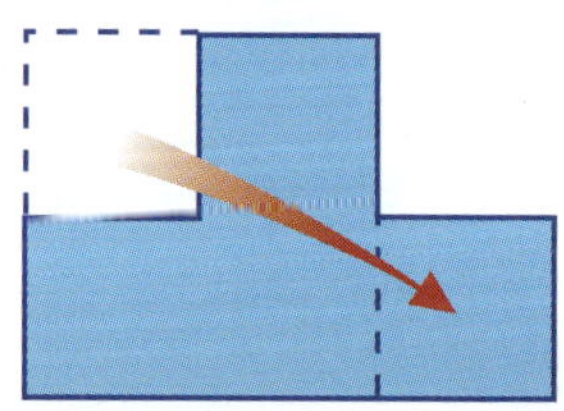

Other Units of Area

FUN SPOT

Investigate other units of area, such as tatami mats (from Japan) and 'squares' for room and house area.

 • *AUSTRALIAN SIGNPOST MATHS 4* • ISBN 978 1 4886 2187 1

3:20 The Square Metre

ACTIVITY

- Make a 1 m^2 template from newspaper, plastic or cloth.
- Estimate, then measure surfaces of area:
  - less than 1 m^2.
  - about 1 m^2.
  - more than 1 m^2.
- Complete the table.

	Surface	
Less than 1 m^2		
About 1 m^2		
More than 1 m^2		

- Use your 1 m^2 template to measure bigger areas.

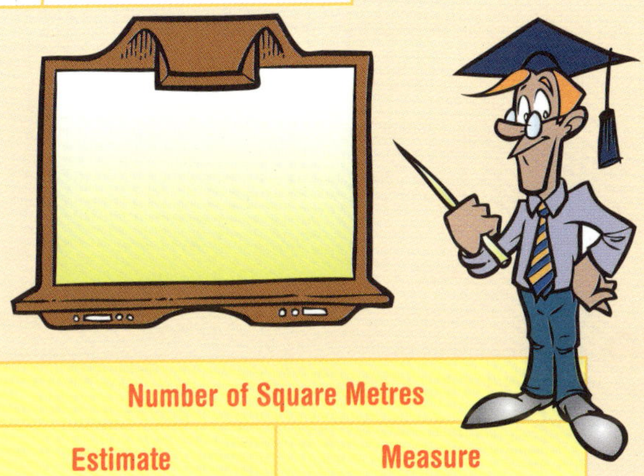

Region or Surface to be Measured	Number of Square Metres	
	Estimate	Measure
Classroom Floor	m^2	m^2
Corridor Floor	m^2	m^2
Whiteboard	m^2	m^2
Section of Path	m^2	m^2

1 Use the short form to write:

a 2 square metres ☐ b 6 square centimetres ☐
c 5 square metres ☐ d 12 square centimetres ☐
e 10 square metres ☐ f 15 square centimetres ☐

2 Which unit of measurement (**cm^2** or **m^2**) would you use to find the area of:

a the top of a brick? ☐ b a car park? ☐ c a DVD cover? ☐
d a long path? ☐ e a playing card? ☐ f a bedroom floor? ☐
g a tennis court? ☐ h a book cover? ☐ i a sheet of notepaper? ☐

3:21 Finding Area

1. Draw these shapes on 1 cm grid paper. Find the area of each one.

a

Area: ______

b
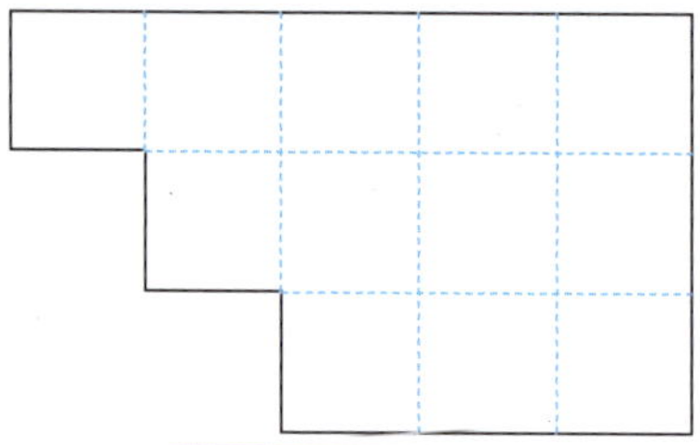
Area: ______

c
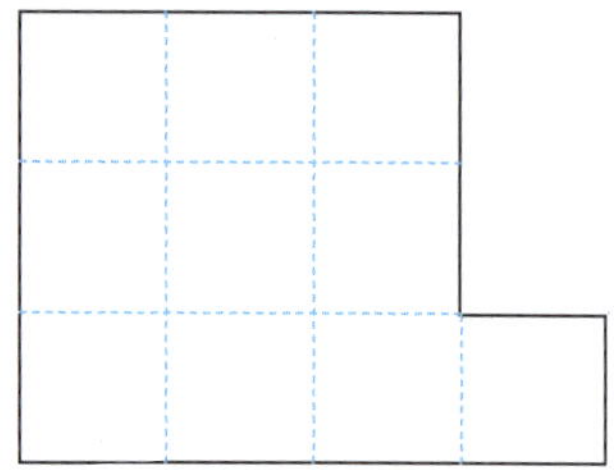
Area: ______

d
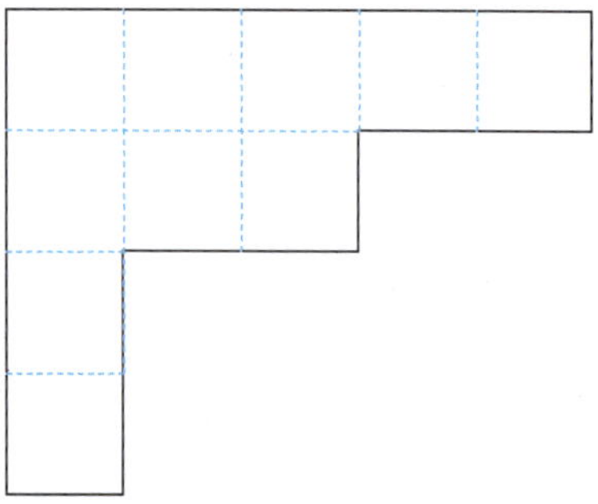
Area: ______

e
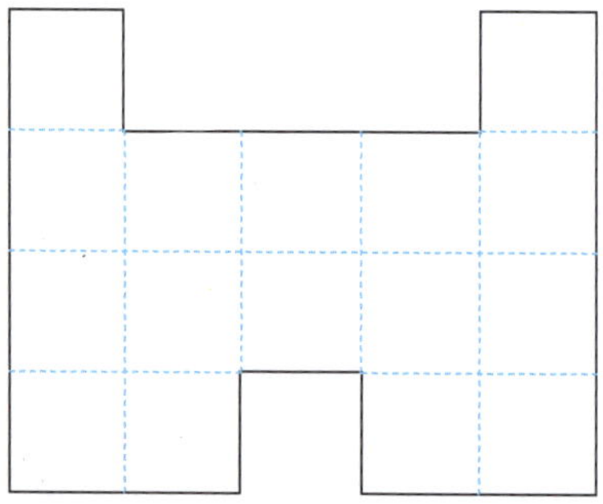
Area: ______

f

Area: ______

2. Overlay these shapes with a 1 cm grid transparency. Find each shaded area.

a

Area: ______

b
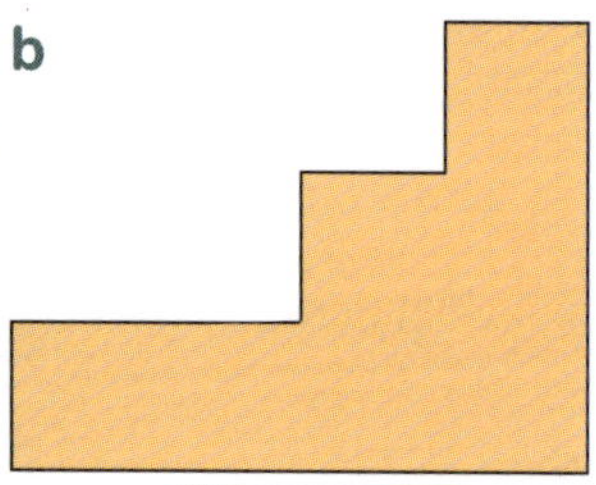
Area: ______

c
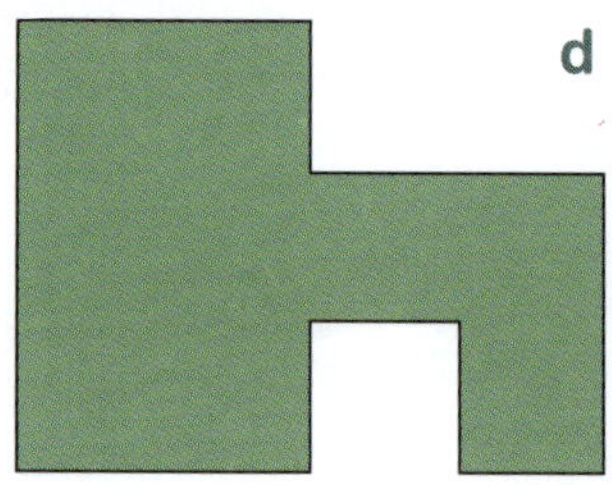
Area: ______

d
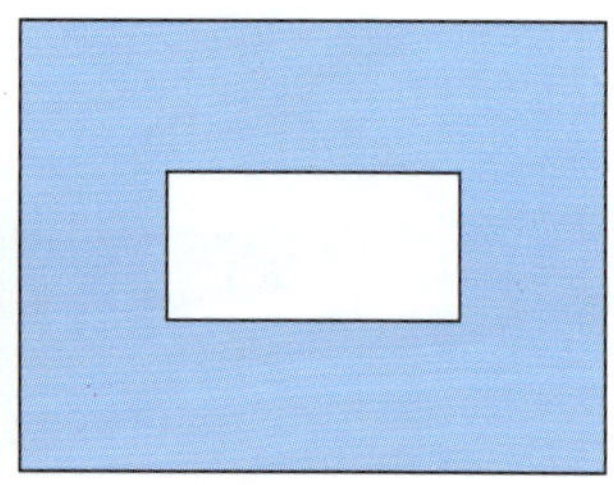
Area: ______

e
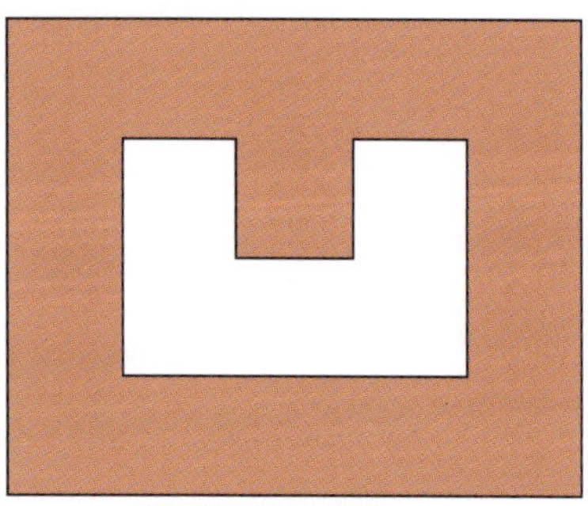
Area: ______

f

Area: ______

g
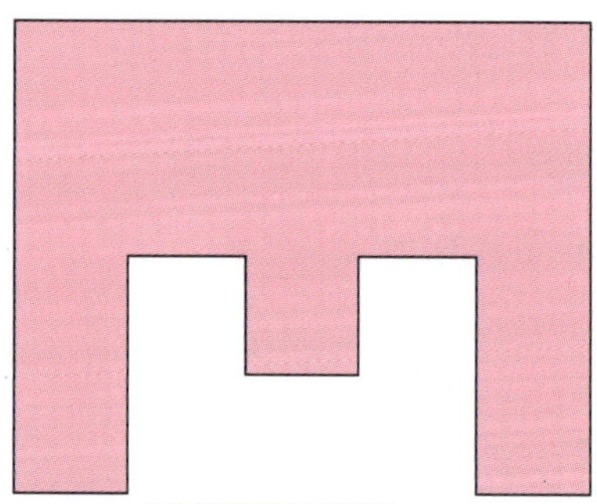
Area: ______

Cut and Find Area

ACTIVITY

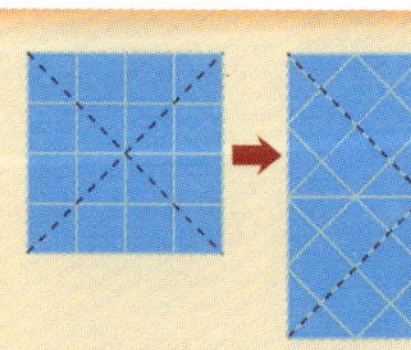

- Cut a square with 4 cm sides from 1 cm grid paper. Find its area.
- Draw in the diagonals and cut along them to form four triangles.
- Rearrange them to make a rectangle (as shown) and find its area. Discuss the results.

 ISBN 978 1 4886 2187 1

3:22 Finding Area

CONCEPT

1 What is the area of each coloured shape?

a

b

c
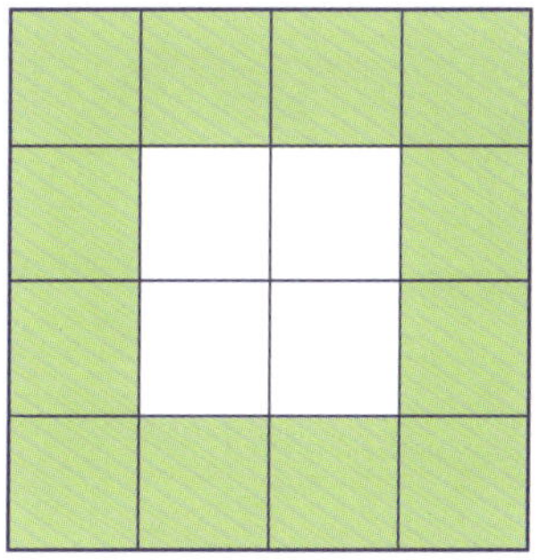

d
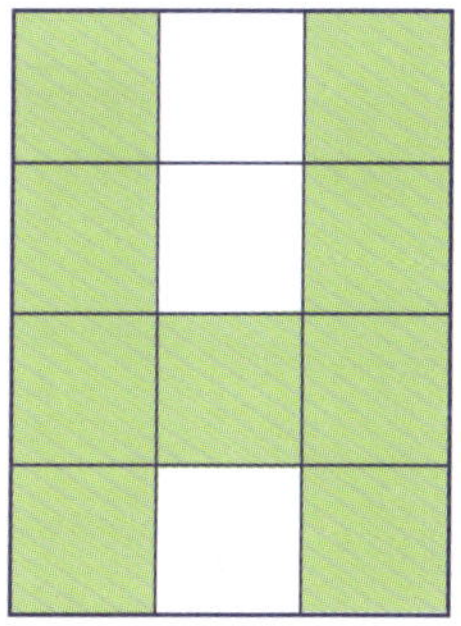

e
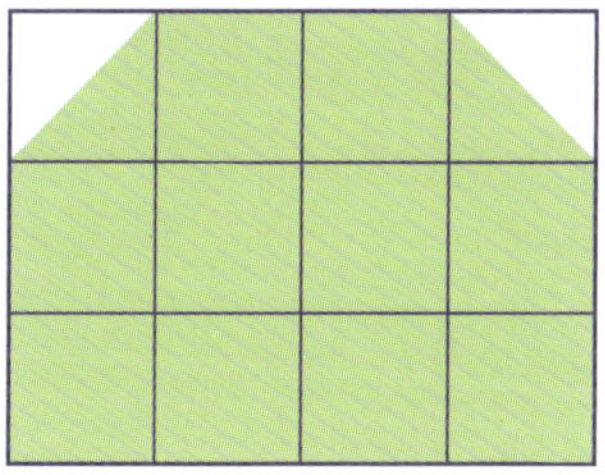

f
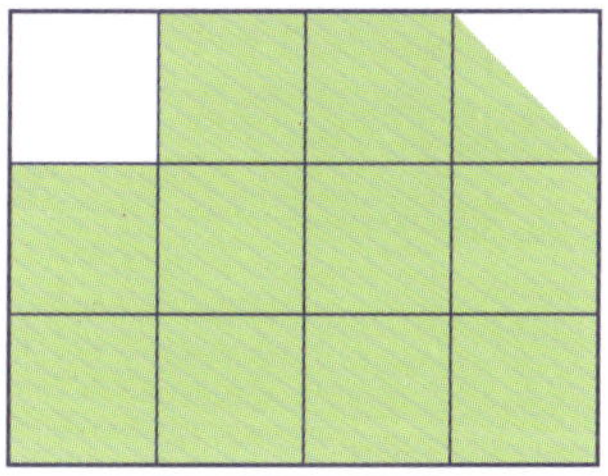

g
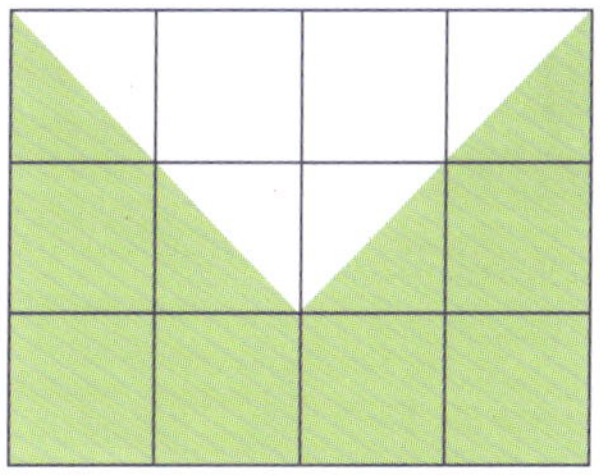

h
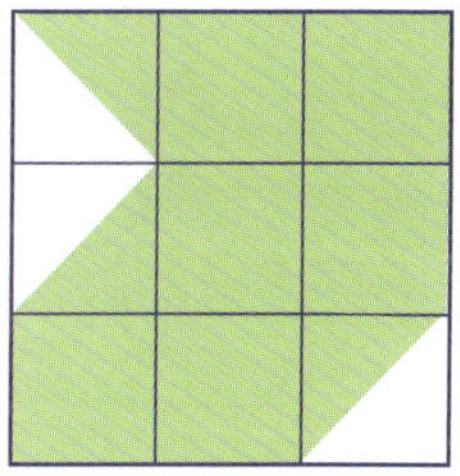

2 Trace these shapes onto 1 cm grid paper and find the area of each.

a
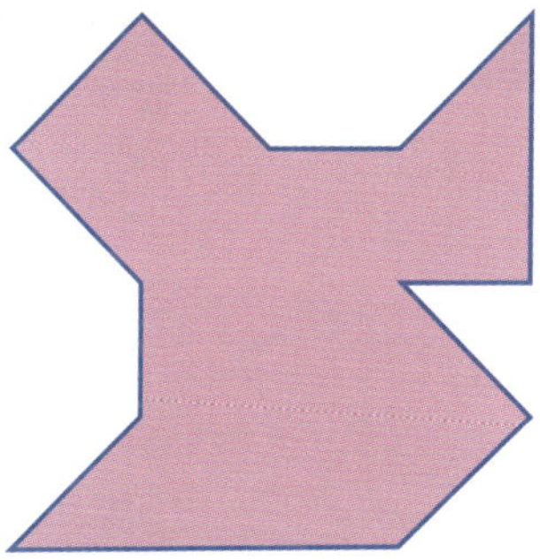

b
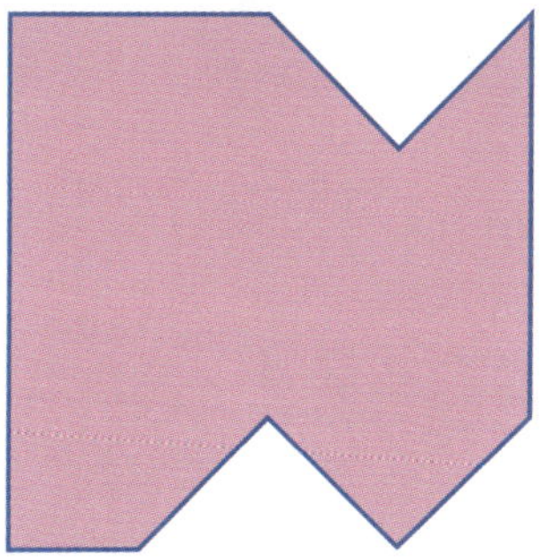

c
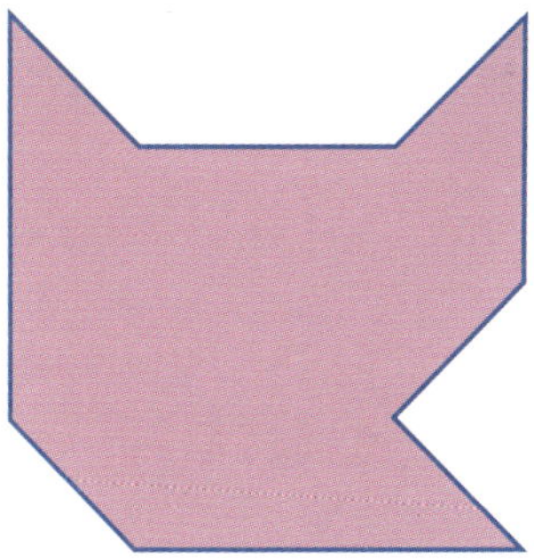

d
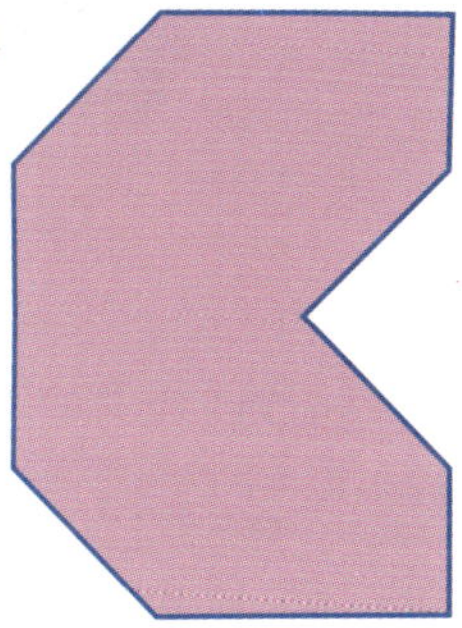

Finding Area

INVESTIGATION

- Collect leaves of different sizes and shapes.
- On 1 cm grid paper trace around each leaf. Count the squares to find each area.

Count each square that has more than half covered.

3:23 Analogue and Digital Time

CONCEPT

The hour hand has passed 2.

The minute hand has gone 5, 10, 15, 16, 17, 18 minutes from 12.

The time is 2:18 or 18 past 2.

1 Complete the labels for each time shown.

a

11 : ___
___ past ___

b

7 : ___
___ past ___

c

4 : ___
___ past ___

d

5 : ___
___ past ___

e

3 : ___
___ past ___

f

12 : ___
___ past ___

g

9 : ___
___ past ___

h

6 : ___
___ past ___

i

10 : ___
___ past ___

j

8 : ___
___ past ___

k

___ past ___

l

___ past ___

m

___ past ___

n

___ past ___

o

___ past ___

2 Write the time that is one minute after:

a 3:16 ___ **b** 2:47 ___ **c** 7:28 ___

d 9:03 ___ **e** 11:41 ___ **f** 4:39 ___

g 12:24 ___ **h** 5:20 ___ **i** 6:55 ___

3 Write the time that is five minutes after:

a 4:13 ___ **b** 9:41 ___ **c** 7:32 ___

d 10:54 ___ **e** 8:16 ___ **f** 5:58 ___

 • *AUSTRALIAN SIGNPOST MATHS 4* • ISBN 978 1 4886 2187 1

3:24 Time

1 On each face show the time given.

a 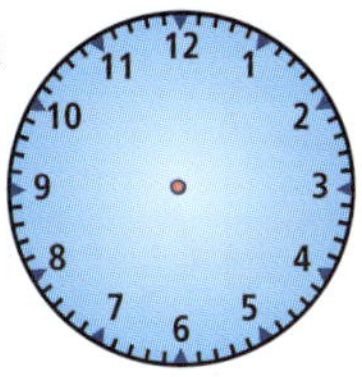
a quarter to 8

b 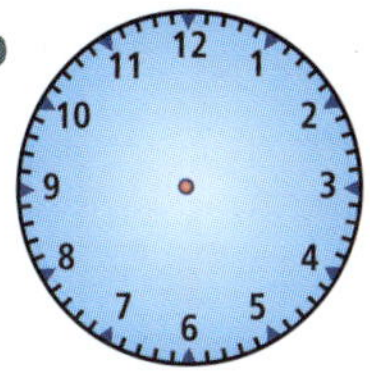
7 past 4

c 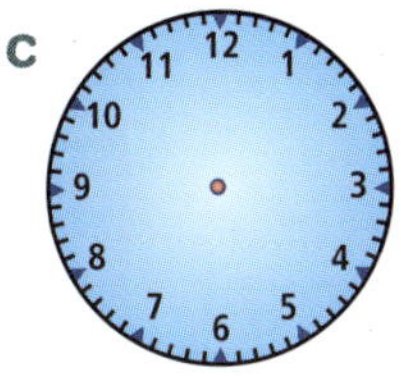
20 to 11

d 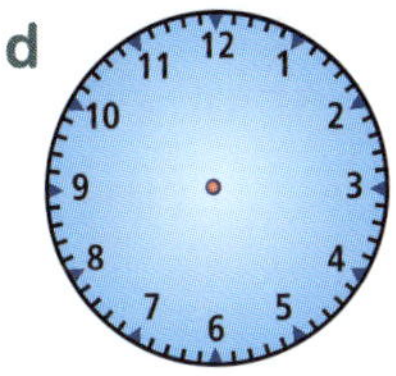
5 to 2

e 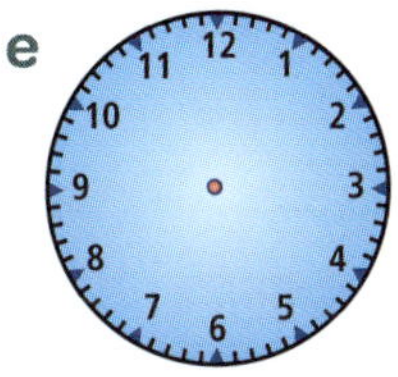
19 past 12

2 Complete the label for each time shown.

a 

☐ to ☐

b 

☐ past ☐

c 

☐ to ☐

d 

☐ past ☐

e 

☐ to ☐

3 How many more minutes will it take for the minute hand to reach the 12?

a 
☐ minutes
The time is: ☐

b 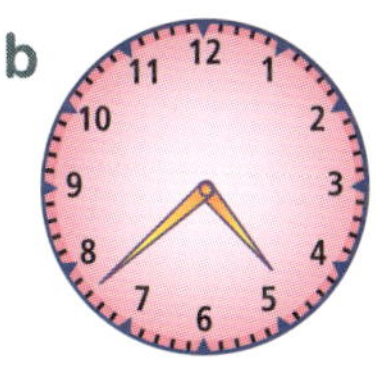
☐ minutes
The time is: ☐

c 
☐ minutes
The time is: ☐

d 
☐ minutes
The time is: ☐

e 
☐ minutes
The time is: ☐

4 Complete:

a 7 weeks = ☐ days
b 3 fortnights = ☐ days
c 28 days = ☐ fortnights
d 3 weeks = ☐ days
e 770 days = ☐ weeks
f 14 weeks = ☐ fortnights
g 21 days = ☐ weeks
h 35 days = ☐ weeks
i 14 days = ☐ weeks

5 Complete:

a 4 consecutive years would include ☐ normal years and ☐ leap year.
b 3 years = ☐ days
c 3 years = ☐ months
d 730 days = ☐ years
e 60 months = ☐ years
f 3 years = ☐ weeks
g 208 weeks = ☐ years

3:25 Volume

Volume is the amount of space within a container or taken up by a model.

To measure volume, we use blocks of equal size. These are placed so that there are no gaps.

Capacity is the amount of liquid a container can hold.

CONCEPT

1. What is the volume of each container?

a

☐ blocks

b
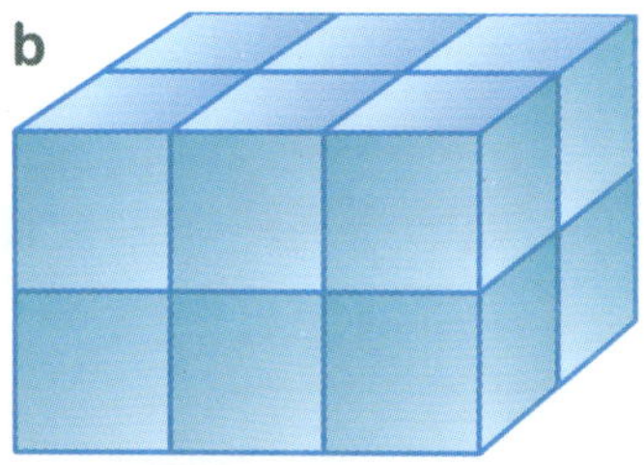
☐ blocks

c
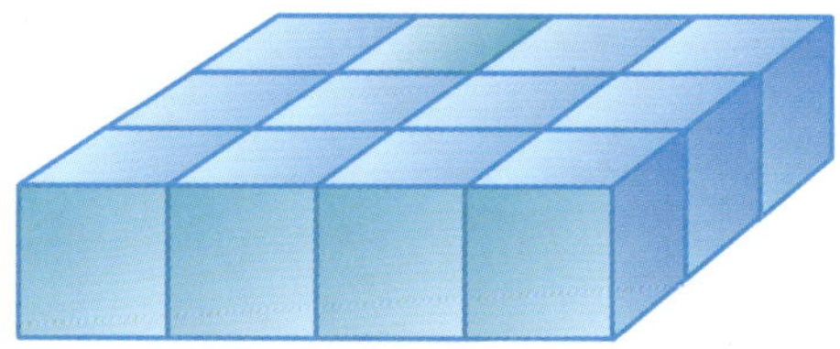
☐ blocks

2. What is the volume of each model?

a

☐ blocks

b

☐ blocks

c

☐ blocks

ACTIVITY

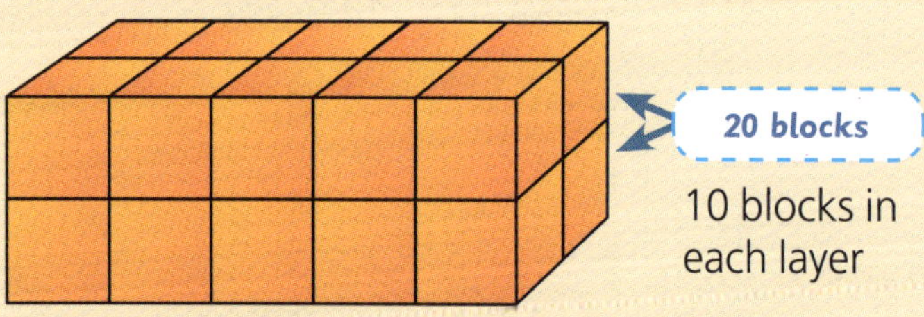

- Pack some small containers with cubic centimetre blocks. Count the number of blocks in each layer. Count the number of layers.
- Describe your packing to a classmate, e.g. 3 layers of 8 cubic centimetre blocks.

AUSTRALIAN SIGNPOST MATHS 4 • ISBN 978 1 4886 2187 1

3:26 Finding Volume

Volume can be the space within a container or the space taken up by an object or model.

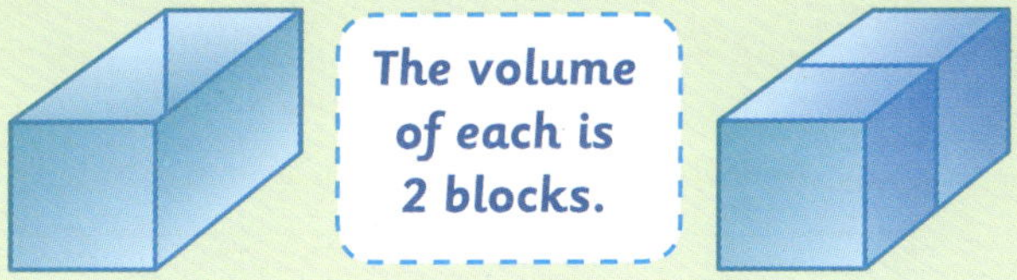

CONCEPT

1. Use place-value blocks to build these models.

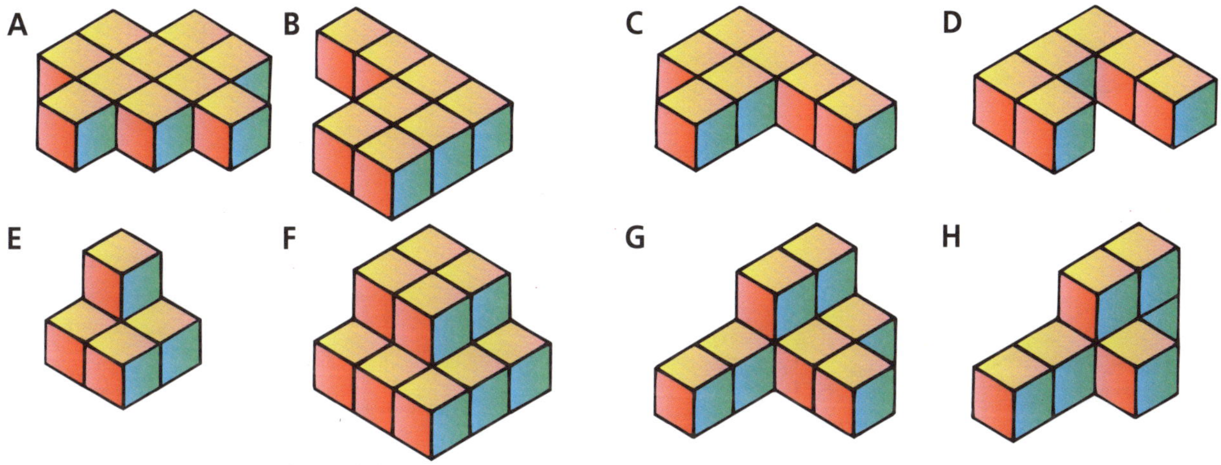

2. What is the volume of model:

   a C? ☐ blocks b E? ☐ blocks c D? ☐ blocks

   d A? ☐ e G? ☐ f B? ☐

3. Which of the models above have a volume of:

   a 8 blocks? ☐ b 6 blocks? ☐ c 9 blocks? ☐

   d 13 blocks? ☐ e 10 blocks? ☐ f 5 blocks? ☐

4. Which two models above have the same volume? ☐

5. Which two models have a total volume of 21 blocks? ☐

Drawing Models

ACTIVITY

- Build different models that are made from four place-value blocks.
- Draw two of these.

3:27 The Cubic Centimetre

CONCEPT

Centicubes and place-value ones are cubes that have a length of 1 centimetre. Their volume is 1 **cubic centimetre.**

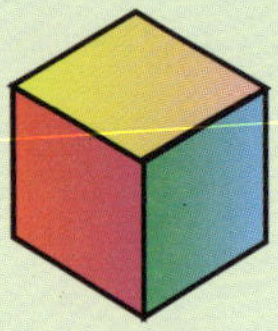

cubic centimetre

Each side is one centimetre in length.

Each face is one square centimetre in area.

The short way to write cubic centimetre is cm^3.

1 How many cubic centimetres are there in each model?

a

☐ cubic centimetres

b

☐ cubic centimetres

c

☐ cubic centimetres

d

☐ cubic centimetres

e

☐ cubic centimetres

f

☐ cubic centimetres

ACTIVITY

- Use cardboard and masking tape to make an open box with 3 cm sides.
- Fill the box with marbles. Pour dry sand into the box until it is completely full.
- Empty the box and re-pack it with place-value ones.

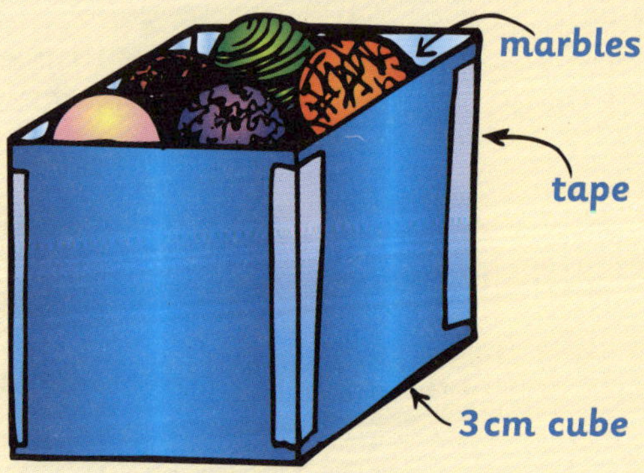

Discussion

- Why did the marbles leave spaces?
- Why did the place-value ones fill all of the space?
- Which one gives the better measure of the space inside the box? Why?

 • *AUSTRALIAN SIGNPOST MATHS 4* • ISBN 978 1 4886 2187 1

3:28 Timetables

Use the timetable to answer the following questions.

Mulga Line Timetable	
Arriving	
Ironbark	6:10
Christmas Ck	7:25
Mt Gordon	7:50
Richmond	8:45
Wandong	10:05
Carbor	11:40
Black Hill	12:30
Mulga	1:15

am means 'before noon' (from midnight until noon).

pm means 'after noon' (from noon until midnight).

1 At what time does the train arrive in:

a Ironbark?
b Richmond?
c Carbor?
d Mt Gordon?
e Wandong?
f Mulga?

2 At which station does the train arrive at:

a 8:45?
b 7:25?
c 12:30?
d 6:10?
e 1:15?
f 10:05?

3 If the train is two hours late, at what time does it arrive in:

a Ironbark?
b Carbor?
c Mulga?
d Wandong?
e Black Hill?
f Richmond?

4 If the train is half an hour early, at what time does it arrive in:

a Black Hill?
b Richmond?
c Carbor?
d Mulga?
e Ironbark?
f Wandong?

5 At which station does the train arrive at:

a half past 12?
b a quarter past 1?
c a quarter to 9?
d 5 past 10?
e 20 to 12?
f 10 to 8?

6 Complete the time line for the Mulga Line train.

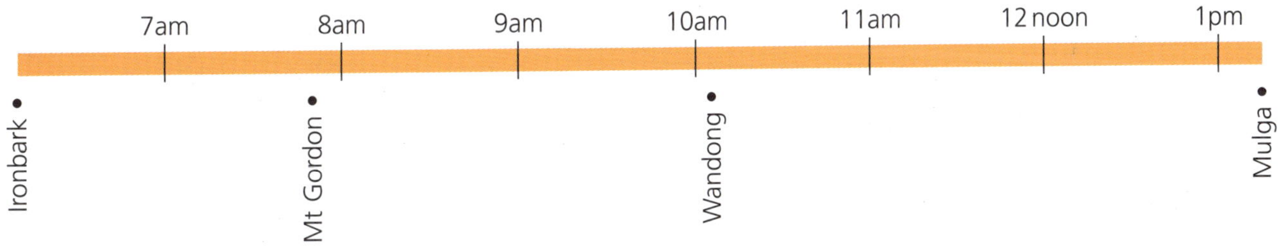

3:29 Comparing Measurements

CONCEPT

- We can compare measurements using subtraction.

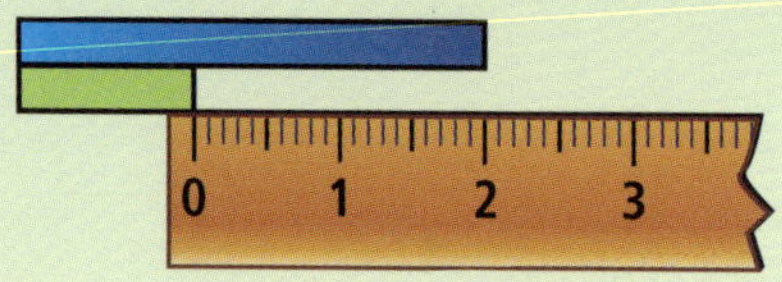

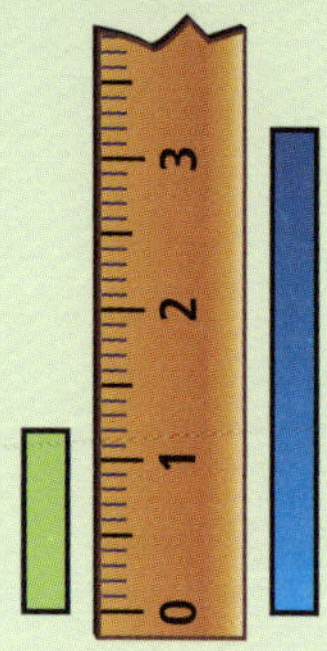

length = 12 mm

length = 32 mm

- We can measure the difference in length.

direct comparison

Difference in Length
= 32 mm – 12 mm
= 20 mm

$$\begin{array}{r} 32 \\ -\,12 \\ \hline 20 \end{array}$$

1. In the example above we compare a blue and a green rod.
   a What is the difference in length of the rods?
   b What is the length of the blue rod?
   c What is the length of the green rod?
   d 32 mm – 12 mm =

2. Measure the length of each rod in millimetres.

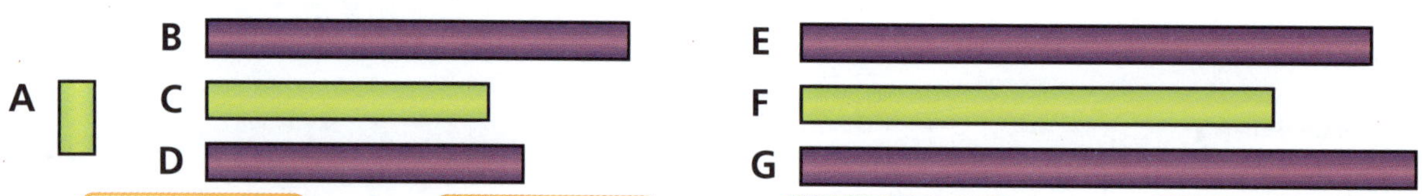

a D　b C　c G　d E
e A　f B　g F

3. On your own paper, calculate the difference in length of:
   a **A** and **B**　b **C** and **D**　c **E** and **F**
   d **G** and **A**　e **E** and **C**　f **D** and **G**
   g **C** and **A**　h **D** and **E**　i **B** and **C**

INVESTIGATION

- Explain how you would find the difference between the distance around a tree and the thickness of the tree.

© PEARSON AUSTRALIA 2018 • *AUSTRALIAN SIGNPOST MATHS 4* • ISBN 978 1 4886 2187 1

3:30 Personal Benchmarks

1 a Draw a 1 metre line on the ground. Practise taking two steps that would cover one metre.

- From a starting point, use the steps you have practised to estimate a length of 5 metres. Use a tape measure to see how good your estimate was. Was your estimate close? ______
- Use this method to estimate other distances.

b Use your practised steps to estimate the length of a building. ______

c Place a 1 metre ruler vertically beside you. Remember how high it reaches on your body. Record this personal measurement for one metre. The metre ruler reached my ______.

2 1 L of water has a mass of 1 kg. Pour two litres of water into a plastic jug.

a Use hefting to find objects that have a mass of about 2 kg.

______

b Use hefting to estimate the mass of a box of toys.

c Try to remember the mass of a litre of water. This will be your personal benchmark for one kilogram.

3 a Fill a bucket with water. Estimate how many 1 litre cartons of water would fill the bucket. ______

- Use a 1 litre carton of milk as your benchmark for a litre.
- Use a 1 litre container to check your estimate of the water in the bucket.

b Partially fill the bucket with water. Estimate how much water you have. ______
Use a 1 litre container to check your estimate.

4 Write your personal benchmarks for length, mass and capacity.

______

3:31 Analogue and Digital Time

1 Complete the label for each time shown.

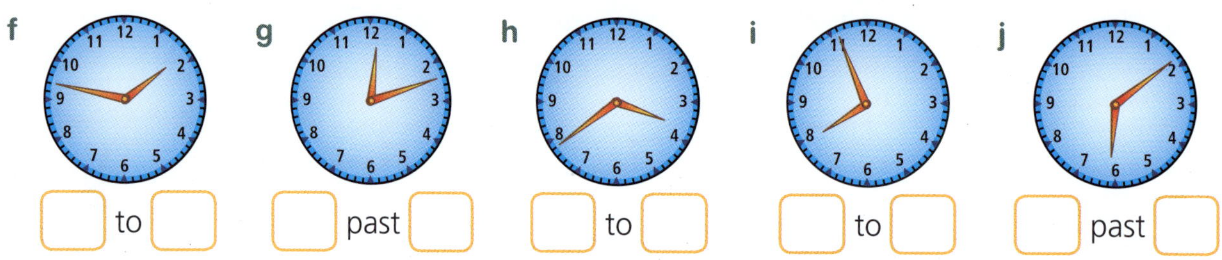

2 Complete the label for each time shown.

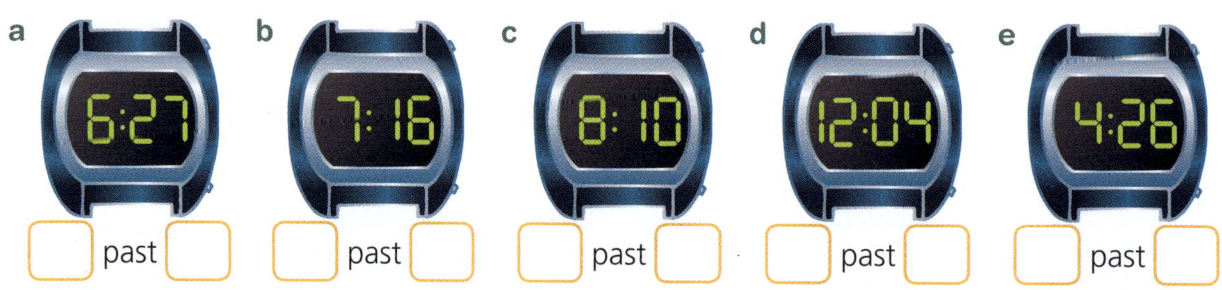

© PEARSON AUSTRALIA 2018 • *AUSTRALIAN SIGNPOST MATHS 4* • ISBN 978 1 4886 2187 1

3:32 am and pm Time

CONCEPT

When we write digital time we usually add **am** or **pm**.

Ante meridiem (am), is Latin for before midday.

7:35 am

Post meridiem (pm), is Latin for after midday.

10:46 pm

The minutes on the digital clock are counted from 00 to 59 before showing 00 again to signal the next hour.

1 Write a digital label for each time shown.

a	b	c	d	e
afternoon	morning	evening	morning	evening
: pm	: am	:	:	:

f	g	h	i	j
afternoon	morning	morning	evening	afternoon
:	:	:	:	:

2 For each digital time, write **before midday** or **after midday**.

a b c d e

f g h i j

3:33 Recording Length

1 Complete this table.

Millimetres	cm and mm	Rounded to the Nearest cm
39 mm		
	6 cm 3 mm	
56 mm		
	7 cm 2 mm	
81 mm		
	4 cm 5 mm	
	9 cm 4 mm	
26 mm		

2 Measure the length of each interval correct to the nearest millimetre.

a ☐ mm

b ☐ mm

c ☐ mm

d ☐ mm

e ☐ mm

f ☐ mm

g ☐ mm

h ☐ mm

3 Find the perimeter of each shape.

a

☐ mm

b
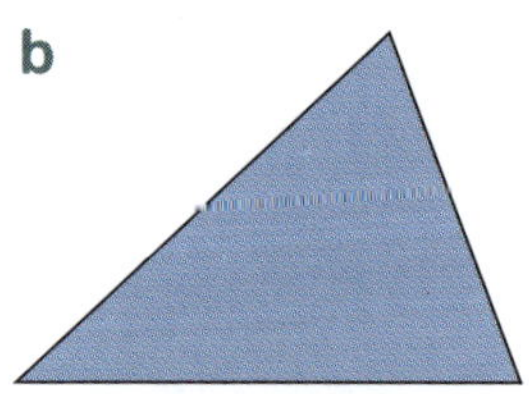
☐ mm

c
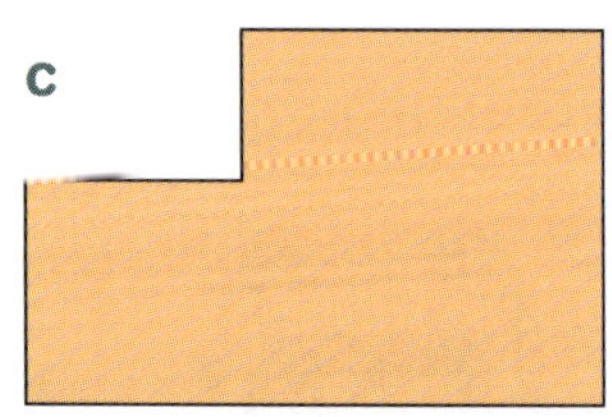
☐ mm

ACTIVITY

- Use a trundle wheel to measure:
  - the length of a path
  - the length of the classroom
  - the length of a fence
  - the length of a hose
  - the length of a car
  - the length of a garden
- Use a tape measure to measure:
  - the height of a door
  - the width of a window

3:34 Using Millilitres

CONCEPT

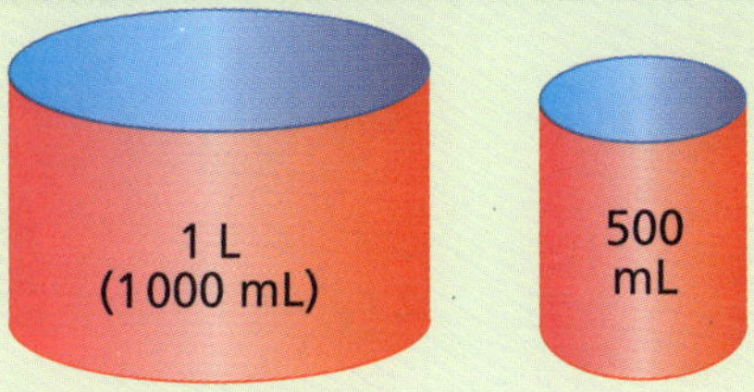

1 litre + 500 millilitres

We can read this as 1500 mL.

1 What is the total capacity of each set of containers?

a
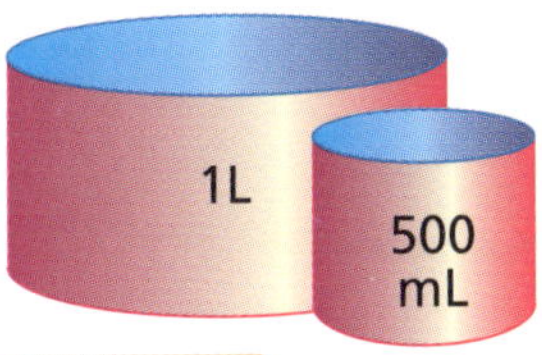

☐ mL

b
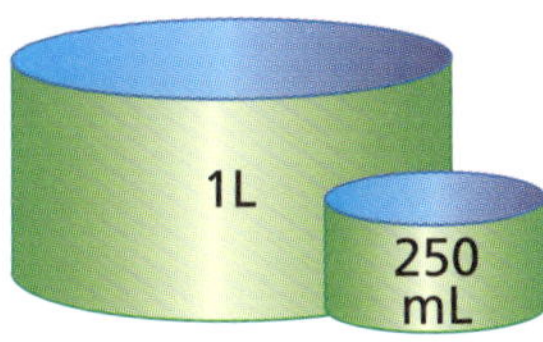

☐ mL

c
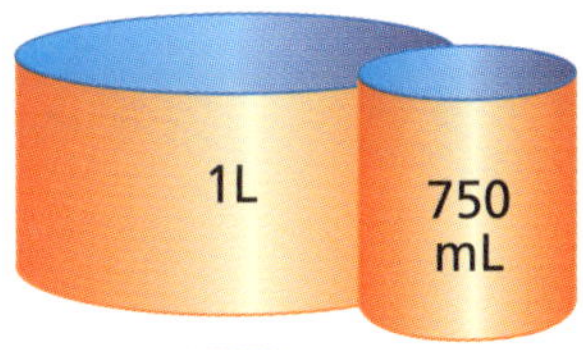

☐ mL

d
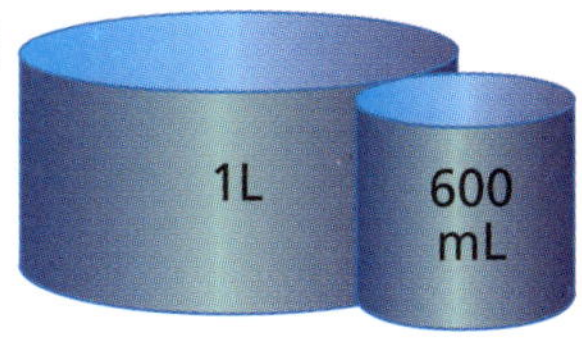

☐ mL

2 Write these as millilitres.

a 1 L 300 mL ☐ b 1 L 490 mL ☐ c 1 L 875 mL ☐

d 1 L 625 mL ☐ e 1 L 750 mL ☐ f 2 L 500 mL ☐

3 Write these as litres and millilitres.

a 1 600 mL ☐ b 1 900 mL ☐ c 1 350 mL ☐

d 1 250 mL ☐ e 1 425 mL ☐ f 2 750 mL ☐

4 Would you use litres or millilitres to measure:

a water in a bucket? ☐ b water in a teacup? ☐

c ice cream in a cone? ☐ d milk in a small carton? ☐

e liquid in an egg? ☐ f petrol in a car? ☐

g medicine in a dessertspoon? ☐ h liquid in an eye-dropper? ☐

5 Which measure (**L** or **mL**) has been left off each label?

a

☐

b

☐

c

☐

d

☐

e

☐

f
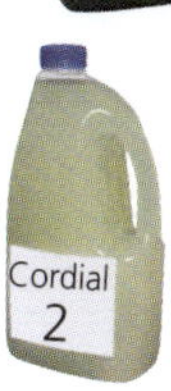

☐

g

☐

h

☐

i

☐

3:35 Using Grams

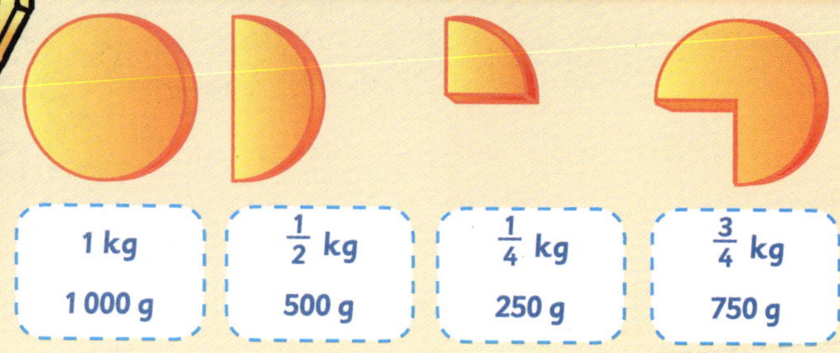

ACTIVITY

Mass	Objects Found
About $\frac{1}{4}$ kg	
About $\frac{1}{2}$ kg	
About $\frac{3}{4}$ kg	
About 1 kg	

1 Write the short form for:

a 100 grams ☐ b 600 grams ☐ c 1 kilogram ☐
d 500 grams ☐ e 3 kilograms ☐ f 300 grams ☐
g 10 kilograms ☐ h 956 grams ☐ i 875 grams ☐

2 How many kilograms are there in:

a 3000 g? ☐ b 7000 g? ☐ c 2000 g? ☐
d 5000 g? ☐ e 9000 g? ☐ f 4000 g? ☐

3 How many grams are there in:

a 2 kg? ☐ b 1 kg 200 g? ☐ c 3 kg? ☐
d 1 kg 600 g? ☐ e 5 kg? ☐ f 1 kg 450 g? ☐
g 1 kg 980 g? ☐ h 1 kg 500 g? ☐ i 7 kg? ☐

4 How many grams are there in:

a $\frac{1}{2}$ kg? ☐ b $\frac{3}{4}$ kg? ☐ c $\frac{1}{4}$ kg? ☐ d $1\frac{1}{2}$ kg? ☐
e $1\frac{3}{4}$ kg? ☐ f $1\frac{1}{4}$ kg? ☐ g $2\frac{1}{2}$ kg? ☐ h $3\frac{1}{2}$ kg? ☐

5 Order these objects from least mass to most mass.

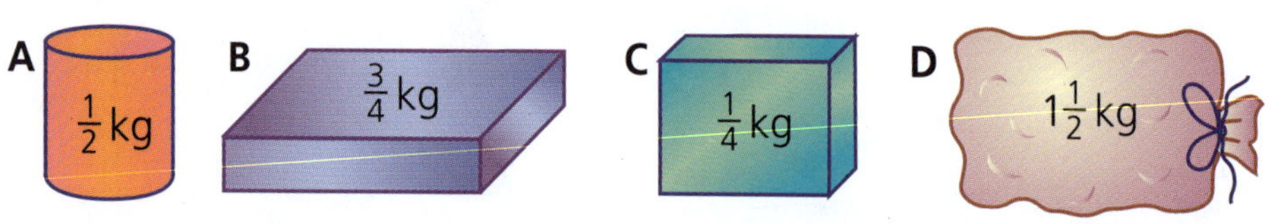

☐

4:01 Flip, Slide and Turn

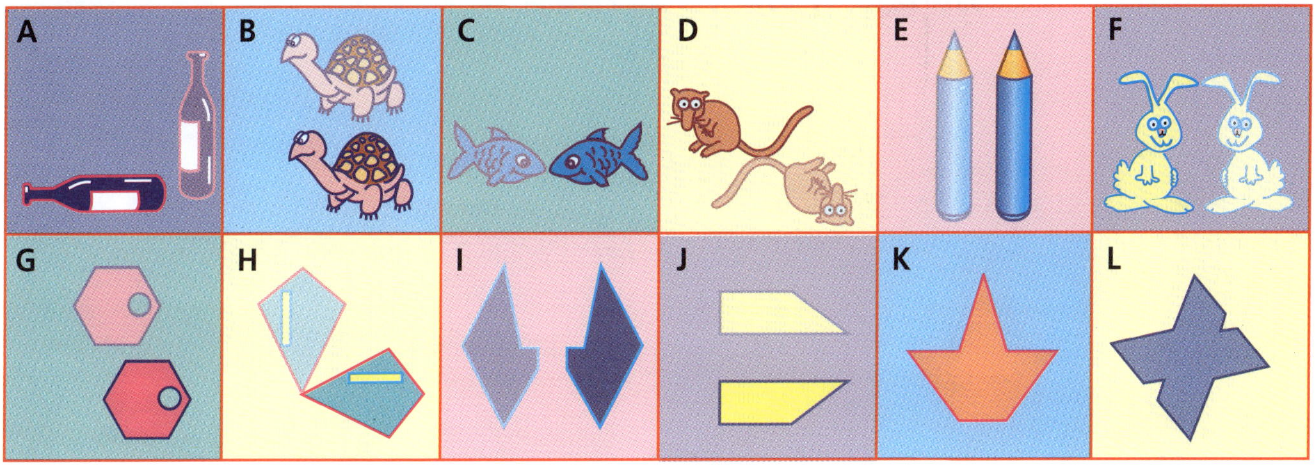

1. Which pictures (**A** to **J**) are samples of:

   a flip (reflection)? ______ **b** slide (translation)? ______

   c turn (rotation)? ______

2. Which picture could be an example of a flip or a slide? ______

3. How many axes of symmetry has picture:

a **K**? ______	**b** **L**? ______	**c** **C**? ______
d **D**? ______	**e** **E**? ______	**f** **F**? ______
g **G**? ______	**h** **H**? ______	**i** **I**? ______

Try to draw pictures of your own that show flip, slide or turn.

Turning Patterns

ACTIVITY

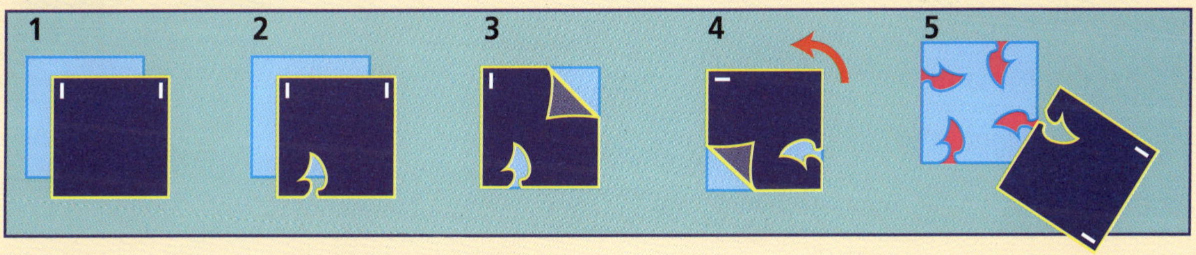

1. Use two identical cardboard squares.
2. From one side of square 1, cut out an interesting shape.
3. Put square 1 on top of the other square and trace the shape you have made.
4. Turn square 1 and trace the shape onto each side of the other square.
5. Colour the pattern you have made.
   Make other patterns starting with:
   a two regular hexagons **b** two regular octagons

 • *AUSTRALIAN SIGNPOST MATHS 4* • ISBN 978 1 4886 2187 1

4:02 Angles and Plane Figures

1

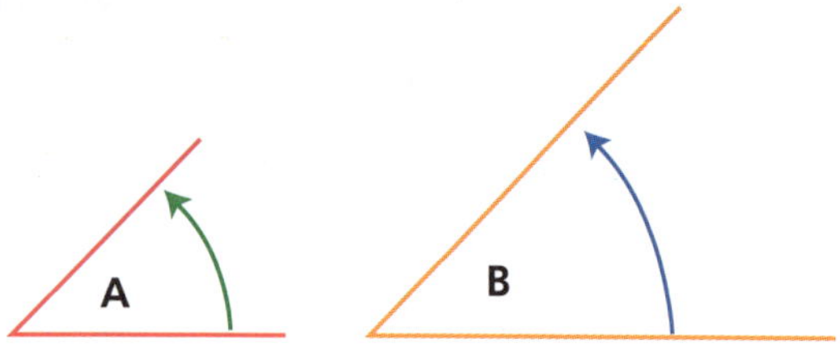

a Are these angles the same size?

b Does an angle get bigger if you make its arms longer?

c The angle at the corner of a page is called a **right angle**. Is angle **B** bigger than a right angle?

2

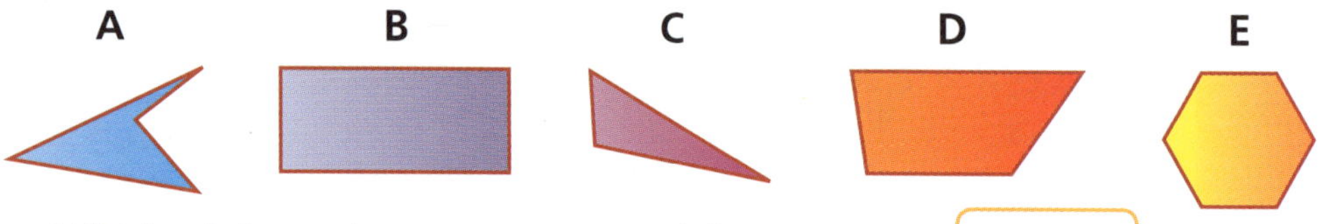

a Which of these shapes are not quadrilaterals?

b Which have parallel sides?

c Which shapes have symmetry?

d How many sides has shape **B**? How many angles?

e How many sides has shape **C**? How many angles?

f How many sides has shape **E**? How many angles?

3 Use the corner of a sheet of paper to test these angles.

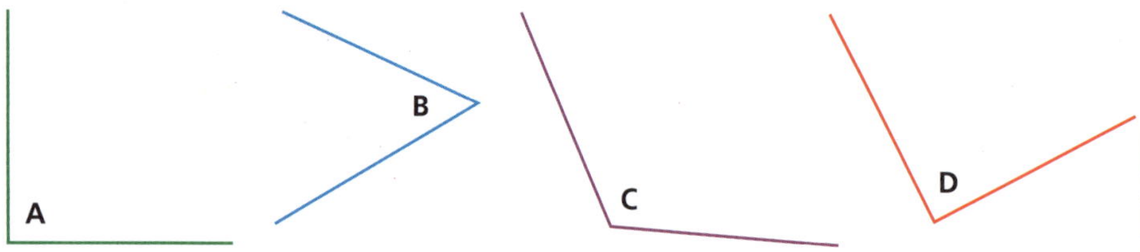

a Which of these angles are right angles?

b Which angle is less than a right angle?

c Which one is bigger than a right angle?

d Are the arms of a right angle perpendicular?

4 On your own paper, use a pencil and ruler to draw or trace:

a a square b a rectangle c a pentagon d a trapezium

e a right angle f an angle smaller than a right angle

5 Which triangle has:

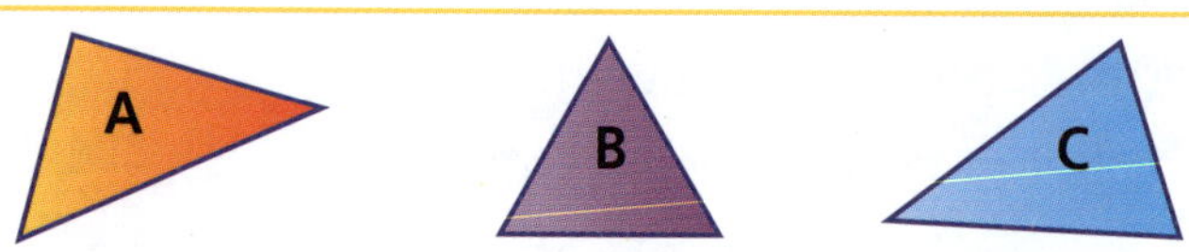

a all sides equal?

b two sides equal?

c no sides equal?

d a right angle?

4:03 Comparing Angles

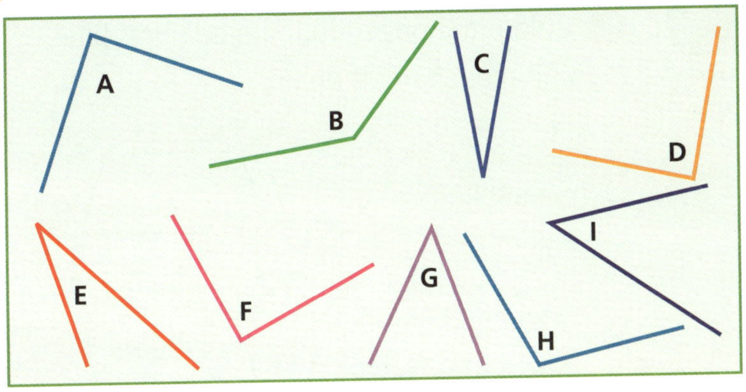

The size of an angle is the amount of turn for one arm to reach the other.

1 Use the corner of a page to test the angles above.

Put the letter of each angle in the correct column.

My Angle (Right Angle)	Smaller	Same as My Angle	Larger
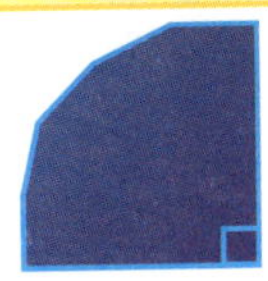			

2 Which angle is smaller:

a **A** or **B**? ______ b **C** or **D**? ______ c **E** or **F**? ______ d **G** or **H**? ______

3 Fold your paper right angle to make half a right angle. Use this new angle to test the angles at the top of the page. Put the letter of each angle in the correct column.

My Angle	Smaller	Same as My Angle	Larger
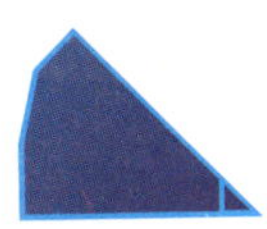			

4 Draw each angle showing both arms and the vertex.

a the opening of a gate

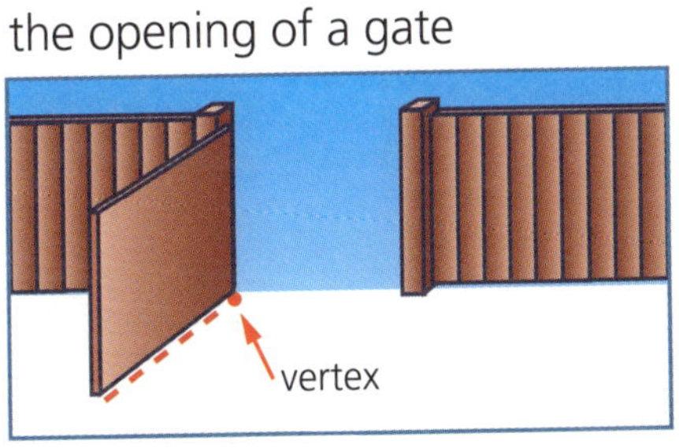

b the slope of a road

c the swing of a bat

ACTIVITY

Name an angle in your classroom that is:

- smaller than a right angle ______
- greater than a right angle ______

Use computer software to create simple shapes that involve direction and angles.

4:04 3D Objects

CONCEPT

Prisms have:
- ends that are the same shape
- sides that are rectangles.

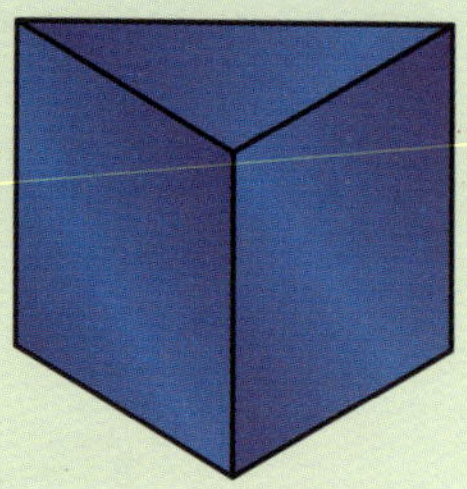

Cylinders have:
- ends that are circles
- one curved surface.

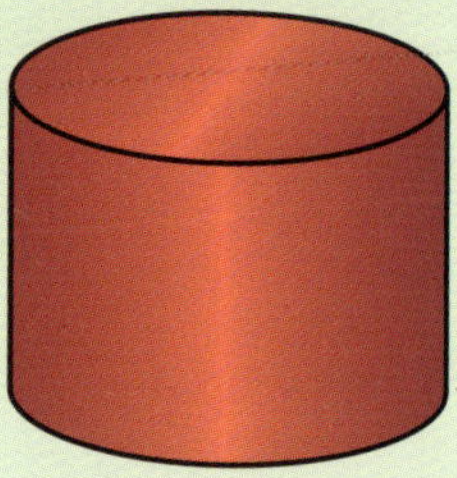

1 Colour the prisms red, the cylinders blue and the pyramids green.

2

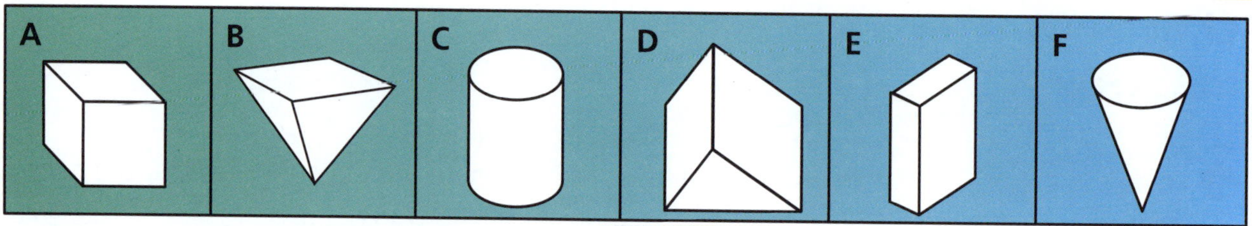

a Which 3D objects have only flat surfaces?

b Which 3D objects have a curved surface?

c Which are prisms? Colour them red.

d Which is a cylinder? Colour it blue.

e Which prism has six corners?

3 Which 3D object in Question 2 has as its faces:

a only squares?

b rectangles and triangles?

c only rectangles?

d triangles and a square?

INVESTIGATION

Make a list of real-life objects that are prisms.

4:05 Prisms and Pyramids

CONCEPT

This pyramid has 4 triangular faces and 1 square face.

This prism has 6 rectangular faces and 2 hexagonal faces.

Prisms and pyramids are named according to the shape of their bases.

1 Which of these 3D objects are:

a prisms? ________

b pyramids? ________

Which shapes have:

c a square base? ________

d a rectangular base? ________

e a triangular base? ________

f a hexagonal base? ________

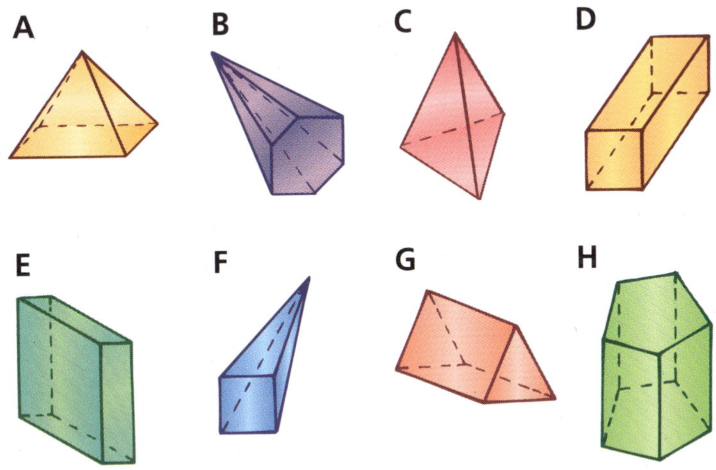

2 These models have been made by class 4B.

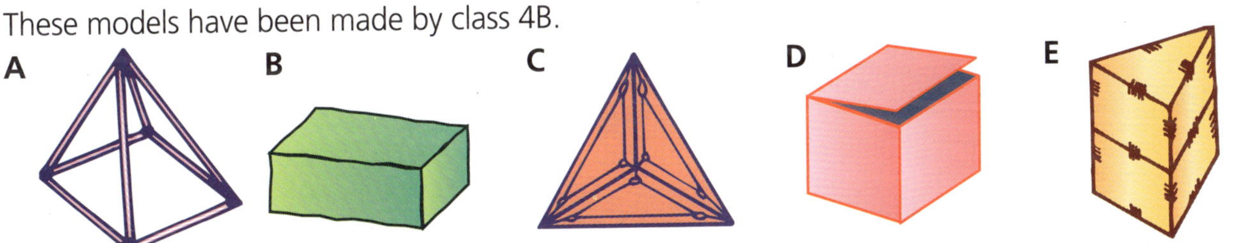

A A skeleton made of straws and Plasticine.

B Plasticine modelled to make the solid.

C Cardboard shapes joined by elastic bands.

D Folded net made of paper.

E Plastic shapes fitted together.

Name the 3D object modelled in:

a A ________ **b** B ________ **c** C ________

d D ________ **e** E ________

ACTIVITY

- Choose at least two of the methods above to model:
  - a cube
  - a pyramid
  - a prism
- Try to make more prisms and pyramids using these methods.

ICT

Can you draw shapes on a computer?

4:06 Surfaces of Prisms and Pyramids

1 Would these faces make a prism (**A**) or a pyramid (**B**)?

a

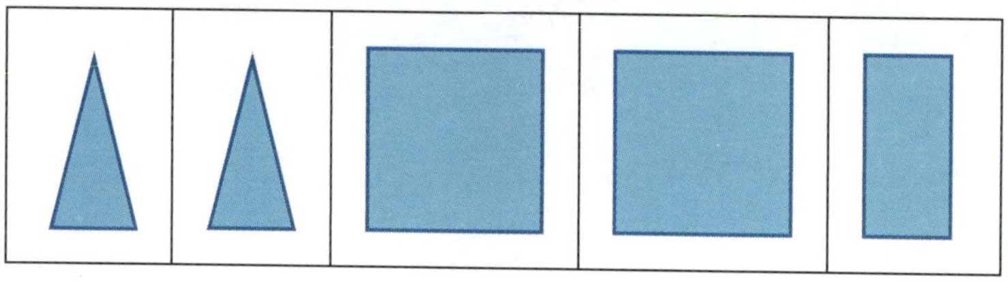

b

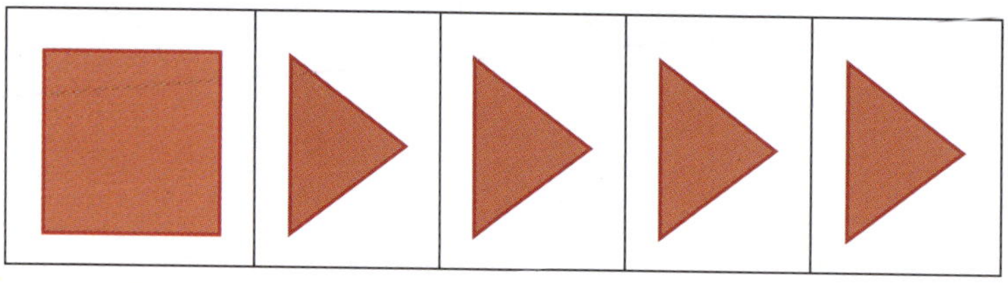

c

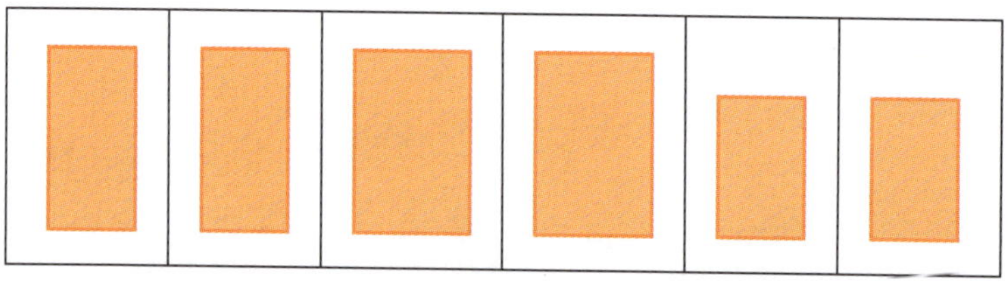

d

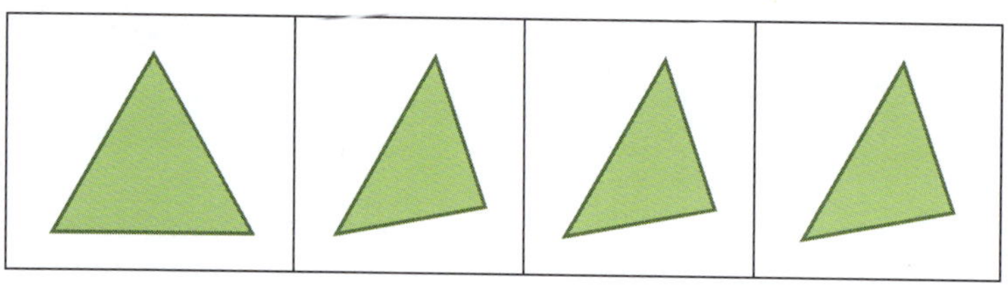

2 Would these nets make a prism or a pyramid? Discuss the reasons for your answers.

a

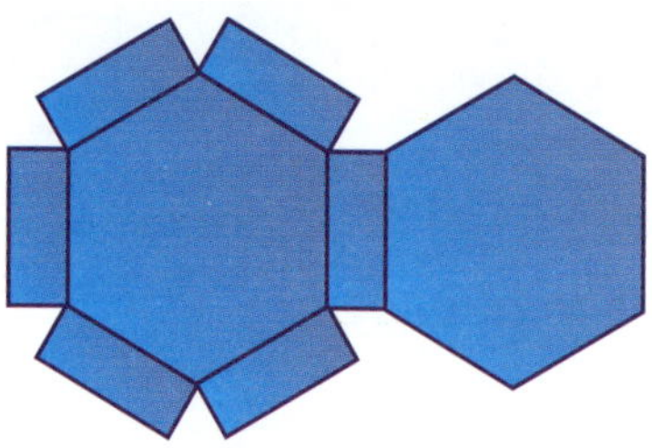

b

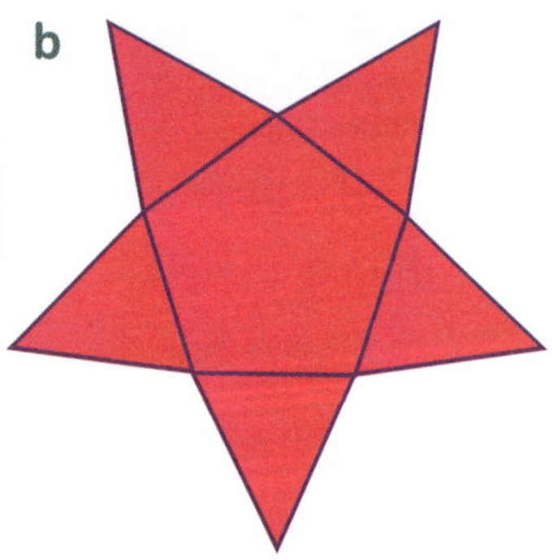

c

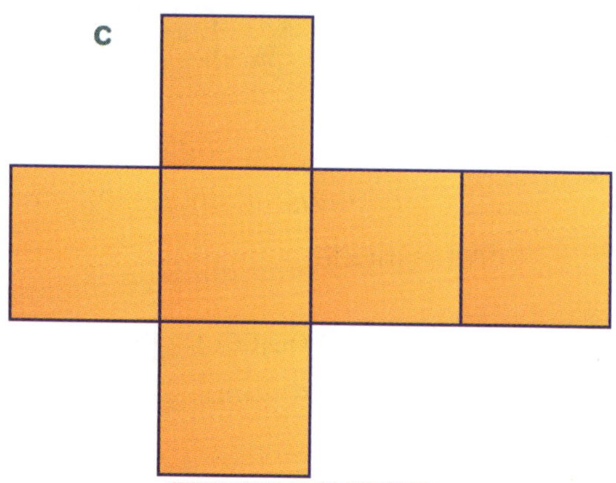

ACTIVITY

Use cardboard shapes and elastic bands or plastic interlocking shapes to make a model of:
- a prism
- a cube
- a pyramid

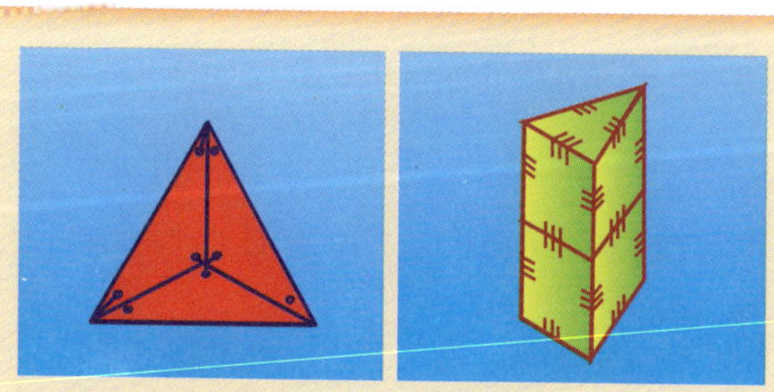

4:07 Drawing Prisms and Pyramids

1 Describe each object.

A
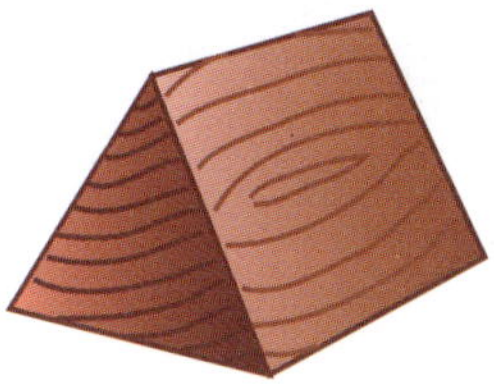

B

C

D
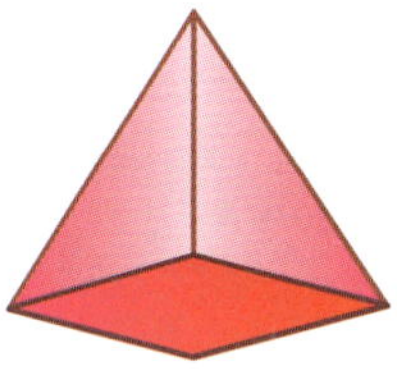

E

F
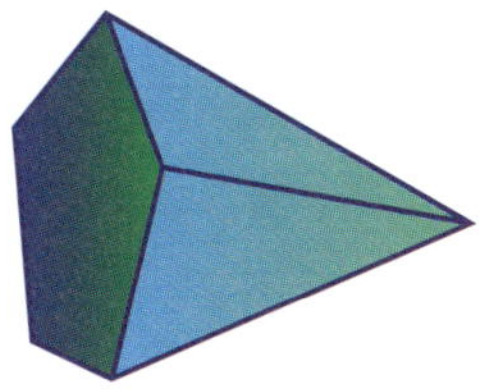

2 Write the name of solid:

a A
b B
c C
d D
e E
f F

3 Trace each solid above onto paper. Under each tracing name the solid.

ACTIVITY

- Some prisms and pyramids can be drawn using parallel lines.

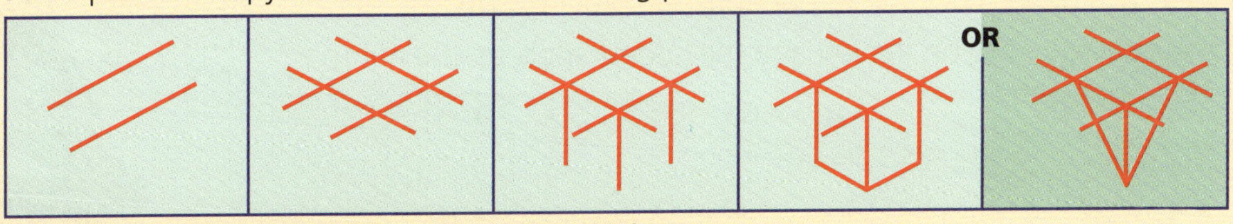

Step 1 Draw two pairs of parallel lines which cross.

Step 2 Draw three lines of equal length straight down from the corners. Join the ends to get a prism.

OR Choose a point in the middle below (or above) the figure. Draw lines from three corners to this point to get the pyramid.

- Use this method to draw: • five prisms • five pyramids

4:08 Drawing Angles

CONCEPT

- Julia took two geostrips and joined them to make a movable angle.
- By moving its arms, she made angles of different sizes.

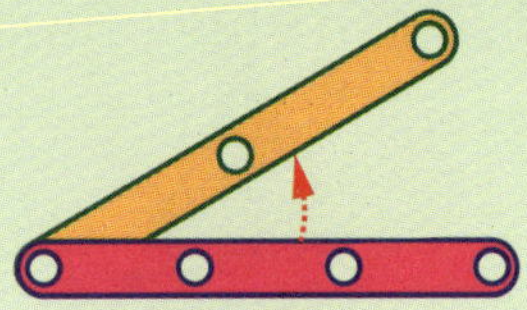

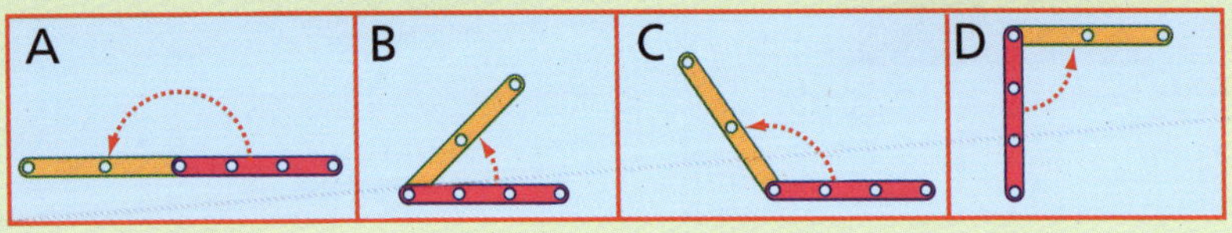

1 Geostrips were used to make the angles in **A**, **B**, **C** and **D** above. Which is the larger angle:

a **A** or **B**? ______ b **B** or **C**? ______

c **C** or **D**? ______ d **A** or **C**? ______

e Which angle is a right angle? ______

The size of an angle is the amount of turning between its arms.

2 Trace these angles.

a the corner of a book

b blades of scissors

c your choice

3 What is a right angle? ______

4 On your own paper, copy these four angles in order of size. Put the smallest angle on the left and the largest angle on the right.

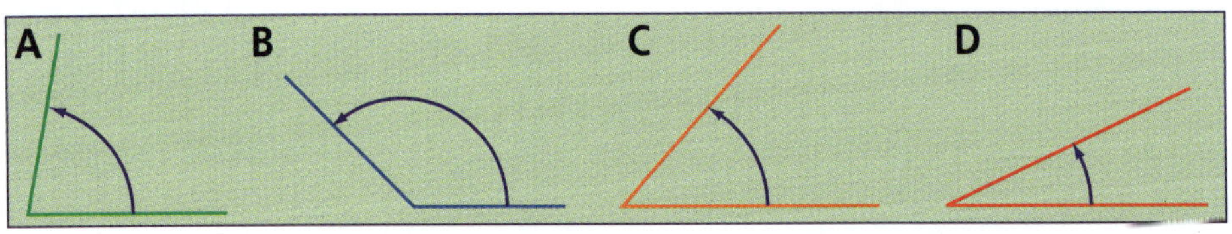

ACTIVITY

- Describe two angles in the room. Identify the smallest and the largest.

______

______

 • *AUSTRALIAN SIGNPOST MATHS 4* • ISBN 978 1 4886 2187 1

4:09 Angles as Half and Quarter Turns

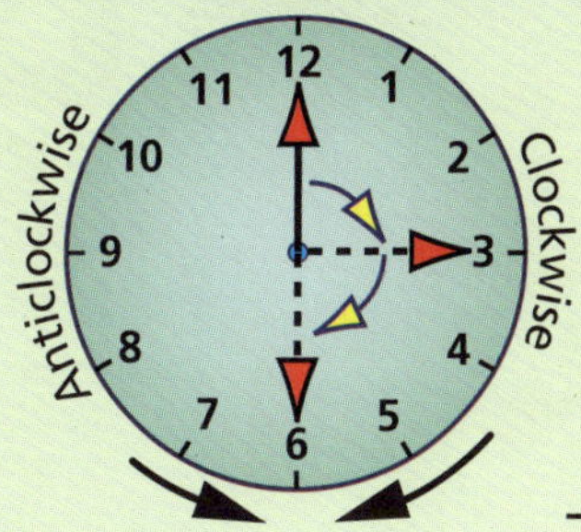

CONCEPT

- A right angle is a quarter turn.
- Two quarter turns make a half turn.
- Two right angles make a straight angle.

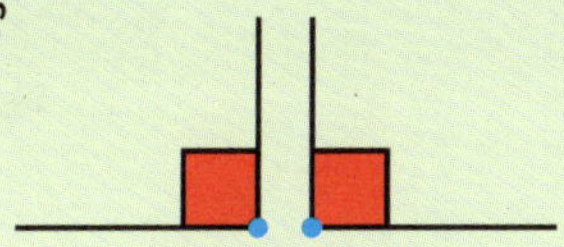
Right Angles

Straight Angle

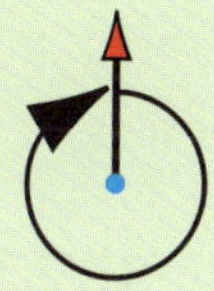
Full Turn

1 Draw an angle that represents:

a quarter turn	a half turn	a full turn
•	•	•

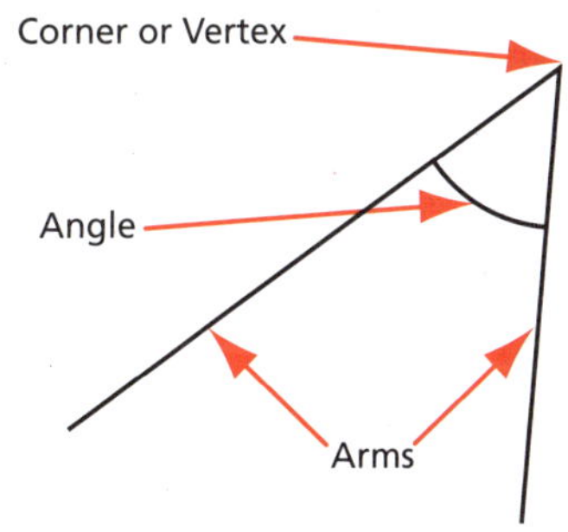

2 When turning the minute hand on a clock:

a a half turn anticlockwise is the same as

b a three-quarter turn anticlockwise is the same as

c two quarter turns clockwise is the same as

3 **a** How many quarter turns make a full turn?

b How many right angles make a full turn?

c How many right angles make a straight angle?

4

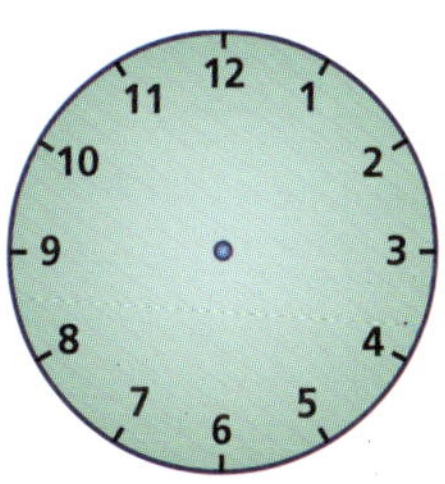

a If each arm must point to a number on the clock, how many different right angles could be drawn?

b Using the centre of the clock as the vertex of the angles, how many right angles can you draw at one time that do not overlap or share an arm?

ACTIVITY

List places in real life that show:

- right angles
- straight angles
- full turns.

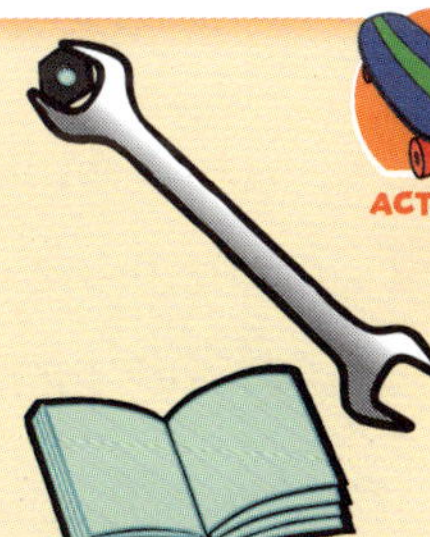

4:10 Investigating Polygons

- A polygon is a figure made of three or more straight sides.
- In a regular polygon, all angles are equal and all sides are equal.
- Diagonals are lines drawn from one corner to another, across a polygon.

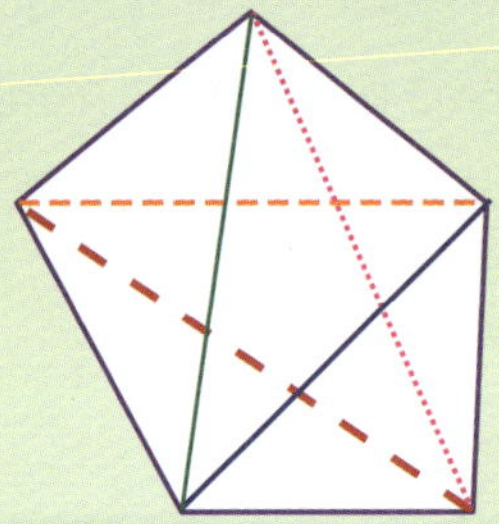

CONCEPT

1 Draw in all the diagonals for each shape below.

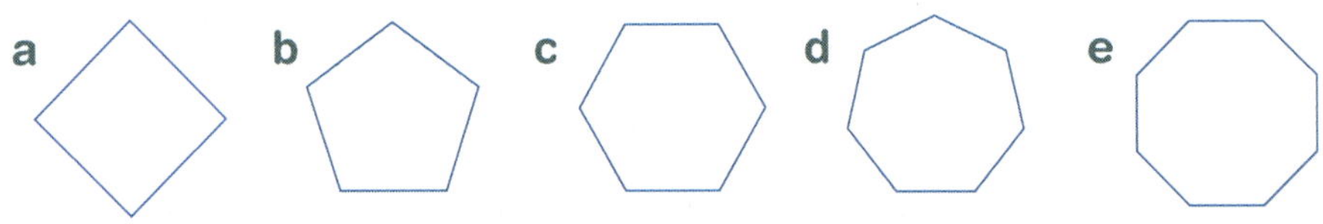

Polygon is a Greek word meaning 'many angles'.

f What are the names of the polygons in **a**, **b** and **c**?

g How many diagonals has a triangle?

h Which of the polygons above has right angles?

i Are the polygons above regular or irregular?

2 Complete this table.

Regular Figure	Number of Sides	Number of Angles	Number of Diagonals
Triangle			
Square			
Pentagon			
Hexagon			
Octagon			

3 Draw in all the axes of symmetry for each regular polygon below.

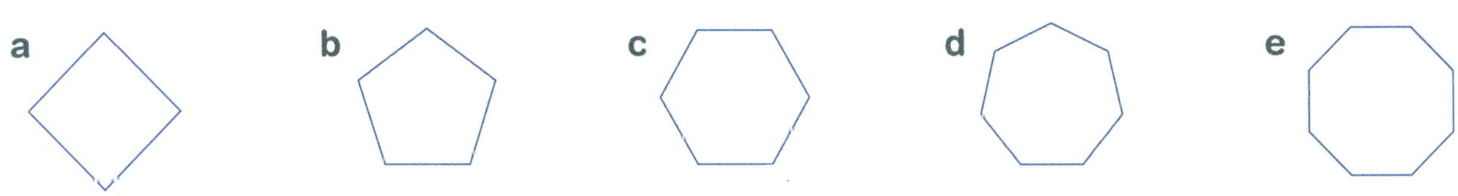

4 Compare the number of sides and the number of axes of symmetry for each shape in Question 3. Predict the number of axes of symmetry for a regular figure with:

a 10 sides   **b** 12 sides   **c** 20 sides

4:11 Visualising Shapes

We can combine shapes in different ways.
Here a square and two triangles have been used.

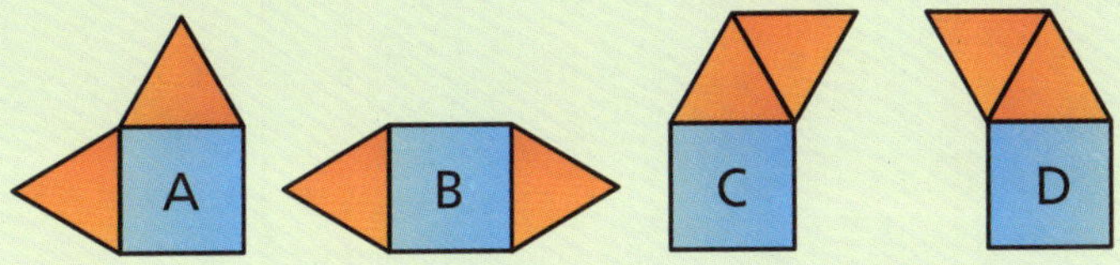

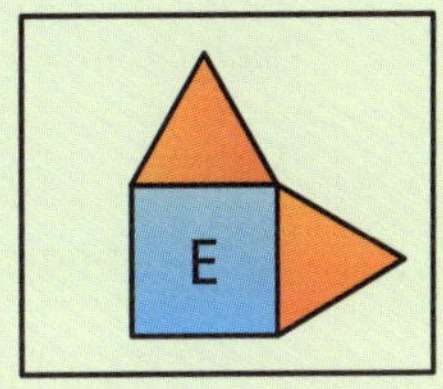

CONCEPT

Conclusion: There are 4 different composite shapes.

1 a Draw as many different composite shapes (combinations) as you can using the two squares and the triangle. A combination is not different if it is turned.

b How many different combinations are possible using just three of the squares? ______

2

a How many different composite shapes are possible using three of these triangles? ______

b Draw four different composite shapes formed by using four of these triangles.

3 **Tetrominoes** are composite shapes made up of 4 identical squares. Using a computer drawing program (or sets of 4 square counters), make the complete set of tetrominoes.

ICT

a Join the two tetrominoes that are the same.

b How many tetrominoes are there altogether? ______

4:12 Maps

- A scale of 1 cm = 1 km means that 1 cm on the map represents 1 km.
- The legend lists the features shown on the map.

Scale: 1 cm = 1 km

Legend

School, Traffic Lights, Hospital, Police

CONCEPT

1 On which street on the map above would you find the:

a school? ______ **b** traffic lights? ______

c hospital? ______ **d** police station? ______

What special feature would you find on:

e Warn St? ______ **f** Joseph St? ______

g Meg St? ______ **h** March St? ______

2 Starting at the red arrow position, where will you be if you drive:

a 1 km forward, then turn left and drive 2 km, then turn right and drive 1 km? ______

b 3 km forward, then turn left and drive 1 km? ______

c 1 km forward, then turn right and drive 2 km, then turn left and drive 2 km? ______

3 Where are the traffic lights? ______

4 **a** How far is it from the school to the police station? ______

b How far is it from the school to the hospital? ______

- Draw a map using the scale 1 cm = 10 m and a legend.

ACTIVITY

1cm = 10m

Scale

______

Legend

	School
	Gym
	Shops

MEASUREMENT & GEOMETRY

4:13 Creating a Map

1 Use the scale 2 cm = 1 m to draw a map of your room. Make up your own legend and icons to represent the features of the room. First record the dimensions of the room.

Scale

1 cm = $\frac{1}{2}$ m

Legend

☒ desk

2 Explain how you made your map.

4:14 Cones, Cylinders and Spheres

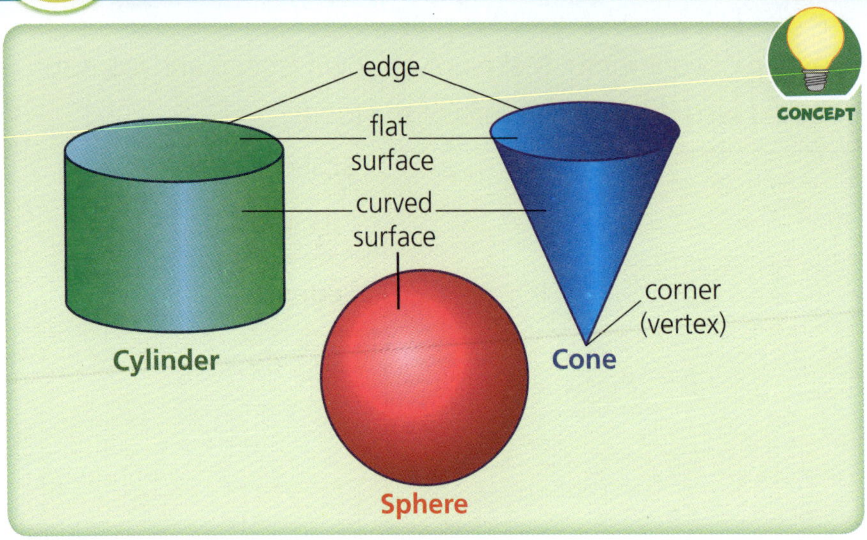

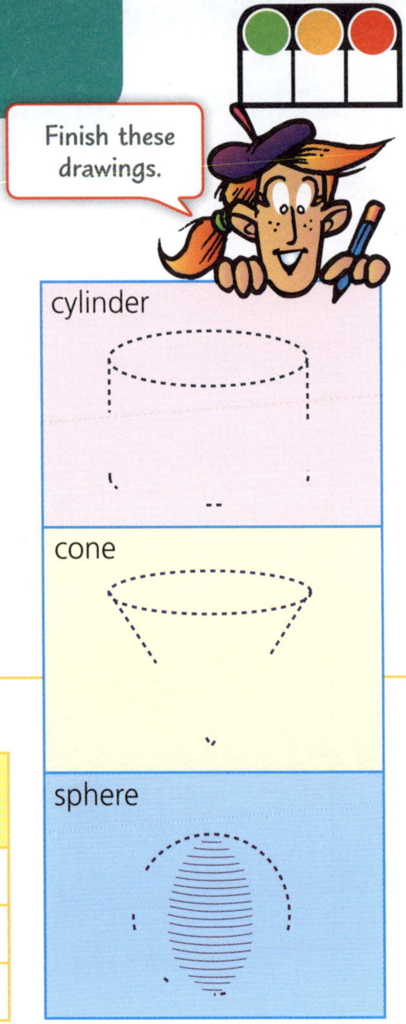

1. Use the pictures above to complete this table.

Shape	Number of Surfaces	Number of Corners	Number of Edges
Cylinder			
Cone			
Sphere			

2. What are some things that look like cylinders? List as many as you can.

3. What are some things that look like cones?

4. What are some things that look like spheres?

5. Write the name of each 3D object under its picture.

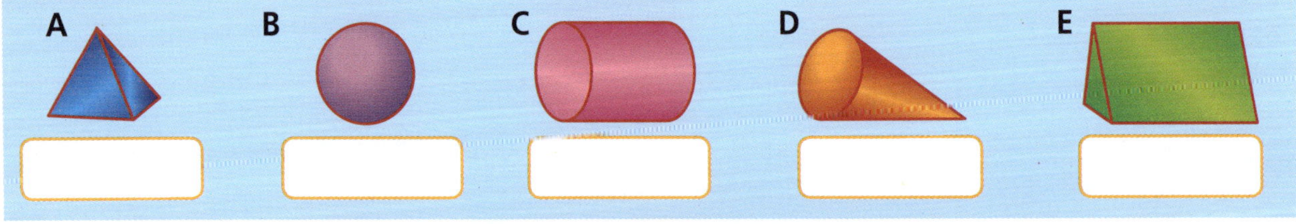

When we cut each 3D object parallel to its base, which of these have:

a uniform cross-sections (always the same)?

b cross-sections that are the same shape but different sizes?

 • *AUSTRALIAN SIGNPOST MATHS 4* • ISBN 978 1 4886 2187 1

4:15 Views of 3D Objects

1 Draw the table and chair from the front, top and side.

Front	Top	Side

2 Join each set of pictures on the left to its 3D object on the right. Name each 3D object.

Top View	Front View	Side View	3D Object	Name
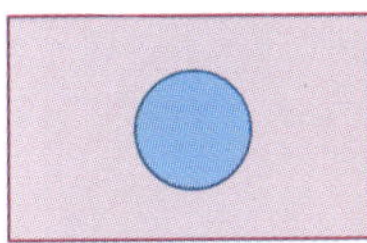	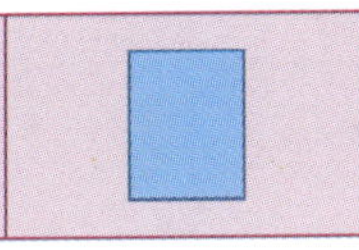	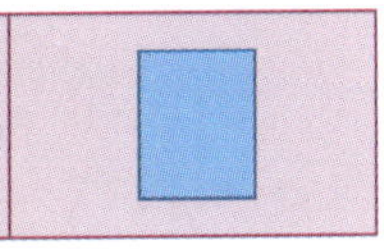	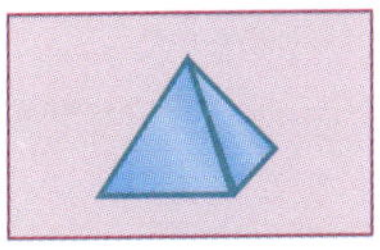	
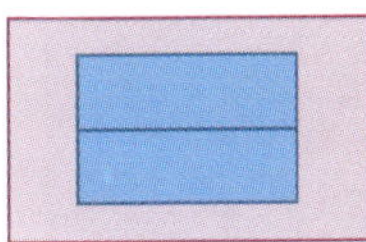	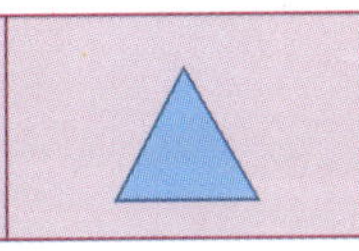	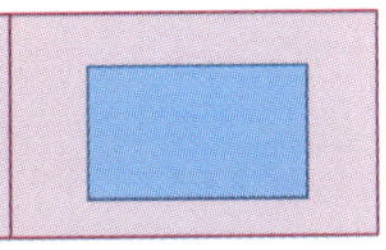	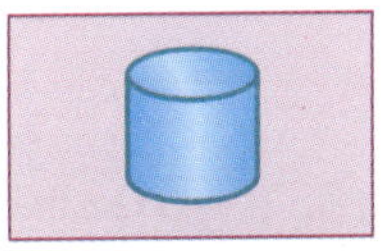	
	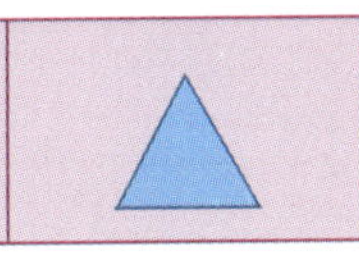	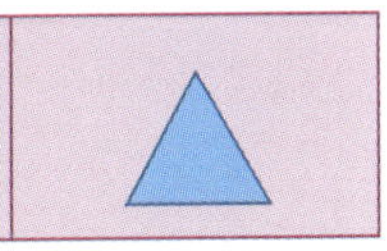	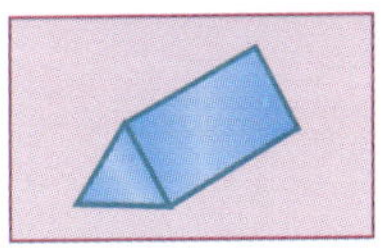	

3 Use blocks to make models of these prisms. Write down the number of blocks used in each model.

a  ☐

b  ☐

c  ☐

d  ☐

ACTIVITY

4 Use blocks to make the model in the picture. Study the model and:

a draw it from the top

b draw it from the side.

How many blocks in the model show:

c 3 faces? ☐

d 2 faces? ☐

e 1 face? ☐

f 4 faces? ☐

4:16 Compass Directions

A compass needle always points north.

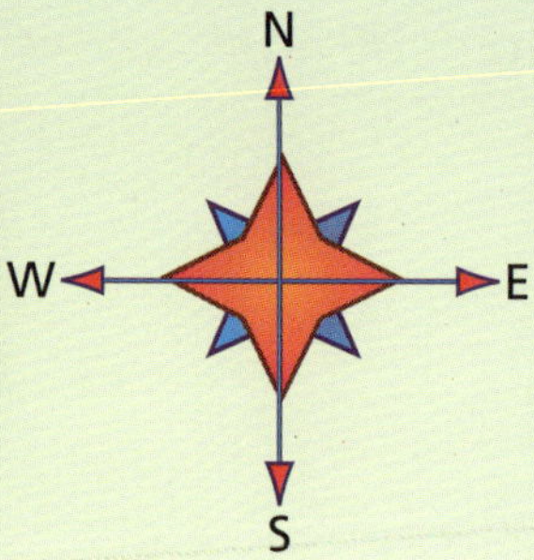

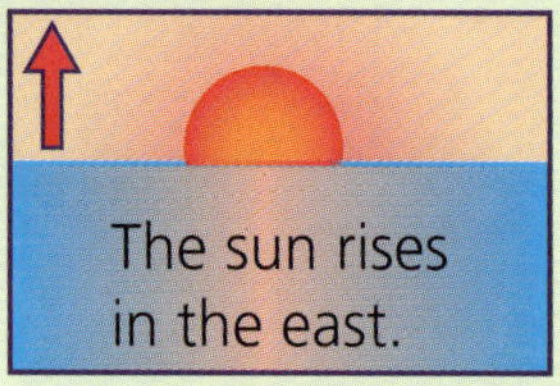

CONCEPT

1 Write **north**, **south**, **east** and **west** around the map.

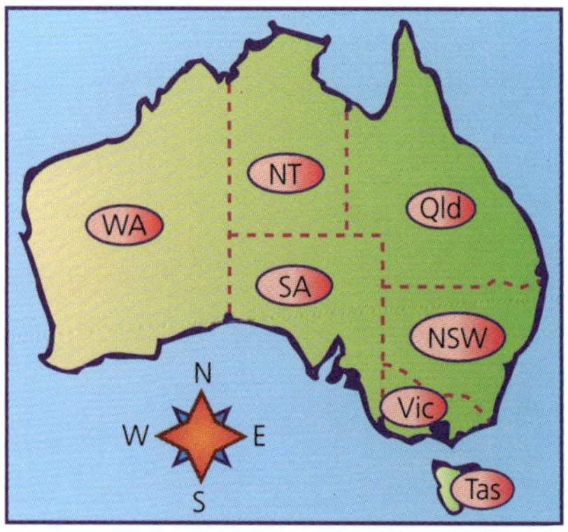

2 Write the full name for:

a WA

b NT

c SA

3 Write the names of the two largest states in the east.

4 Which landmark is:

a north of Bart's Cave?

b west of Queen Hill?

c south of Black Rock?

d west of Red Cove?

e south of Jake's Grave?

f east of Bart's Cave?

g north of Queen Hill?

5 Use the map in Question 4 to answer these questions.

a The treasure is buried east of Land's End and south of Queen Hill. Mark this spot on the map with an **X**.

b Our ship sank north of Land's End and west of Jake's Grave. Mark this spot on the map with an **S**.

c The captain broke his leg south of Bart's Cave and east of Land's End. Mark this spot on the map with a **C**.

4:17 Compass Directions

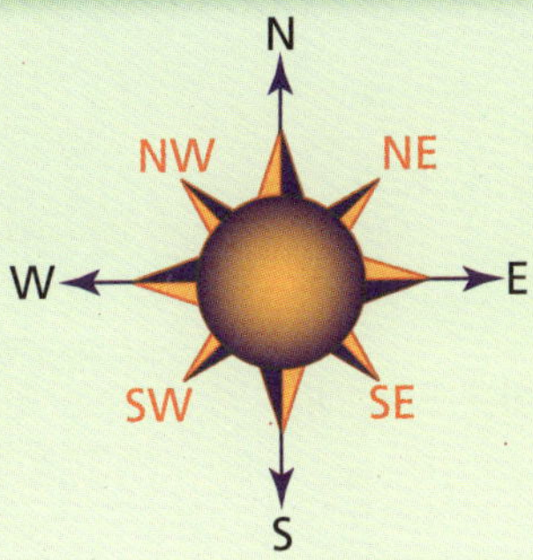

- North-east is half-way between north and east.
- North-west is half-way between north and west.
- South-east is half-way between south and east.
- South-west is half-way between south and west.

The needle of a compass points towards north.

CONCEPT

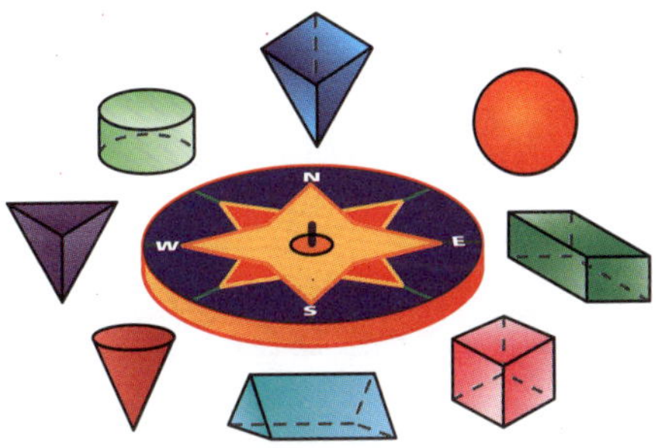

1. Around this compass, find which 3D object is:
   - **a** east
   - **b** west
   - **c** south
   - **d** north
   - **e** north-east
   - **f** south-west
   - **g** south-east
   - **h** north-west

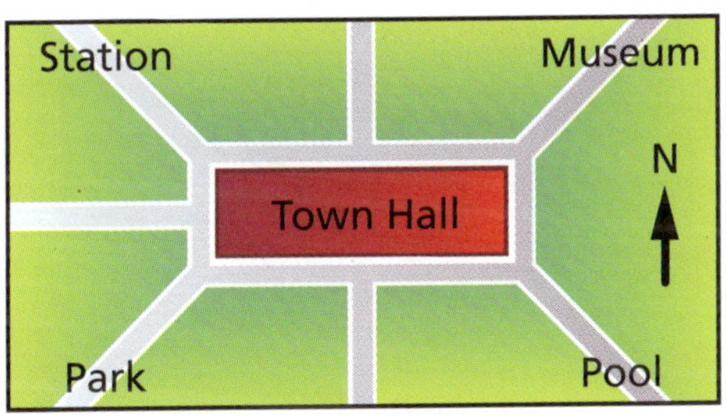

2. On the map above, what place is:
   - **a** north-east of Town Hall?
   - **b** south-east of Town Hall?
   - **c** south-west of Town Hall?
   - **d** north-west of Town Hall?

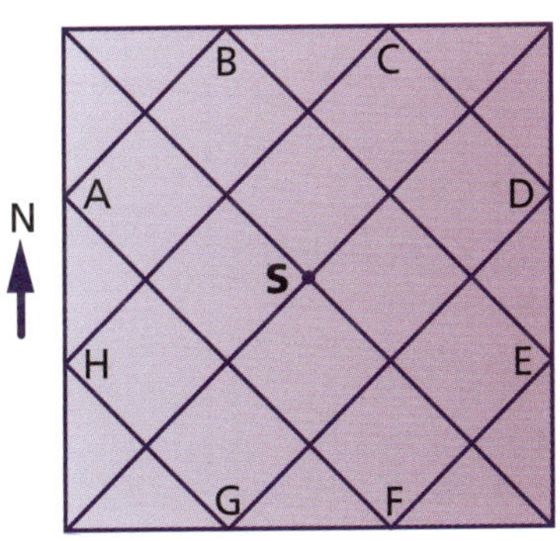

North, East, West, South = NEWS

3. Starting at **S**, where do you finish, if you travel:
   - **a** 2 units north-west, then 1 unit south-west?
   - **b** 1 unit south-east, then 2 units south-west?
   - **c** 2 units north-east, then 1 unit south-east?
   - **d** 1 unit south-west, then 2 units north-west?
   - **e** 2 units north-west, then 1 unit south-west, then 3 units south-east, then 3 units north-east?

 • • ISBN 978 1 4886 2187 1

4:18 Describing Position

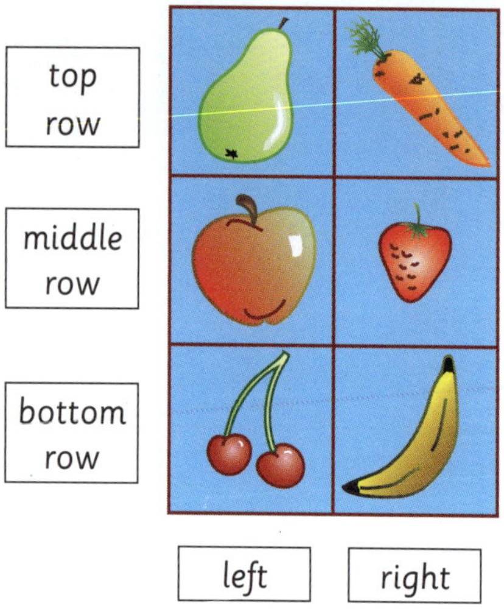

1 Name the fruit in the box that is:

a left, bottom ____ b right, top ____
c left, middle ____ d right, bottom ____
e left, top ____ f right, middle ____

Write down the position of the:

g apple ____ h strawberry ____
i cherries ____ j pear ____
k carrot ____ l banana ____

2 These are desks in a classroom.
Name the person in:

blue	Tom	Sophie	Rex	Peter
red	Ruby	Kate	Ethan	Chris
green	Georgia	Rocco	Ben	Oscar
	1	2	3	4

a column one, red row ____
b column three, blue row ____
c column one, blue row ____
d column four, red row ____
e 2, blue ____ f 3, red ____ g 1, green ____ h 4, green ____

3 Write down the column number and row letter for the:

a rectangle ____
b square ____
c hexagon ____
d triangle ____
e circle ____
f prism ____
g pyramid ____
h cone ____
i net ____
j pentagon ____
k rhombus ____
l trapezium ____

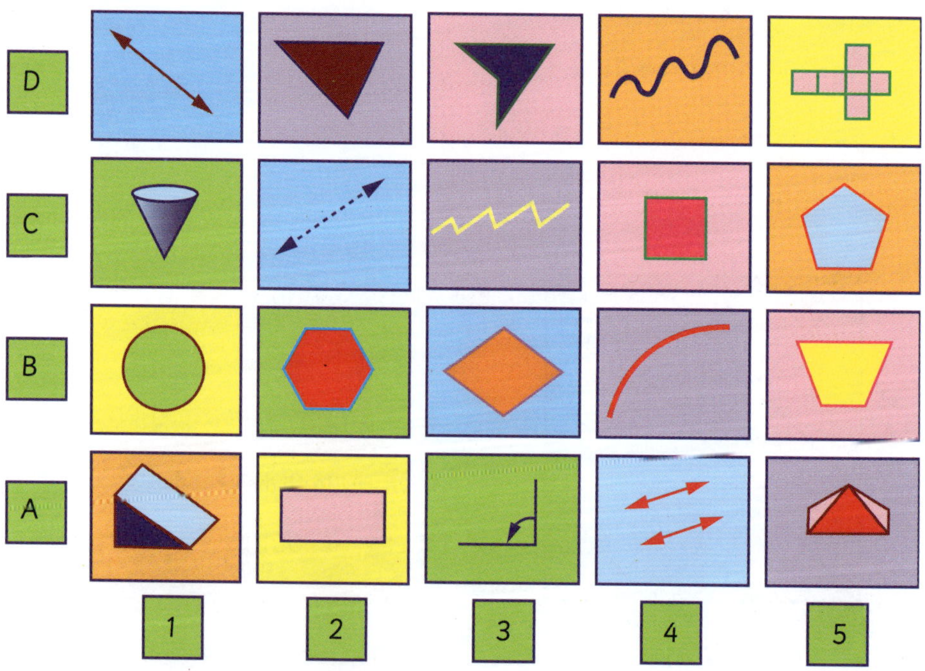

m right angle ____ n parallel lines ____

4:19 Using Position in Maps

1 What town can be found in:

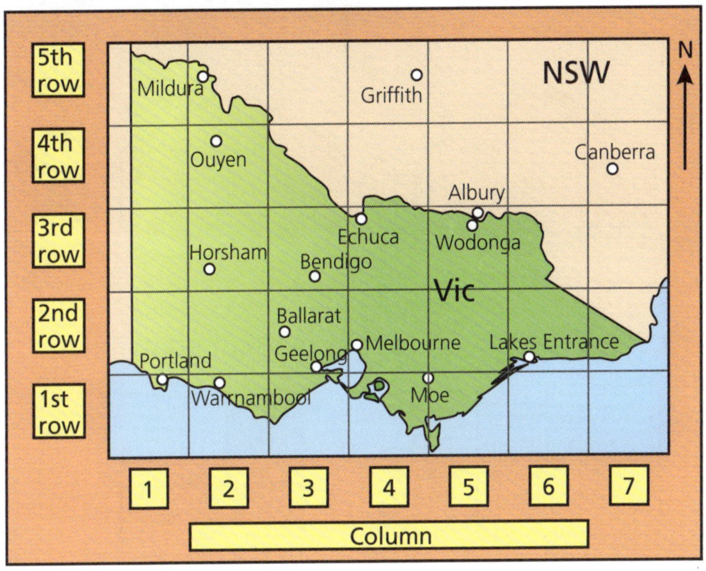

a column 2, 5th row?

b column 6, 2nd row?

c column 4, 5th row?

d column 2, 1st row?

e column 2, 3rd row?

2 Using the map in Question 1, give the position of the box where you find:

a Melbourne

b Echuca

c Geelong

d Ballarat

e Portland

f Wodonga

g Bendigo

h Ouyen

i Which boxes show part of Moe?

j Which town is west of Wodonga and north of Melbourne?

k Which town is east of Echuca and in New South Wales?

l Which town is west of Griffith but in Victoria?

3 On the grid to the right draw:

a a rectangle in C2

b an oval in A3

c a square in A1

d a triangle in C3

e a rhombus in B2

f an octagon in B1

g a trapezium in A2

h a parallelogram in B3

i a pentagon in C1

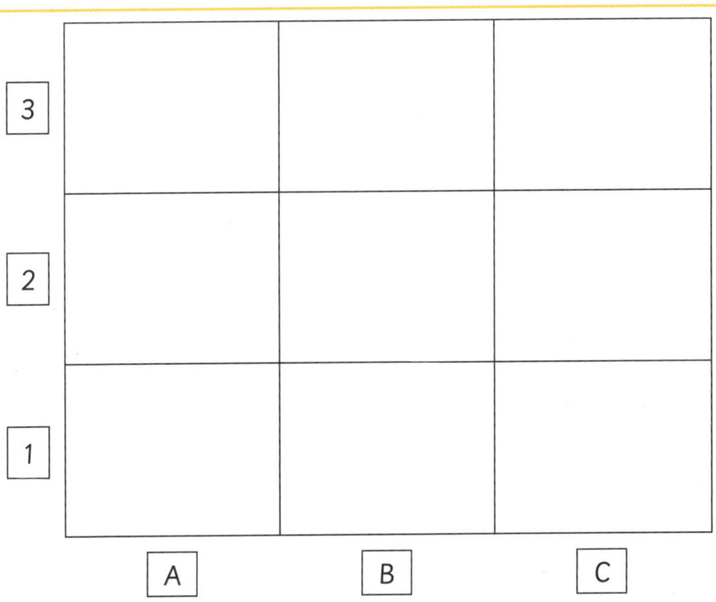

4:20 Visualising Shapes

CONCEPT

A shape can often be cut into other shapes.

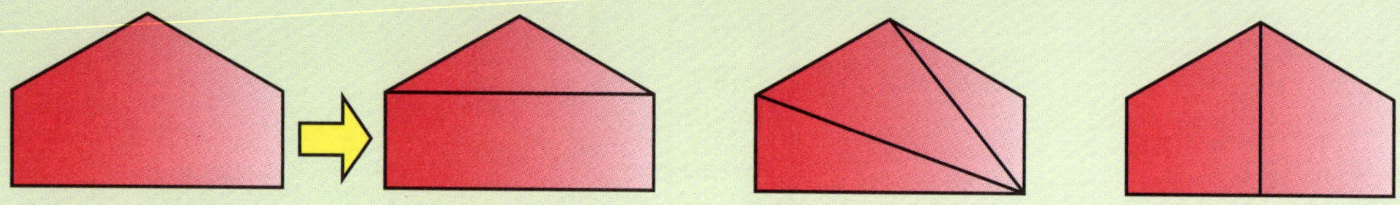

1. Show how the shape could be cut into:

2 triangles | a square and a rectangle | a triangle and a rectangle | 5 squares

2. Draw lines only from the vertices (corners) to make 3 triangles.

3. Show how the shape could be cut into:

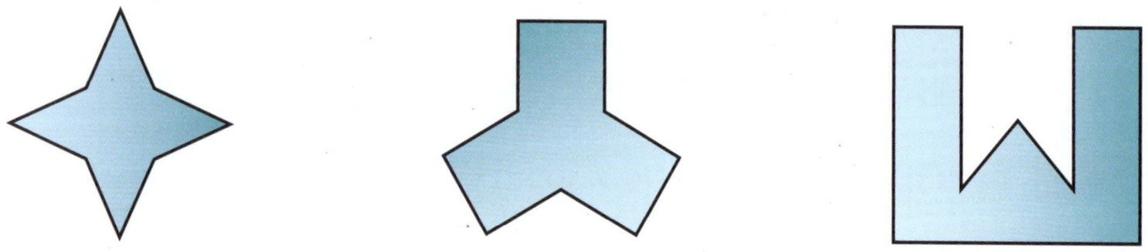

4 triangles and a square | a triangle and 3 squares | 3 rectangles and a triangle

ACTIVITY

4. On your own paper, copy the shapes above that have line symmetry.

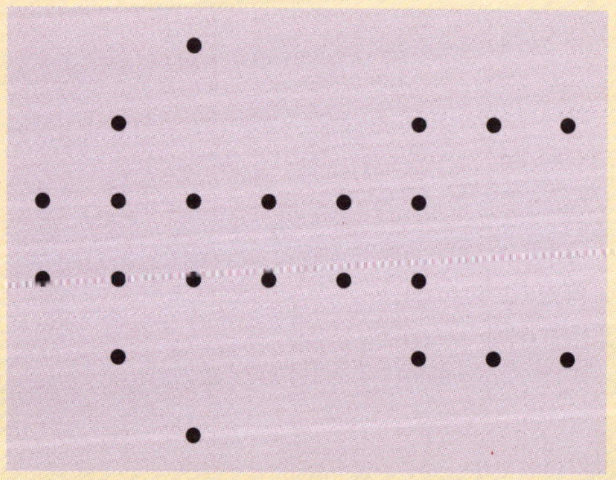

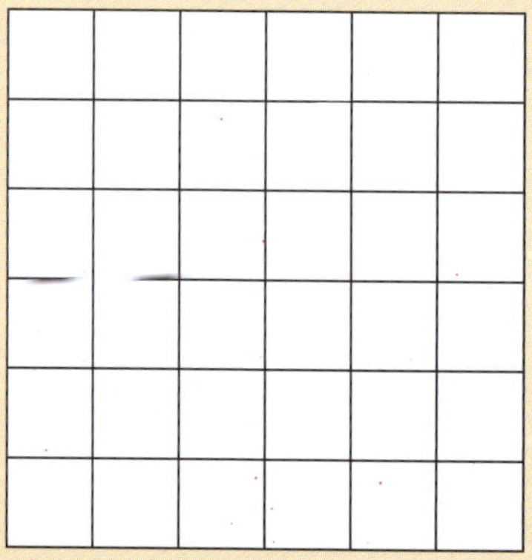

- Use a grid to draw a symmetrical pattern using large dots like the picture above.

4:21 Acute and Obtuse Angles

CONCEPT

acute

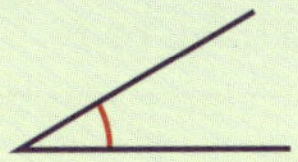

obtuse

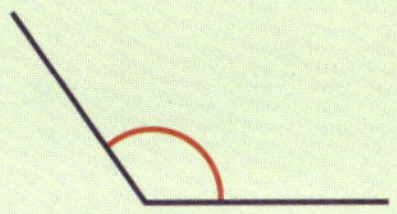

- An angle that is less than a right angle is called an **acute angle**.
- An angle that is larger than a right angle but smaller than a straight angle is called an **obtuse angle**.

1 Which of these angles are:

a right angles?

b acute angles?

c obtuse angles?

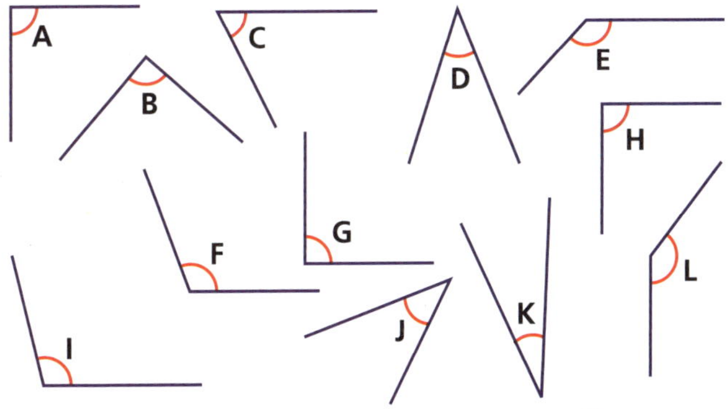

2 Place in order from smallest angle to largest angle:

a **A**, **B** and **C**

b **D**, **E** and **G**

c **F**, **J** and **H**

d **B**, **K** and **L**

e **C**, **G** and **K**

f **G**, **F** and **L**

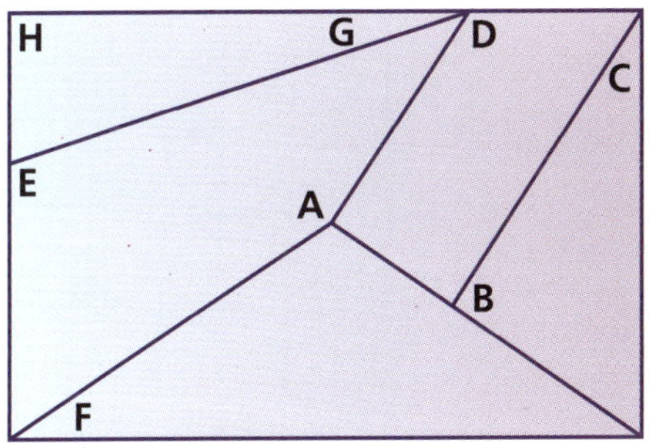

3 Name each angle as acute, right or obtuse.

a **A**

b **B**

c **C**

d **D**

e **E**

f **F**

g **G**

h **H**

INVESTIGATION

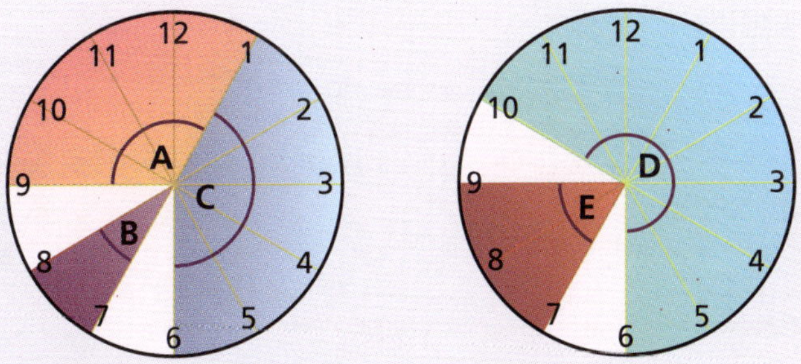

4 Which two angles have a total size equal to:

a a right angle?

b a straight angle?

c a full turn?

 • *AUSTRALIAN SIGNPOST MATHS 4* • ISBN 978 1 4886 2187 1

4:22 Horizontal and Vertical

CONCEPT

Surfaces that go straight up and down are **vertical**.

Surfaces that go straight across are called **horizontal**.

1 Naomi and Luke were building a birdcage.

- Some pipes had to be horizontal, and others vertical. To test these they used a spirit level.

- When the bubble is in the middle of the glass tube, the spirit level is horizontal (or vertical).

Which pipes in this cage are:

a horizontal?

b vertical? **c** sloping?

2 **a** Why would triangles be used in the cage?

b Which three pipes are parallel to **H**?

c Which three are parallel to **J**?

d Which three are parallel to **A**?

e List examples of parallel lines in our community.

3 Which pipes have been used to make:

a triangles?

b rectangles?

4 **a** Are any of the shapes used in the birdcage regular?

b List the pipes that are perpendicular to pipe **G**.

ACTIVITY

Write down the way you would test a surface to see if it is:

- horizontal
- vertical

Test as many surfaces as you can.

4:23 Tessellations

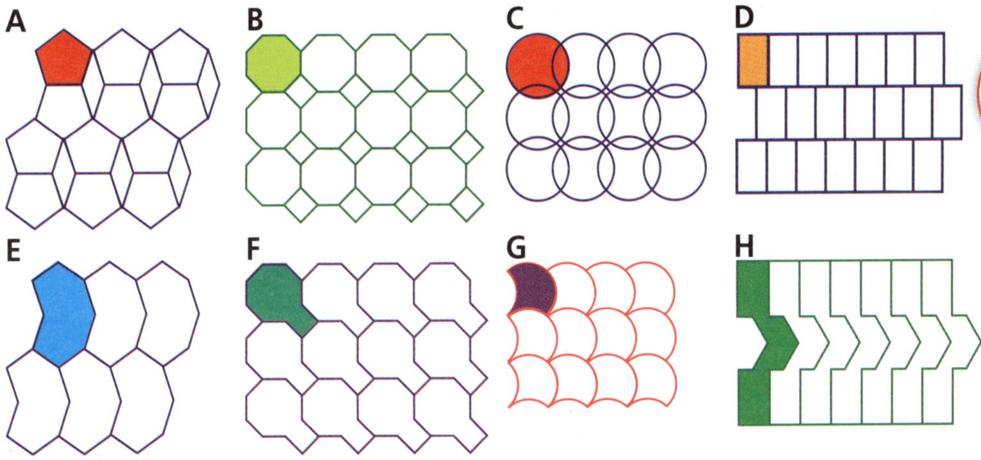

1 a In which pictures above does the coloured shape tessellate?

b In **A**, **B** and **C**, colour any gaps or overlaps black.

Identical shapes tessellate if they fit together without gaps or overlaps.

2 Peter wanted to tile his parents' bathroom. He used dot paper to draw the two tile patterns below.

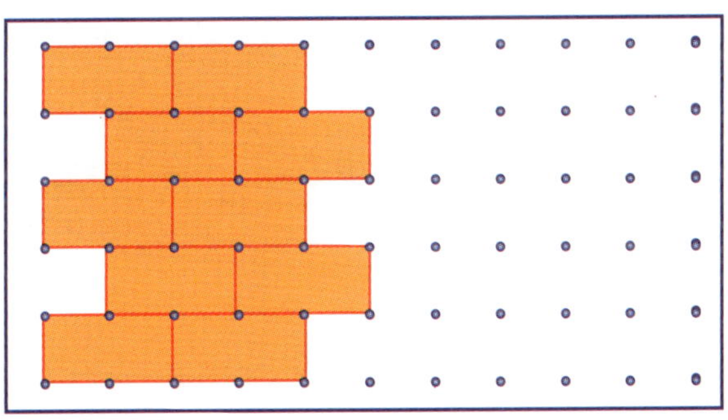

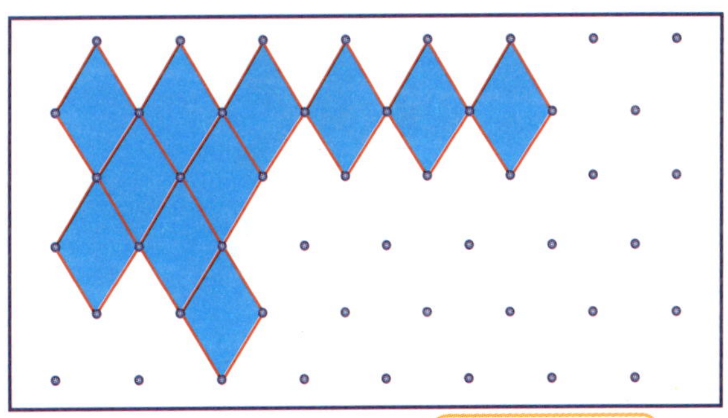

a Draw ten more tiles on each pattern.

b Do the tiles tessellate?

- Alan used this pattern to draw a tessellating bird.
- Complete the tessellation by drawing more birds. Dots have been drawn to help you.
- Use computer drawing tools to create a tessellating design by copying, pasting and rotating regular shapes.

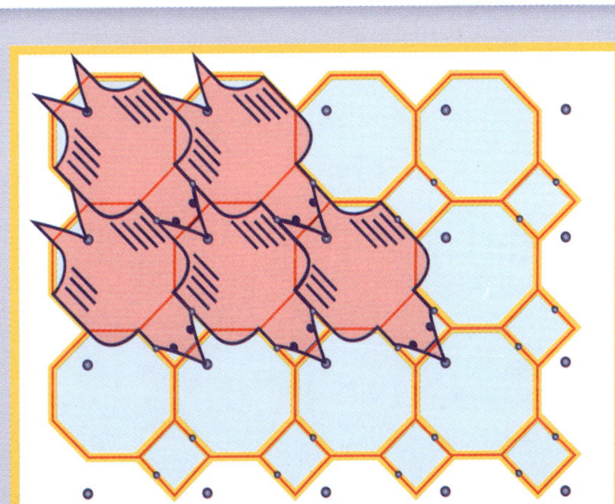

ICT

4:24 Tangrams

1

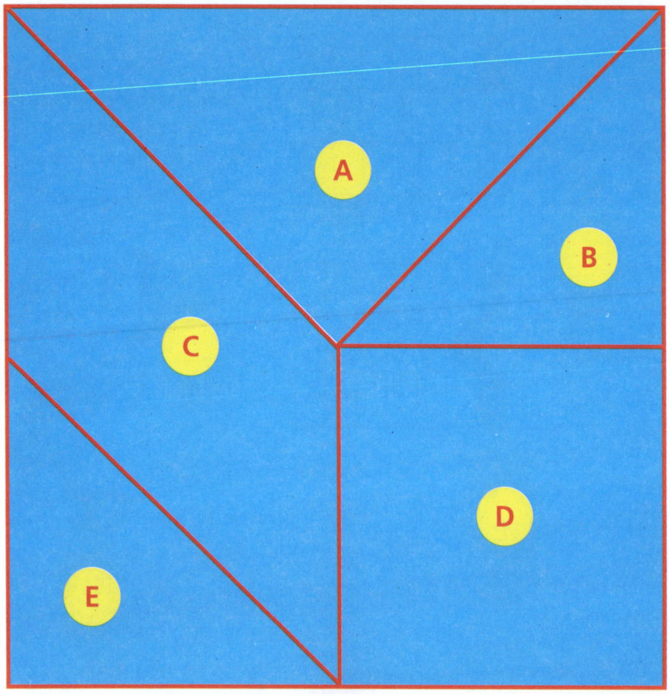

a Trace this tangram onto paper, cut it into 5 pieces and label them **A** to **E**.

b Cover the tangram shown on the left with your 5 pieces.

c Without looking at this page, arrange your 5 pieces to make a square.

d Use your 5 pieces to make these pictures.

e Make patterns of your own.

f Paste your best pattern onto cardboard or coloured paper.

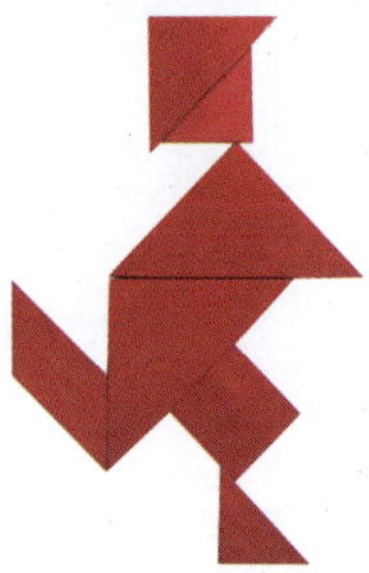

2

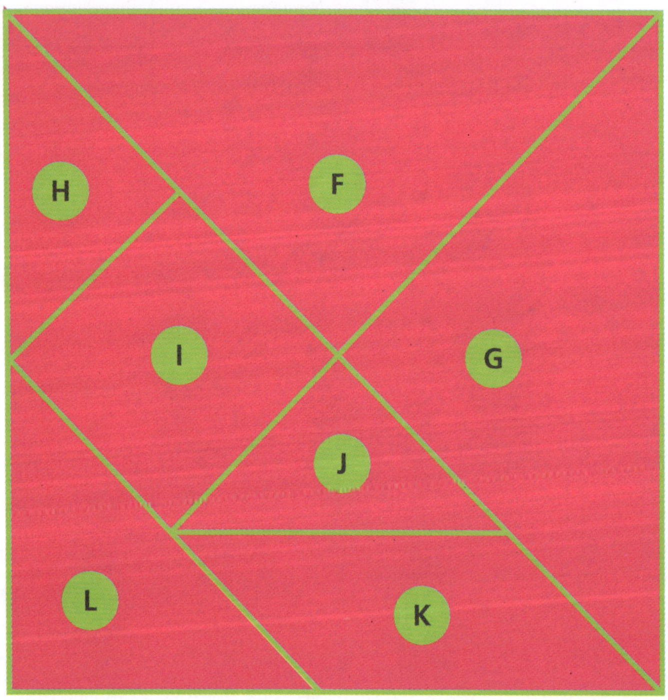

a Trace this tangram onto paper, cut it into 7 pieces and label them **F** to **L**.

b Cover this square with your 7 pieces.

c Without looking at this page, arrange your 7 pieces to make a square.

d Make a square using pieces **F** and **G**.

e Using pieces **H**, **I** and **J**, make these shapes:

5:01 Drawing Tables

1. Use the picture to fill in the two-way table.

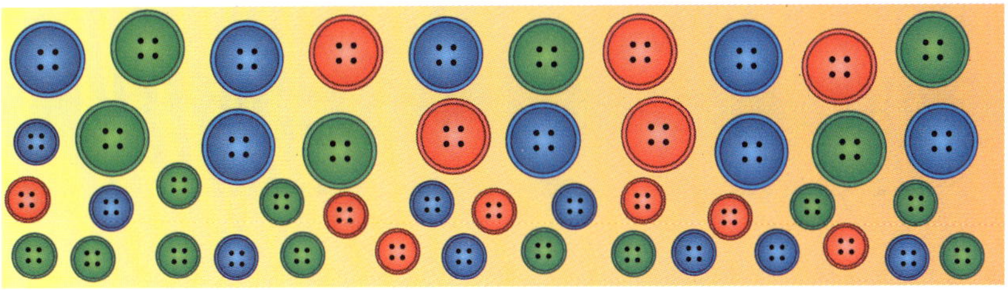

Buttons			
	Red	Blue	Green
Large			
Small			

2. Use the table above to fill in this simpler table.

Buttons		
Red	Blue	Green

3. Use the table above to fill in this simpler table.

Buttons	
Large	
Small	

4. Show the data from Question 1 on a graph.

5. Fill in this table using data about your family and friends.

Name	Age	Male or Female
Me		

6. Use the dice to complete the table.

Number of Times Thrown					

Graphing the Weather

ACTIVITY

- For the next month record the weather like this.
- At the end of the month cut up the calendar. Glue the pieces to make a graph like this.
- Use the graph to describe the month's weather.

 • *AUSTRALIAN SIGNPOST MATHS 4* • ISBN 978 1 4886 2187 1

5:02 Chance

1 Without looking, Min must choose one card.

a Is she more likely to choose red or black? ______

b Is she certain to choose an ace? ______

c Is there an even chance to choose red? ______

d The chance of choosing black is 1 out of ______.

e The chance of choosing red is 2 out of ______.

2 The spinner can land on any colour.

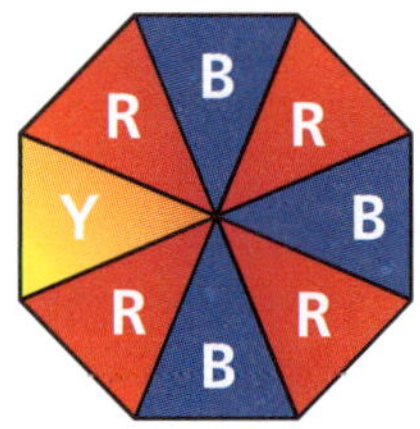

a Which colour is it most likely to land on? ______

b Which colour is it least likely to land on? ______

c Which colour has an even chance of occurring? ______

d What is the chance of spinning pink? ______

e The chance of landing on red is ______ out of 8.

f The chance of landing on blue is ______ out of 8.

g The chance of landing on yellow is ______ out of 8.

3 Order these events from 1 (least likely) to 4 (most likely).

- The principal will visit our classroom tomorrow. ______
- It will be sunny tomorrow. ______
- Our class will use books tomorrow. ______
- The next person to visit our room will be a female. ______

4 Write three things that might happen tomorrow.
Order each event from 1 (least likely) to 3 (most likely).

______ ______

______ ______

______ ______

Heads or Tails?

ACTIVITY

Toss a coin 50 times and keep a tally of the number of heads and tails.
Discuss the experiment and which results you think are most likely.

5:03 Chance

CONCEPT

Here, each ball has the same chance of being selected.

1 Write these events in order from least likely to most likely.

○ ○ ○ ○ ○

least likely — **most likely**

A The principal will come to school on Monday with no shoes on.

B It will be sunny tomorrow.

C If I drop a glass it will break.

D If I toss a coin it will be a head.

E Our next teacher will be from Mars.

2

To decide who works together, **two numbers** will be taken from the jar at the same time.

Does each number have the same chance of being chosen? ☐

*In a **random selection** each number has the same chance of being chosen.*

Carry out this experiment in your classroom.

Record the results: 6 is matched with ☐, ☐ with ☐ and ☐ with ☐.

3 Choose a label to answer each question.

In Question 2, what is the chance that:

a the numbers 6 and 7 are picked? ☐

b the numbers 1 and 2 are picked? ☐

c two numbers less than 10 are picked? ☐

d when Alf is chosen, he will be working with a girl? ☐

e at least one of the pairs will be two girls? ☐

impossible | **not likely** | **even chance** | **very likely** | **certain**

5:04 Using Graphs

1 Rhonda graphed the number of calls made each day last week by her family.

a How many calls were made on Monday?

b On what day was the greatest number of calls made?

c On what day was the least number made?

d What is the difference between the number of calls made on Saturday and the number made on Thursday?

e What is the total number of calls made for the week?

f Why do you think more calls were made on Sunday and Saturday?

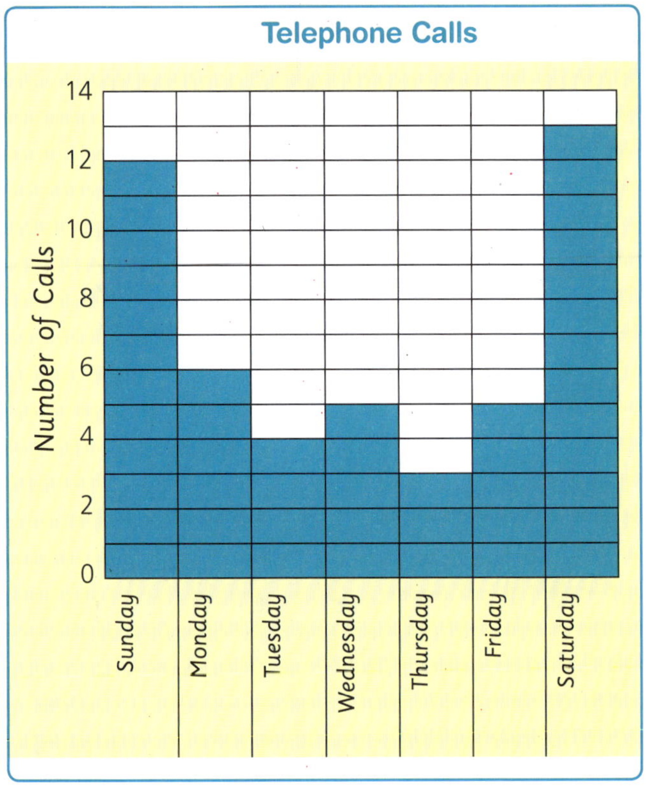

2 Rhonda kept this tally of calls made by her children. Use the tally to draw the graph.

Name	Number of calls
Alana	卌 卌 ||||
Rachel	卌 卌
Naomi	||||
Luke	卌 卌 ||
Heather	|||

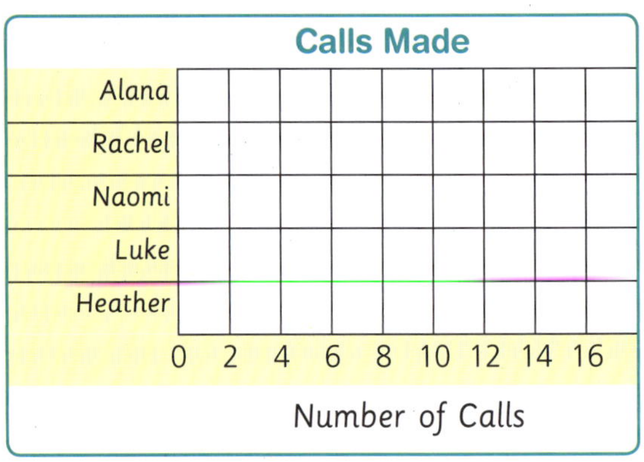

3 Trees were planted in a row. A rabbit sat in each space.

How many rabbits would there be between:

a 3 trees?

b 7 trees?

c 10 trees?

Draw a picture.

INVESTIGATION

 • *AUSTRALIAN SIGNPOST MATHS 4* • ISBN 978 1 4886 2187 1

5:05 Reading Graphs

1

Bottles Brought to School

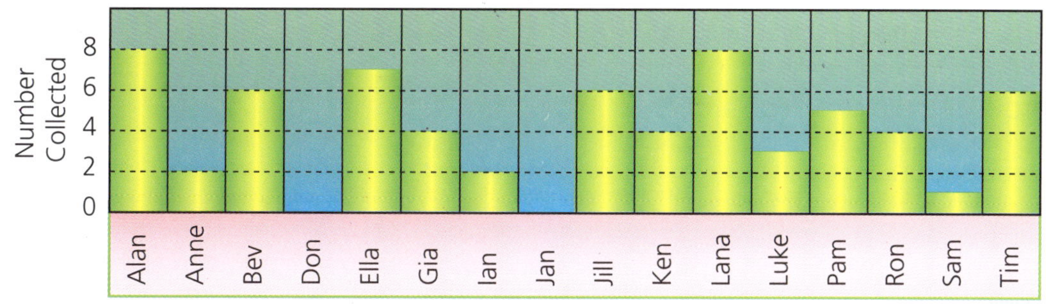

a How many bottles were brought to school by:

i Alan? ii Luke? iii Jan?

b How many people brought bottles?

c How many bottles altogether were brought by Alan, Anne and Bev?

d How many bottles were brought to school altogether?

e If all these bottles were brought to school during the first week, how many bottles do you think would be brought during the first two weeks? How did you get your answer?

2

Sport Chosen by Class 4R

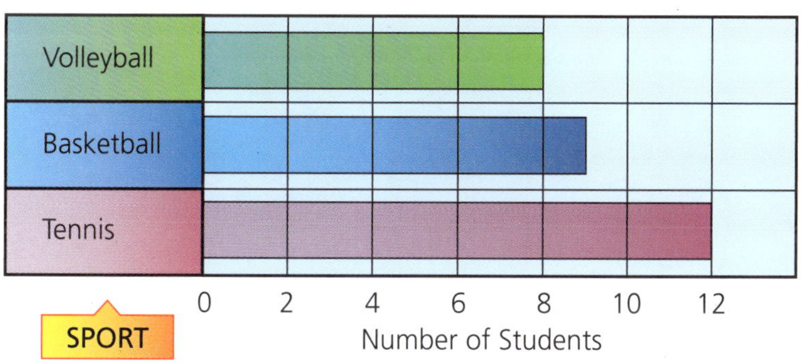

a Which was the most popular sport?

b How many students made a choice?

c How many sports were chosen?

d What is the title of this graph?

Tally Marks

ACTIVITY

Tally marks used in South America and Asia also have five lines.
With a partner, create your own five-line tally marks.

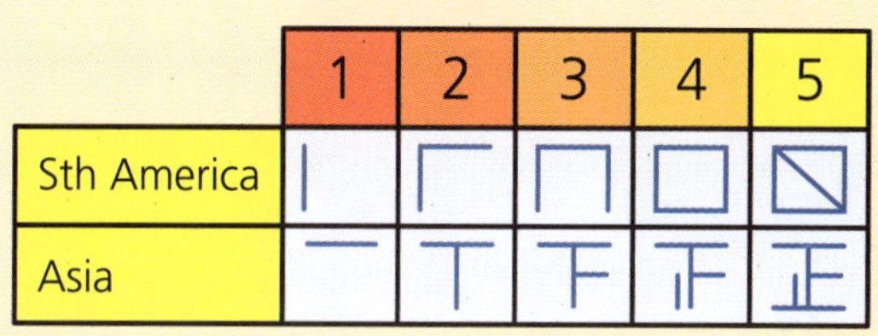

Discuss: When would tally marks be useful?

5:06 Ordering Events

1 Choose the label you think best for each part below.

impossible | **unlikely** | **even chance** | **very likely** | **certain**

a My first throw of the dice will be a one.

b I will throw a zero on a dice.

c When I throw a dice the number will be greater than one.

In this jar are 3 red and 3 yellow marbles. Blindfolded, I draw one out.

Which label from above is best for each of these statements?

d I will draw out a coloured marble.

e I will draw out a blue marble.

f I will draw out a yellow marble.

2 Ms Banner placed two red and two blue counters in a hat. She had two more red and two more blue counters. Which two counters must she put into the hat so that she would be:

a more likely to draw out a red than a blue counter?

b more likely to draw out a blue than a red counter?

c equally likely to draw out a blue or a red counter?

3

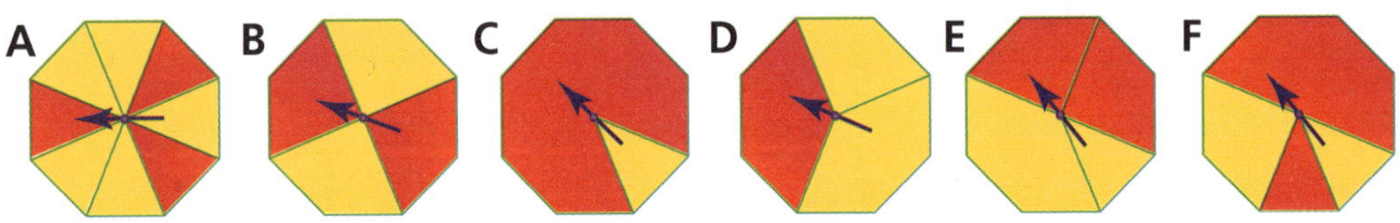

The spinners above stop on either red or yellow.

a Which spinners are equally likely to stop on red as on yellow?

b Which spinners are more likely to stop on red than on yellow?

c Which spinners are more likely to stop on yellow than on red?

4 Here is the head and tail of an old Australian penny.

a If I toss a coin in the air, would a head be more likely to show than a tail?

b If two coins are tossed, what could result?

 ISBN 978 1 4886 2187 1

5:07 Chance Used in Games

1 Caroline tossed a coin many times. She kept this tally of the results.

Tossing a Coin		
Tossed	**Tally**	**Total**
Heads	卌 ||||	
Tails	卌 卌 |	

a How many times did she toss the coin? ______

b What fraction of the time did she toss:

i a head? ______ **ii** a tail? ______

2 When playing cricket, we toss a coin to see who will bat first. How would you describe the chance that our team will win the toss? ______

3 100 tickets were sold in a raffle. John bought 70 of the tickets, Peta bought 25, Jeremy bought 2 and Alan bought none.

Choose a label that describes the chance that the winner will be:

impossible	even chance
very unlikely	likely
unlikely	certain

a John ______ **b** Peta ______

c Alan ______ **d** Jeremy ______

e none of the four people mentioned ______

4 A dice is rolled. Jack wins if an odd number is rolled.

Emma wins if a 6 is rolled and Luke wins if a 2 or 4 is rolled.

a Write their names in order from most likely to win, to least likely to win.

______

b Which person has an even chance of winning? ______

What's My Age?

- Ask friends to write the month in which they were born. (1 for January, 2 for February, 3 for March, etc.)
- Tell them to multiply this number by 5, then add 2, then multiply by 20, then add their age in years, then subtract 40.
- Ask the person their answer.

Solution

- The last two digits of the answer give you the age. The other digits give you the month of birth. For example, 714 means the person is 14 and was born in July (the 7th month).

  ISBN 978 1 4886 2187 1

5:08 Surveys

CONCEPT

- **Surveys** are used to discover information, to give us a general view of a situation or to check predictions.
- If everyone is surveyed from the group we wish to study, we have carried out a **census**.

1 Jared chose as his survey topic: *The time students go to bed*. He asked everyone in his class.

Jared's Survey

Tick the box next to the time you went to bed last night.

- ☐ before 8 pm
- ☐ between 8 pm and 9 pm
- ☐ between 9 pm and 10 pm
- ☐ between 10 pm and 11 pm
- ☐ after 11 pm

Jared's Results

before 8 pm	|
between 8 pm and 9 pm	~~||||~~ ~~||||~~ |||
between 9 pm and 10 pm	~~||||~~ ~~||||~~
between 10 pm and 11 pm	||||
after 11 pm	||

How many students went to bed:

a before 8 pm? ______ **b** between 9 pm and 10 pm? ______

c after 11 pm? ______ **d** before 9 pm? ______ **e** before 11 pm? ______

f What was the most popular bed time? ______

g How many students were surveyed altogether? ______ Was this a census? ______

h Make up a question of your own that could be answered using these results.

______

2 Write down a topic that you would like to use in a survey of your class.

______

Write two questions you would like to ask about this topic.

a ______

b ______

5:09 Tally Marks

Tally the Colours

ACTIVITY

- Start on any white square.
- Throw a dice and move that many spaces clockwise.
- Each time record the colour you land on, as a tally mark.
- Throw the dice 50 times.

Before you start, try to guess which colour will appear the most.

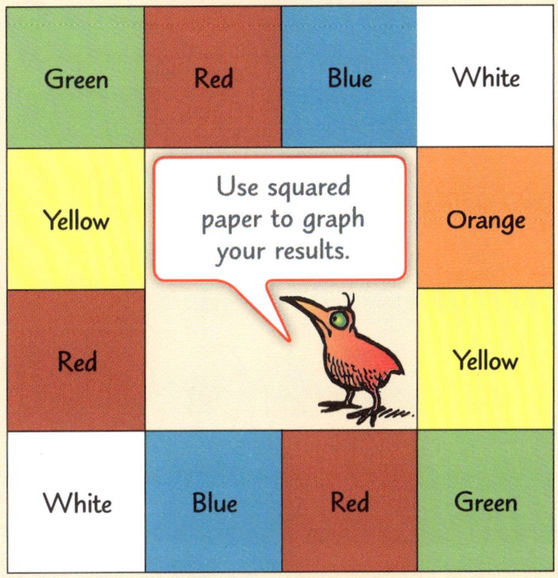

Colour	Tally	Total
White		
Blue		
Red		
Green		
Yellow		
Orange		

1 Eric asked 300 people to name their favourite sport. He made this tally of their answers.

Sport	Tally	Total
Tennis	卌 卌 卌 卌 卌 卌 卌 卌 卌	
Athletics	卌 卌 卌 卌 卌 卌 卌	
Swimming	卌 卌 卌 卌 卌 卌 卌 卌 |||	
Hockey	卌 卌 卌 卌 卌 |||	
Basketball	卌 卌 卌 卌 卌 卌 卌 |||	
Cricket	卌 卌 卌 卌 卌 卌 卌 卌 ||	
Football	卌 卌 卌 卌 卌 卌 卌 卌 卌		
Other	卌 卌 卌 卌 ||||	

a Complete the **Total** column.

b What was the most popular sport? ______

c How many more people chose swimming than athletics? ______

d Altogether, how many chose athletics and tennis? ______

e Name two sports which might have been included in *other*. ______

INVESTIGATION

- Ask 50 people to name their favourite sport.
- Keep a tally, like the one above, of their answers.
- Before you begin, estimate how many out of 50 will choose each sport.

Anyone for tennis?

5:10 Constructing Spinners

INVESTIGATION

Aim: Construct and test spinners with 3, 4, 5, 6 and 8 sides.

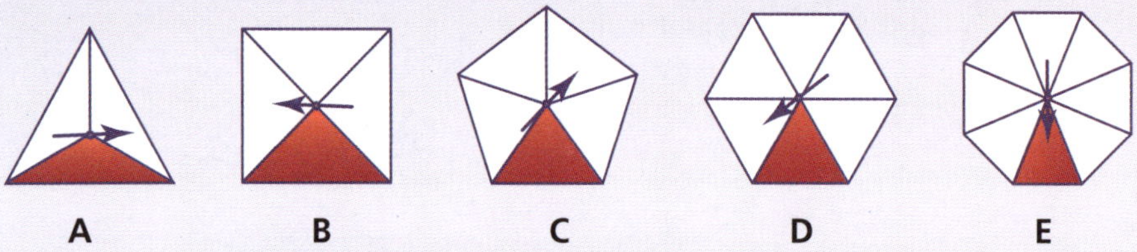

Step 1: Use a program such as Microsoft® Word® to create these shapes. (Go to **Insert**, then **Shapes**, select the shape and hold the shift key to give the shape equal sides.)

Step 2: Print these shapes onto cardboard. Divide the spinners into sections (as above) and colour one section of each.

Step 3: Make these shapes into spinners with one section coloured.

1 Experiment

Spin each spinner 20 times. Keep a tally of coloured spins for each.

a Which spinner do you think will have most coloured spins?

b Which spinner do you think will have least coloured spins?

c Carry out the experiments.

Spinner	A	B	C	D	E
Tally of Coloured Spins					

d Which spinner had most coloured spins?

e Which spinner had least coloured spins?

From your results, on which spinner is a colour:

f most likely to be spun?

g least likely to be spun?

h If we combined the results of many students, would we have a better idea of the answers to parts **f** and **g**? Why or why not?

i Put **A** to **E** above in order, from most likely to spin a colour to least likely to spin a colour.

5:11 Unequal Outcomes

CONCEPT

- Each side of these squares has the same chance of being spun.
- The two colours usually have a different chance of being spun.

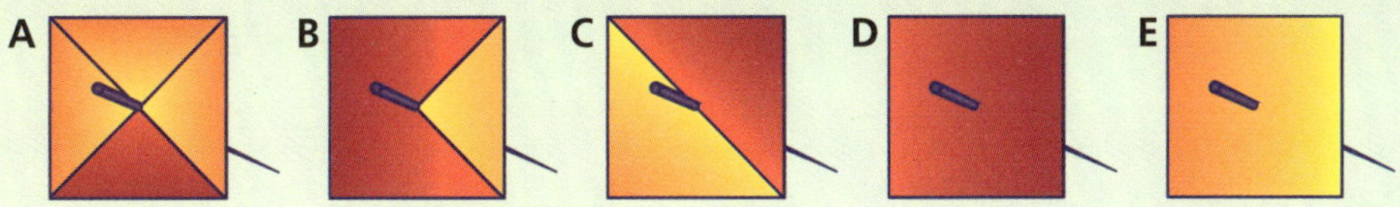

1 For which of the spinners above is the chance of spinning red:

a impossible? ______ **b** certain? ______ **c** an even chance? ______

d unlikely? ______ **e** likely? ______

2 Put the spinners above in order, least likely to most likely of spinning red. Discuss your answer. ______

3 Put the spinners in order, least likely to most likely of spinning red.

a 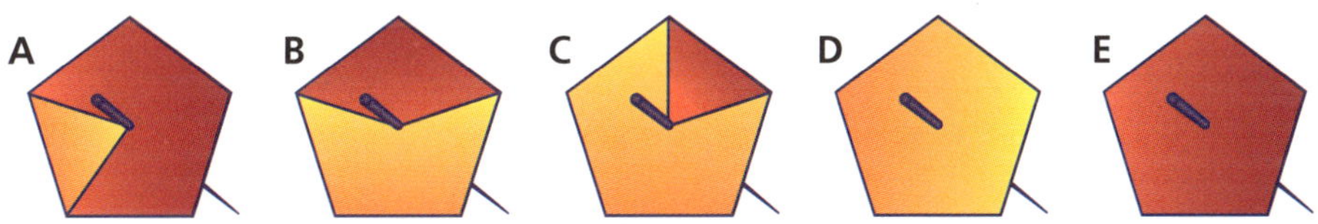

b 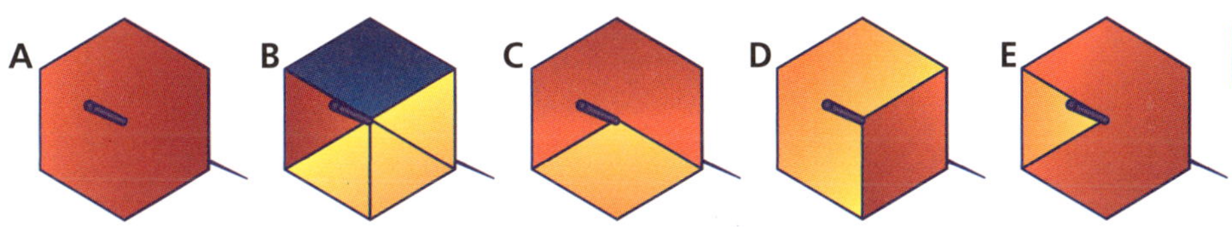

c 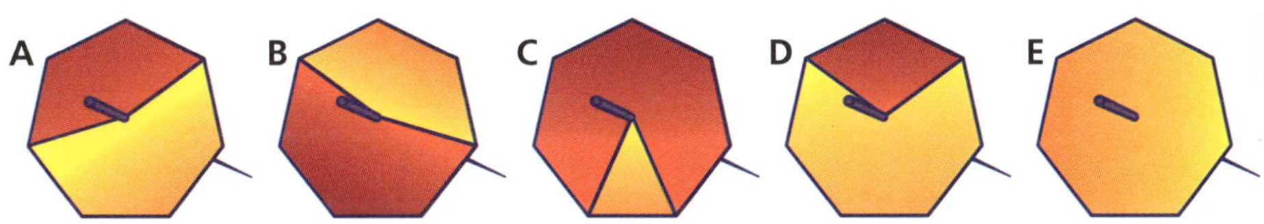

INVESTIGATION

4 Put the numbers 1 to 6 on the faces of a matchbox.
Are the faces equally likely to land face up when the box is tossed? ______
Carry out this experiment 50 times, keeping a tally of the results.
What did you discover? ______

Tally					
1	2	3	4	5	6

 • • ISBN 978 1 4886 2187 1

5:12 Collecting Information

INVESTIGATION

You are to measure the hand length of 12 boys and 12 girls, in centimetres (to the nearest cm).

Compare your results with the results of others.

1 Who will you measure?

2 How will you measure hand length?

I will use

3 Make some predictions.

Who will have the longest hand, a boy or a girl?

Who will have the shortest hand, a boy or a girl?

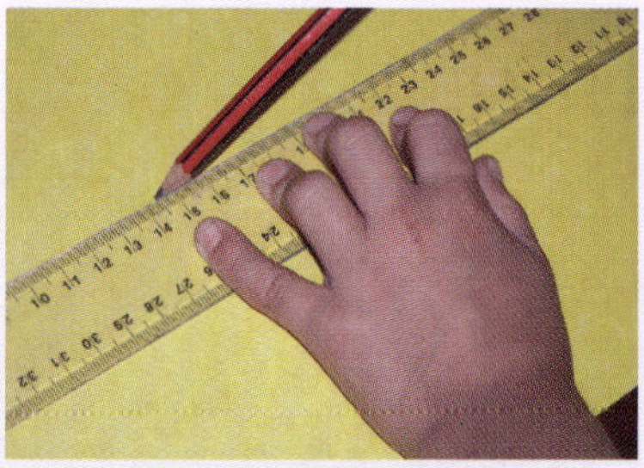

4 Measure a few hands before you decide which measurements you will use in the left column.

5 Use tallies to record your results below.

Hand Lengths of Boys and Girls

Measurement in Centimetres	Boys' Tally	Total	Girls' Tally	Total

6 **Results**

The longest hand belonged to a ______. The shortest hand belonged to a ______.

7 Make a strip of paper that has the same length as:

a the shortest male hand

b the shortest female hand

c the longest male hand

d the longest female hand

8 **Your comments**

 ISBN 978 1 4886 2187 1

5:13 Graphing Data

1 a How many trains were sold on Monday?

b On which day were the most trains sold?

c How many trains were sold altogether?

d If there were 200 trains for sale, how many trains were not sold?

e Why do you think no trains were sold on Thursday?

f If the trains were sold in boxes of 10, how many boxes were sold?

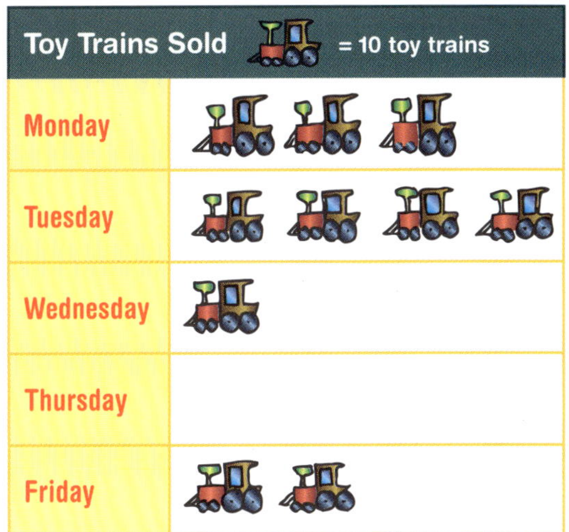

2 Katie surveyed students at her school to find out how they travelled to school. She recorded her results (to the nearest 10).

Bus	30
Walk	60
Car	70
Bike	40

a Which option was most common?

b How many students use a bus or car?

c Approximately how many students did she survey?

d Should Katie expect the same results every day?

e Complete the picture graph to display her results.

f Why is it useful to make [student icon] = 10 students? Discuss.

Key: [student icon] = 10 students	
Bus	
Walk	
Car	
Bike	

3 Peter rolled two dice 25 times and found each total.

He used a column graph to record his results.

INVESTIGATION

a Carry out this activity and record your results.

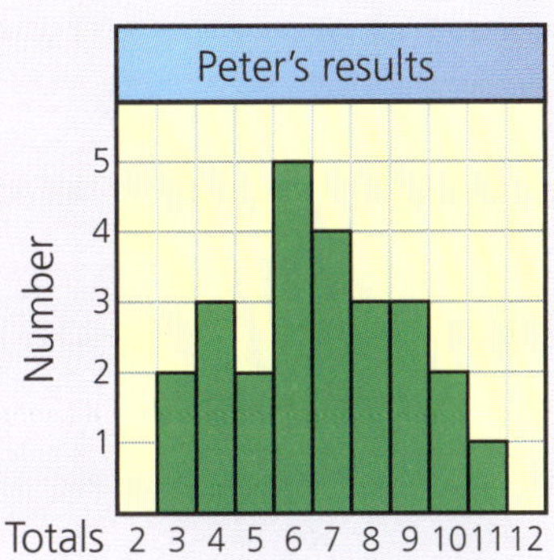

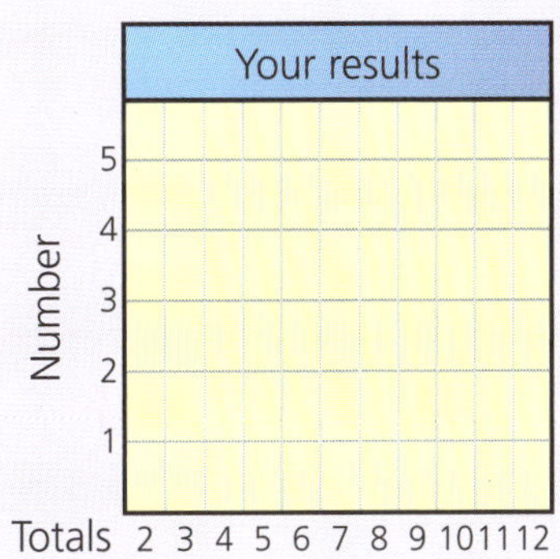

b Compare your results with Peter's results.

5:14 Chance Experiments

1. Put the same number of red and green counters in a container.
One counter will be drawn out and then replaced.
   a How many red counters would you expect to get in 10 turns?
   b Give a reason for your answer.
   c Carry out this experiment. The number of red counters chosen = ______. Discuss.
   d Carry out this experiment a second time. Were your results the same?

2. 

- Take a counter at random from a bag containing 5 red, 3 blue and 1 yellow counter. Record a tally mark for that colour.
- Return the counter to the bag.
- Repeat the experiment 50 times.

	Tally	
Red		
Blue		
Yellow		

3.

Total	Tally
2	
3	
4	
5	
6	
7	
8	
9	
10	
11	
12	

Throw two dice, total the numbers and tally the result. Repeat 100 times.
   a Which total occurred most?
   b Which total occurred least?
   c List the totals in order from 'occurred least' to 'occurred most'.

Use **never**, **sometimes**, **highly unlikely** or **always** to describe the chance of tossing a total of:
   d 7
   e 1
   f a total less than 14
   g Make up a sentence using one of these words.

4. Write **equally likely** (**E**) or **not equally likely** (**N**) in each case.

   a It will land on 1, 2 or 3.
   b The drawing pin will land face up or face down.

   c The next child born will be a girl or a boy.

 • *AUSTRALIAN SIGNPOST MATHS 4* • ISBN 978 1 4886 2187 1

5:15 Carry Out Your Own Survey

INVESTIGATION

1 Work in small groups to plan and carry out a survey.

Step 1

Choose a topic to survey.

Topic:

Step 2

Write a question to ask people about your topic.

Question:

Step 3

Decide how you will record the answers (called **data**).

Step 4

Decide on the best way to collect the data.

Guess (predict) what the data will show.

I think

Step 5

Represent your results using tables and graphs either hand drawn or on computer.

I found

Step 6

Write predictions or conclusions drawn from your data.

2 Discuss with your family the uses made of surveys.

3 Find a survey in a newspaper, magazine or at the library. Share this survey with your class.

5:16 Chance Experiments

1 Matt tossed a coin 10 times and made this graph.

a Which outcome occurred more often?

b Do these results mean that heads is more likely to be tossed than tails?

Why or why not?

2 **a** If you tossed a coin 10 times, how many heads would you predict?

b Toss a coin 10 times. Colour a circle after each toss.

Heads	○	○	○	○	○	○	○	○	○	○
Tails	○	○	○	○	○	○	○	○	○	○

c Do you think heads or tails is more likely, or are they equally likely?

d If many students combined their results would we have a better idea of whether heads and tails are equally likely?

e Why or why not?

3 **a** If we toss two coins we could toss **2 heads**, **2 tails** or **1 head and 1 tail**. Which of these do you think is most likely?

b Toss two coins 50 times and make a tally of the results.

c Which outcome occurred most often?

d Did this agree with your prediction?

e Compare your results with others in the class.

f What do you think is the likelihood of the three results occurring? Report your findings here.

Tally		
2 heads	**1 head 1 tail**	**2 tails**

 ISBN 978 1 4886 2187 1

4 Answers

1:01

1 a $\frac{1}{12}$ b $\frac{2}{3}$ c $\frac{2}{6}$ d $\frac{3}{12}$ e $\frac{7}{12}$
f $\frac{5}{12}$ g $\frac{3}{6}$ h $\frac{11}{12}$

2 a $\frac{11}{12}$ b $\frac{1}{3}$ c $\frac{4}{6}$ d $\frac{9}{12}$ e $\frac{5}{12}$
f $\frac{7}{12}$ g $\frac{3}{6}$ h $\frac{1}{12}$

3 a b c d e f g h

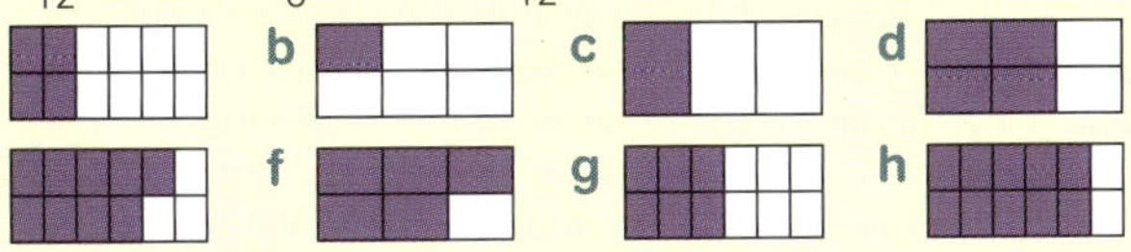

4 a $\frac{5}{6}$ b $\frac{3}{12}$ c $\frac{5}{12}$ d $\frac{11}{12}$ e $\frac{2}{6}$ f $\frac{9}{12}$

5 a $\frac{1}{6}$ b $\frac{9}{12}$ c $\frac{7}{12}$ d $\frac{1}{12}$ e $\frac{4}{6}$ f $\frac{3}{12}$

1:02

1 a $\frac{10}{100}$ b $\frac{16}{100}$ c $\frac{32}{100}$ d $\frac{48}{100}$
e $\frac{57}{100}$ f $\frac{69}{100}$ g $\frac{83}{100}$ h $\frac{91}{100}$

2 a b c d

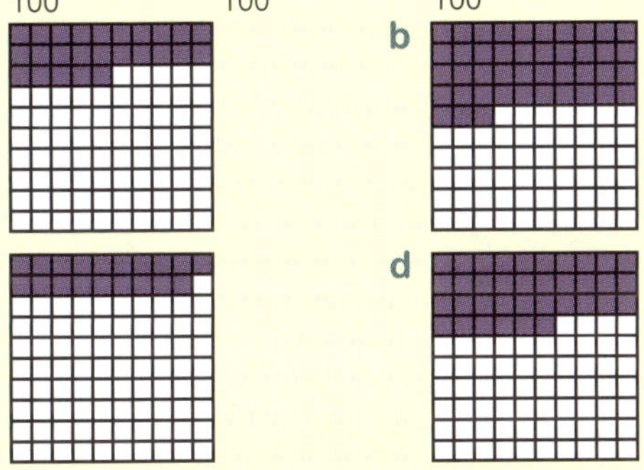

3 Estimates will vary. $\frac{30}{100}$ has been covered.
Pictures will vary.

1:03

1 100

2 a 16 out of 100 $\frac{16}{100}$ 0·16
b 27 out of 100 $\frac{27}{100}$ 0·27
c 18 out of 100 $\frac{18}{100}$ 0·18

3 0·39

4 a 0·36, $\frac{36}{100}$ b 0·79, $\frac{79}{100}$
c 0·57, $\frac{57}{100}$ d 0·85, $\frac{85}{100}$
e 0·27, $\frac{27}{100}$ f 0·18, $\frac{18}{100}$
g 0·62, $\frac{62}{100}$ h 0·44, $\frac{44}{100}$

1:04

1 a 1324

one thousand three hundred and twenty-four

b 1413

one thousand four hundred and thirteen

c 1522

one thousand five hundred and twenty-two

d 1179

one thousand one hundred and seventy-nine

2 a 3 b 4 c 4 d 3
e 4 f 3 g 2 h 4

3 a 260 b 152 c 1040 d 7018
e 5179 f 2634 g 8568

1:05

1 a 1002 b 1005 c 1027
d 999 e 1020 f 1017

2 a 1008, 1009 b 3008, 3010
c 5025, 5030 d 1020, 1030
e 1017, 1016 f 7000, 6990

3 a 1016 b 1020 c 1027 d 5010
e 8018 f 2012 g 7024 h 6001
i 9436 j 4965 k 3689

4 a four thousand and twenty-three
b nine thousand and thirty
c seven thousand five hundred
d two thousand nine hundred and one

5 a 1007 — one thousand and four
1017 — one thousand and seventeen
1004 — one thousand and seven
b 1003 — one thousand and thirteen
1013 — one thousand and five
1005 — one thousand and three

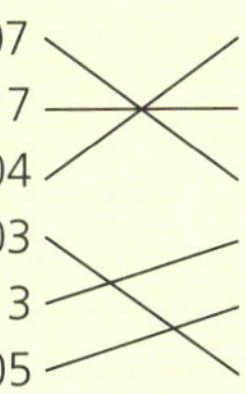

1:06

1 a 10 b 10 c 10 d 100
e 100 f 1000

2 a A b D c F d H

3 a E b A and B c E and F d A and B
e 10 times as big

1:07

1 a

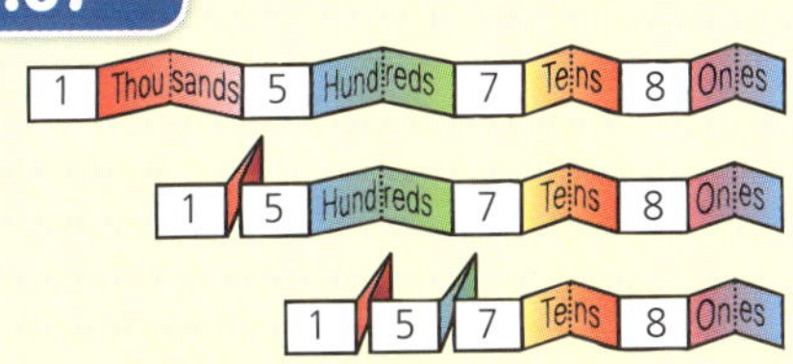

b

c 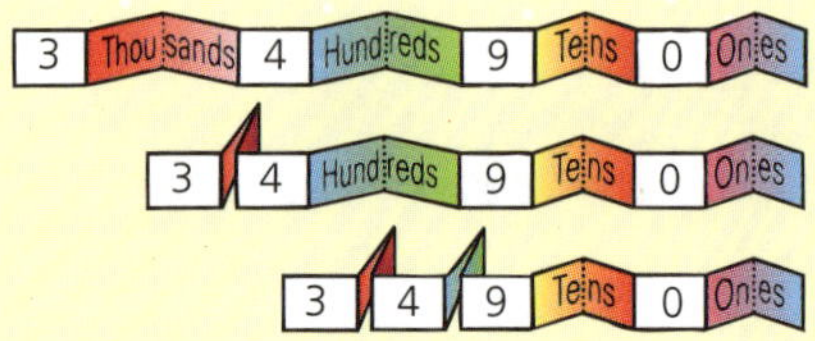

d 

2 a 17 b 19 c 15 d 26 e 39
f 62 g 96 h 71

3 a 162 b 147 c 219 d 347 e 538
f 736 g 457 h 825

4 a 30 tens b 80 tens c 300 tens d 700 tens
e 2 thousands + 3 hundreds f 1 ten thousand

1:08

1 a 3700 b 4200 c 1400 d 9300
e 6500 f 6700 g 9000 h 5900

2 a 32000 b 83000 c 11000 d 57000
e 23000 f 52000 g 47000 h 69000

3 a 2694 2781 2639
2721 2675 2642
2765 2748 2683
b 53640 52967 52849
52621 52076 53297
53599 53346 52374

4 a true b false c false d true

5 a 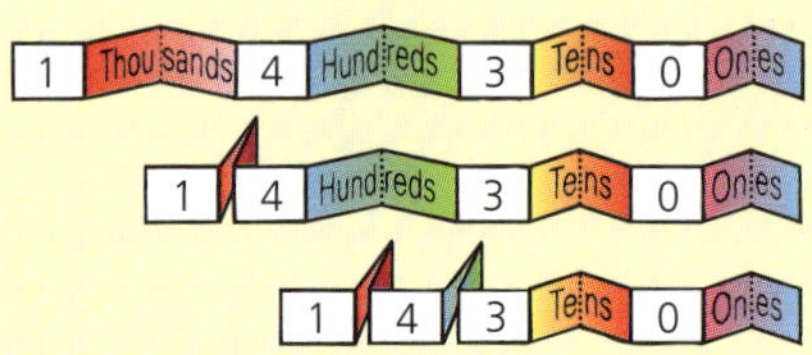

b 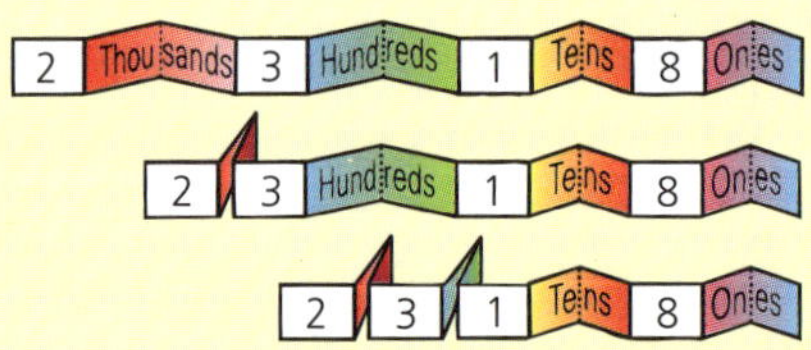

1:09

1 a 2749 b 7124 c 4592 d 6432
e 36836 f 83575 g 12570 h 50746

2 a 4000 + 200 + 80 + 6 b 9000 + 300 + 10 + 7
c 3000 + 700 + 50 + 2 d 6000 + 400 + 80 + 9
e 70000 + 1000 + 400 + 20 + 3
f 20000 + 9000 + 400 + 60

3 a 5774 b 6347 c 9875 d 8483
e 4657 f 3473 g 7498 h 8758

4 a 7 tens (70) b 2 units (2)
c 5 thousands (5000) d 5 hundreds (500)
e 4 thousands (4000) f 1 hundred (100)
g 3 hundreds (300) h 4 tens (40)

1:10

1 a $\frac{1}{2}$ b $\frac{1}{4}$ c $\frac{1}{4}$ d $\frac{1}{4}$ e $\frac{1}{2}$ f $\frac{1}{8}$

2 a 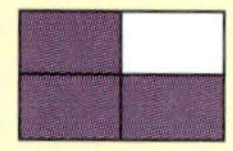 b 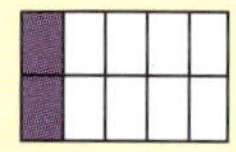 c 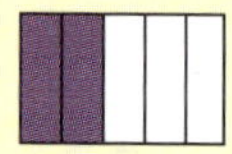
d 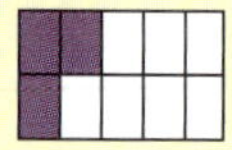 e 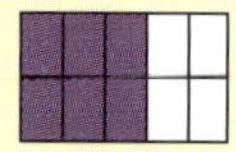 f 
g 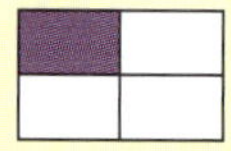 h 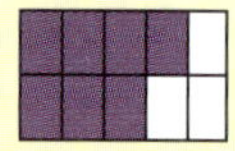

3 a true b false c true d true e false f true

4 a $\frac{1}{2}$ b $\frac{1}{5}$ c $\frac{1}{10}$ d $<$
e $>$ f $<$ g $<$ h $>$
i $<$ j $>$ k $>$

1:11

1 a $\frac{2}{10}$ b $\frac{2}{5}$ c $\frac{8}{10}$ d $\frac{3}{5}$
e $\frac{1}{2}$ f $\frac{4}{10}$ g $\frac{5}{10}$ h 1

2 a true b true c true d true

3 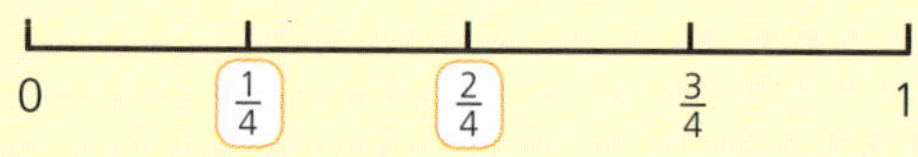

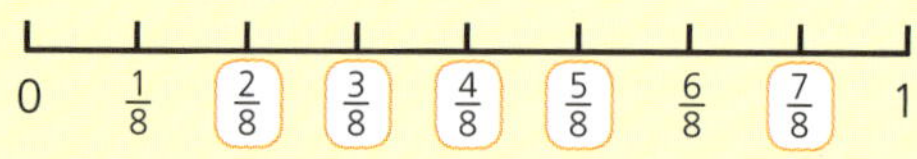

4 a $\frac{2}{8}$ b $\frac{2}{4}$ or $\frac{1}{2}$ c $\frac{1}{2}$ or $\frac{4}{8}$ d $\frac{2}{4}$ or $\frac{4}{8}$
e $\frac{3}{4}$ f $\frac{2}{2}$ or $\frac{4}{4}$ or $\frac{8}{8}$

5 a true b true c true d true

1:12

1 a 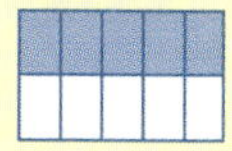 b 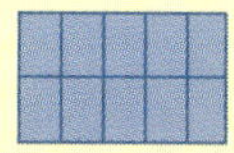 c 
d  e 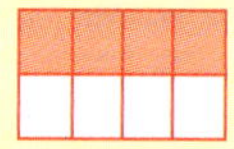 f 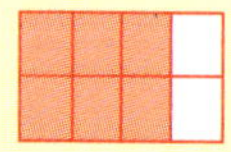
g 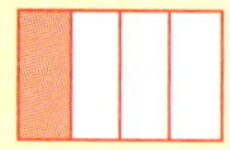 h 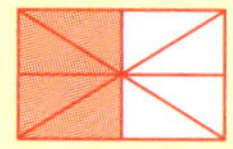

2 a $\frac{1}{8} < \frac{3}{8}$ b $\frac{3}{10} > \frac{1}{10}$ c $\frac{3}{4} > \frac{1}{4}$
d $\frac{3}{10} < \frac{7}{10}$ e $\frac{6}{10} < \frac{9}{10}$ f $\frac{5}{8} > \frac{2}{8}$
g $\frac{4}{5} > \frac{3}{5}$ h $\frac{9}{10} > \frac{8}{10}$

3 a $\frac{2}{4}$ b $\frac{4}{5}$ c $\frac{1}{2}$ d $\frac{1}{2}$
e $\frac{3}{5}$ f $\frac{1}{4}$ g $\frac{3}{4}$ h $\frac{2}{5}$

4 a 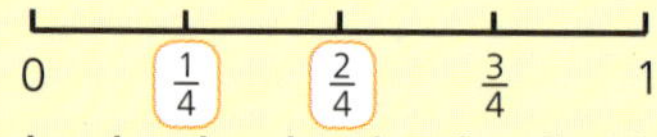

b 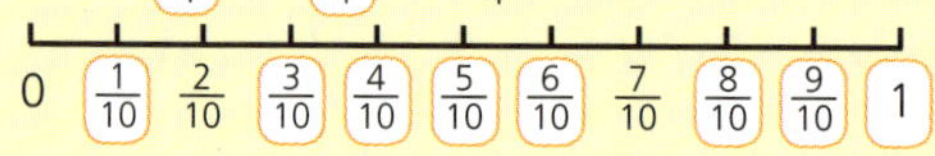

c 0, $\frac{1}{8}$, $\frac{2}{8}$, $\frac{3}{8}$, $\frac{4}{8}$, $\frac{5}{8}$, $\frac{6}{8}$, $\frac{7}{8}$, 1

d 0, $\frac{1}{5}$, $\frac{2}{5}$, $\frac{3}{5}$, $\frac{4}{5}$, 1

5 **a** $2\frac{1}{2}$ **b** $2\frac{1}{4}$ **c** $1\frac{1}{2}$
d $1\frac{1}{4}$ **e** $2\frac{1}{4}$ **f** $1\frac{3}{4}$

1:13

1 **a** $\frac{1}{2}$ or $\frac{4}{8}$ **b** $\frac{1}{2}$ or $\frac{2}{4}$ **c** $\frac{3}{4}$
d 1 or $\frac{2}{2}$ or $\frac{8}{8}$ **e** $\frac{2}{2}$ or $\frac{4}{4}$ or $\frac{8}{8}$ **f** $\frac{2}{4}$ or $\frac{4}{8}$

2 **a** $\frac{1}{3}$ **b** $\frac{2}{3}$ **c** 1 or $\frac{2}{2}$ or $\frac{3}{3}$
d 1 or $\frac{2}{2}$ or $\frac{6}{6}$ **e** $\frac{2}{6}$ **f** $\frac{4}{6}$

3 **a** T **b** T **c** T **d** F **e** T **f** T

4 **a** T **b** T **c** T **d** F **e** F **f** T

5 **a** $\frac{4}{3}, \frac{5}{3}, \frac{6}{3}, \frac{7}{3}$ **b** $\frac{4}{4}, \frac{5}{4}, \frac{6}{4}, \frac{7}{4}, \frac{8}{4}$
c $\frac{4}{6}, \frac{5}{6}, \frac{6}{6}, \frac{7}{6}, \frac{8}{6}, \frac{9}{6}$ **d** $\frac{4}{8}, \frac{5}{8}, \frac{6}{8}, \frac{7}{8}, \frac{8}{8}, \frac{9}{8}, \frac{10}{8}$

1:14

1 **a** $\frac{5}{4}$ **b** $\frac{4}{3}$ **c** $\frac{7}{6}$ **d** $\frac{3}{2}$ **e** $\frac{7}{4}$ **f** $\frac{11}{8}$

2 **a** $1\frac{3}{4}$ **b** $2\frac{2}{3}$ **c** $2\frac{1}{6}$ **d** $2\frac{1}{2}$ **e** $3\frac{3}{4}$ **f** $2\frac{3}{5}$

3 0, $\frac{1}{4}$, $\frac{2}{4}$, $\frac{3}{4}$, 1, $\frac{5}{4}$, $\frac{6}{4}$, $\frac{7}{4}$, 2
0, $\frac{1}{4}$, $\frac{2}{4}$, $\frac{3}{4}$, 1, $1\frac{1}{4}$, $1\frac{2}{4}$, $1\frac{3}{4}$, 2 (continues below)

$\frac{9}{4}$, $\frac{10}{4}$, $\frac{11}{4}$, 3, $\frac{13}{4}$, $\frac{14}{4}$, $\frac{15}{4}$, 4, $\frac{17}{4}$
$2\frac{1}{4}$, $2\frac{2}{4}$, $2\frac{3}{4}$, 3, $3\frac{1}{4}$, $3\frac{2}{4}$, $3\frac{3}{4}$, 4, $4\frac{1}{4}$

4 **a** $1\frac{3}{4}$ **b** $1\frac{4}{5}$ **c** $2\frac{3}{4}$ **d** $2\frac{1}{5}$
e $1\frac{3}{5}$ **f** $3\frac{1}{4}$ **g** $2\frac{4}{5}$ **h** $4\frac{1}{4}$

5 **a** $\frac{3}{2}$ **b** $\frac{11}{5}$ **c** $\frac{14}{4}$ **d** $\frac{9}{5}$
e $\frac{11}{4}$ **f** $\frac{13}{4}$ **g** $\frac{14}{3}$ **h** $\frac{21}{8}$

1:15

1 **a** $1\frac{1}{2}$ **b** $1\frac{1}{4}$ **c** $3\frac{1}{2}$ **d** $2\frac{1}{4}$

2 **a** 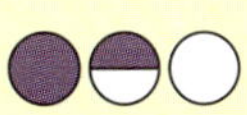

b 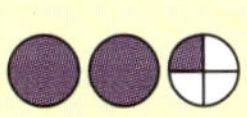

c 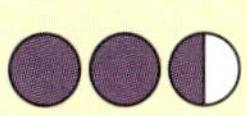

d 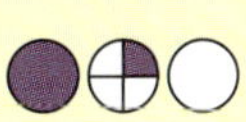

e 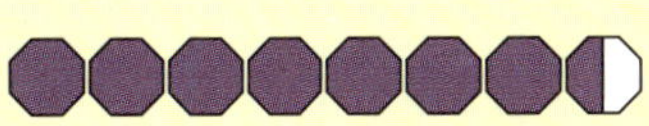

3 **a** 3, $3\frac{1}{2}$, 4, $4\frac{1}{2}$
b $\frac{6}{4}, \frac{7}{4}, \frac{8}{4}, \frac{9}{4}$

1:16

1 **a** 49 734 **b** 68 392 **c** 56 438 **d** 32 584

2

	T Thous	Thous	Hunds	Tens	Ones
a	2	6	3	2	4
b	3	5	1	6	2
c	8	2	9	7	0
d	5	2	8	1	4
e	7	4	2	6	0

3 **a** 34 528 **b** 67 934 **c** 58 462
d 92 748 **e** 82 359 **f** 48 673

1:17

1 **a** 25 748 **b** 30 957 **c** 73 808

2

	T Thous	Thous	Hunds	Tens	Ones
a	6	4	3	7	0
b	7	0	7	3	9
c	4	1	6	0	2
d	8	5	0	7	2
e	3	9	5	0	8
f	5	3	0	2	9

3 **a** 96 248 **b** 79 531 **c** 47 093
d 62 856 **e** 28 460 **f** 15 704

4 **a** 8 thousands (8 000) **b** 4 units (4)
c 1 thousand (1 000) **d** 6 tens of thousands (60 000)
e 8 tens (80) **f** 9 tens (90)
g 9 hundreds (900) **h** 6 hundreds (600)

1:18

1 **a** $\frac{1}{3}$ **b** $\frac{2}{3}$ **c** 1 **d** $\frac{4}{6}$

2 **a** 0, $\frac{1}{4}$, $\frac{2}{4}$, $\frac{3}{4}$, 1
0, $\frac{1}{8}$, $\frac{2}{8}$, $\frac{3}{8}$, $\frac{4}{8}$, $\frac{5}{8}$, $\frac{6}{8}$, $\frac{7}{8}$, 1

b 0, $\frac{1}{3}$, $\frac{2}{3}$, 1
0, $\frac{1}{6}$, $\frac{2}{6}$, $\frac{3}{6}$, $\frac{4}{6}$, $\frac{5}{6}$, 1

3 **a** T **b** T **c** T **d** F **e** F **f** T
g T **h** T **i** F **j** F **k** F **l** T

4 **a** 6 **b** 3 **c** 8 **d** 4

1:19

1 **a** $\frac{2}{4}, \frac{3}{4}$
b $\frac{2}{8}, \frac{4}{8}, \frac{6}{8}, \frac{8}{8}$ or 1

c $\frac{2}{3}$, $\frac{3}{3}$ or 1, $1\frac{2}{3}$ or $\frac{5}{3}$, $1\frac{3}{3}$ or 2

d $\frac{2}{6}$ or $\frac{1}{3}$, $\frac{4}{6}$ or $\frac{2}{3}$, $\frac{6}{6}$ or 1, $1\frac{2}{6}$ or $1\frac{1}{3}$, $1\frac{4}{6}$ or $1\frac{2}{3}$, $1\frac{6}{6}$ or 2

e $\frac{1}{5}$, $\frac{3}{5}$,

f $\frac{2}{10}$ or , , $\frac{4}{10}$ or $\frac{2}{5}$, or $\frac{1}{2}$, $\frac{6}{10}$ or $\frac{3}{5}$, or 1

2 0, $\frac{1}{4}$, $\frac{2}{4}$, $\frac{3}{4}$, $\frac{4}{4}$, $\frac{5}{4}$, $\frac{6}{4}$, $\frac{7}{4}$, $\frac{8}{4}$, $\frac{9}{4}$, $\frac{10}{4}$, $\frac{11}{4}$, $\frac{12}{4}$

0, $\frac{1}{4}$, $\frac{2}{4}$, $\frac{3}{4}$, 1, $1\frac{1}{4}$, $1\frac{2}{4}$, $1\frac{3}{4}$, 2, $2\frac{1}{4}$, $2\frac{2}{4}$, $2\frac{3}{4}$, 3

3 **a** $1\frac{2}{4}$ or $1\frac{1}{2}$ **b** $1\frac{3}{4}$ **c** 3

1:20

1 **a**

Ones	Tenths	Hundredths
0 ·	4	6

b

Ones	Tenths	Hundredths
0 ·	5	5

c

Ones	Tenths	Hundredths
1 ·	2	8

2 **a** 3 Tenths 1 Hundredths / 3 1 Hundredths
b 6 Tenths 8 Hundredths / 6 8 Hundredths
c 9 Tenths 5 Hundredths / 9 5 Hundredths
d 4 Tenths 3 Hundredths / 4 3 Hundredths
e 7 Tenths 2 Hundredths / 7 2 Hundredths
f 8 Tenths 6 Hundredths / 8 6 Hundredths

3 **a** $\frac{35}{100}$ — 0·35; $\frac{59}{100}$ — 0·59; $1\frac{22}{100}$ — 1·22
b $\frac{41}{100}$ — 0·41; $2\frac{97}{100}$ — 2·97; $\frac{63}{100}$ — 0·63
c $\frac{89}{100}$ — 0·89; $1\frac{56}{100}$ — 1·56; $3\frac{38}{100}$ — 3·38

1:21

1 **a** 0·5 **b** 0·3 **c** 0·6 **d** 0·1 **e** 0·8
f 0·9 **g** 0·2 **h** 0·4 **i** 1·1 **j** 1·3
k 1·9 **l** 1·8

2 **a** $\frac{4}{10}$ — 0·4; $\frac{9}{10}$ — 0·9; $\frac{7}{10}$ — 0·7
b $\frac{3}{10}$ — 0·3; $\frac{1}{10}$ — 0·1; $\frac{2}{10}$ — 0·2
c $1\frac{9}{10}$ — 1·9; $1\frac{8}{10}$ — 1·8; $1\frac{6}{10}$ — 1·6

3 **a** 0·5 **b** 0·3 **c** 0·9 **d** 0·7 **e** 0·4
f 0·8 **g** 0·6 **h** 1·0 **i** 0·1 **j** 0·8
k 0·5 **l** 1·9 **m** 1·3 **n** 1·0

4 **a** 0, $\frac{1}{10}$, $\frac{2}{10}$, $\frac{3}{10}$, $\frac{4}{10}$, $\frac{5}{10}$, $\frac{6}{10}$, $\frac{7}{10}$, $\frac{8}{10}$, $\frac{9}{10}$, 1
0, 0·1, 0·2, 0·3, 0·4, 0·5, 0·6, 0·7, 0·8, 0·9, 1·0

b 0, $\frac{1}{10}$, $\frac{2}{10}$, $\frac{3}{10}$, $\frac{4}{10}$, $\frac{5}{10}$, $\frac{6}{10}$, $\frac{7}{10}$, $\frac{8}{10}$, $\frac{9}{10}$, 1
0, 0·1, 0·2, 0·3, 0·4, 0·5, 0·6, 0·7, 0·8, 0·9, 1·0

1:22

1 **A** 0·34 **B** 0·50 **C** 0·30
D 0·06 **E** 0·03

2 **a** C **b** A **c** A **d** D **e** B
f B **g** C **h** B **i** B **j** A

3 **a** C **b** A **c** B **d** A **e** B

4 **a** 0·28 **b** 0·85 **c** 0·49 **d** 0·37

5 1·3, 1·4, 1·5, 1·6, 1·7, 1·8, 1·9, 2·0, 2·1, 2·2, 2·3, 2·4, 2·5, 2·6, 2·7, 2·8, 2·9, 3·0, 3·1, 3·2, 3·3, 3·4

6

0·01	0·02	0·03	0·04	0·05	0·06	0·07	0·08	0·09	0·1
0·11	0·12	0·13	0·14	0·15	0·16	0·17	0·18	0·19	0·2
0·21	0·22	0·23	0·24	0·25	0·26	0·27	0·28	0·29	0·3
0·31	0·32	0·33	0·34	0·35	0·36	0·37	0·38	0·39	0·4
0·41	0·42	0·43	0·44	0·45	0·46	0·47	0·48	0·49	0·5
0·51	0·52	0·53	0·54	0·55	0·56	0·57	0·58	0·59	0·6
0·61	0·62	0·63	0·64	0·65	0·66	0·67	0·68	0·69	0·7
0·71	0·72	0·73	0·74	0·75	0·76	0·77	0·78	0·79	0·8
0·81	0·82	0·83	0·84	0·85	0·86	0·87	0·88	0·89	0·9
0·91	0·92	0·93	0·94	0·95	0·96	0·97	0·98	0·99	1·00

1:23

1 **a** 0·3 **b** 0·06 **c** 0·09 **d** 0·5 **e** 0·9
f 0·02 **g** 0·4 **h** 0·1 **i** 0·05 **j** 0·7

2 **a** 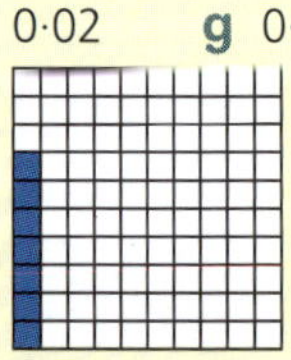 **b** 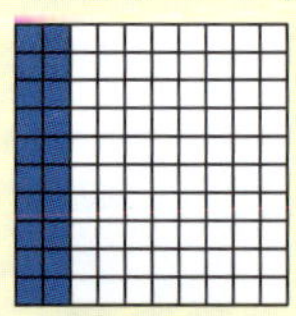 **c** 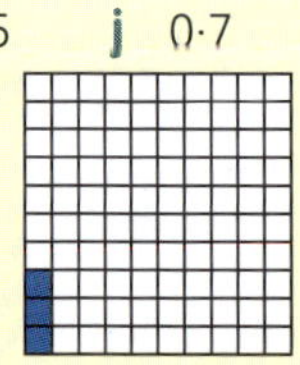
d 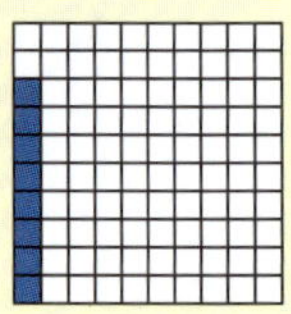 **e** 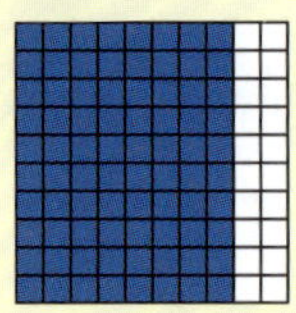

1:24

1 a 49768 b 37281 c 84971 d 70923
2 a 26301 b 69309 c 46526
3 a 28747 b 50378 c 82409 d 75682 e 44936
4 a 26493 b 85374 c 95651 d 48769
5 a forty-two thousand six hundred and three
b seventy thousand nine hundred and forty-one
c twenty-seven thousand and eighty-six
d sixty-three thousand five hundred and fifty

1:25

1 a 49769 b 75238 c 80652 d 39043
2 a 10000 + 6000 + 300 + 20 + 7
b 50000 + 2000 + 800 + 50 + 1
c 60000 + 7000 + 500 + 9
d 90000 + 300 + 20 + 5
3 a 4 thousands (4000) b 8 tens (80)
c 5 hundreds (500) d 9 tens of thousands (90000)
e 2 tens (20) f 2 thousands (2000)
g 6 tens of thousands (60000) h 6 hundreds (600)
4 a 34529 < 35014 b 62916 < 63845
c 18397 > 17624 d 71310 < 76243
e 27592 > 26991 f 49620 > 48375
5 a true b true c false d false e true f false
6 a 86394 b 52731

1:26

1 a 1·48 b 1·46 c 1·62 d 1·59 e 1·61
2

	Units		Tenths	Hundredths
a	2	·	5	9
b	6	·	7	3
c	9	·	4	1
d	5	·	2	6
e	8	·	4	8
f	7	·	1	3
g	6	·	7	9

3 a 2 hundredths b 4 tenths c 3 units
d 8 hundredths e 1 unit f 5 tenths
g 2 units h 2 hundredths i 6 tenths
j 1 hundredth

1:27

1 a 53·64 b 75·46 c 95·48 d 68·96
2 a 19·75 b 40·91 c 67·82 d 30·06
e 52·14 f 27·43
3 a 8 tenths and 6 hundredths
b 0 tenths and 9 hundredths
c 6 tenths and 0 hundredths
4 a 629·37 b 463·09 c 731·6 d 957·42
e 5·6 f 10·3 g 2·8 h 7·9
5 a 6 tens b 9 units c 8 hundreds
d 4 tenths e 2 hundredths f 3 tenths
g 1 ten h 5 tens i 8 tenths
j 9 hundredths k 6 tenths l 5 hundredths

1:28

1 a 57804 b 84045 c 60127 d 32060
2 a 9000 + 400 + 70 + 1 b 60000 + 1000 + 300 + 8
c 20000 + 9000 + 80 + 2 d 30000 + 600 + 50 + 7
e 80000 + 3000 + 200 + 40 + 1
f 60000 + 5000 + 400 + 2
3 a 357497 b 785724 c 297045
d 144385 e 645784 f 573057
4 a 1 hundred = 10 tens
b 1 thousand = 10 hundreds
c 1 ten thousand = 10 thousands
d 1 hundred thousand = 10 ten thousands
5 a 759364 b 532849

1:29

1

	H Thous	T Thous	Thous	Hunds	Tens	Ones
a		3	6	9	7	4
b	5	7	5	3	8	0
c	9	1	5	4	2	9

2 a 64972 b 176356 c 383749 d 75923
3 a hundreds b thousands c tens d thousands
4 a five hundred and three thousand, two hundred and seventy-seven
b two hundred and thirty thousand, four hundred and twenty-six
5 a 262144 b 482354 c 364688
d 639486 e 875038 f 595178
g 310607 h 1526426

1:30

1 a 21000 + 586 b 85000 + 142
2 a 4028 (grams) b 7352 (grams)
3 a $5000 b $10000 c $20000
d $50000 e $70000 f $100000

1:31

1 a 135000 b 458000 c 680000
d 263000 e 917000 f 543000
g 795000 h 382000 i 828000
2 a 250000 b 980000 c 530000 d 790000
e 140000 f 370000
3 a 64527 b 39278 c 48981 d 98356
e 71064 f 50761
4 a 80000 + 3000 + 100 + 60 + 2
b 30000 + 5000 + 200 + 70 + 5
c 40000 + 6000 + 900 + 10 + 3
d 70000 + 8000 + 500 + 40 + 8
e 90000 + 2000 + 300 + 1
f 60000 + 800 + 20 + 5
5 a 758497 b 438327 c 143927

6 a 2 hundreds (200)
b 6 tens (60)
c 4 hundred thousands (400 000)
d 5 ten thousands (50 000)
e 3 hundred thousands (300 000)
f 9 ten thousands (90 000)
g 8 thousands (8 000)
h 7 hundred thousands (700 000)
i 3 ten thousands (30 000)
j 9 thousands (9 000)
k 6 hundreds (600)
l 8 hundred thousands (800 000)

1:32

1 a = 1 b $\frac{3}{2} = 1\frac{1}{2}$ c $\frac{4}{2} = 2$ d $\frac{5}{2} = 2\frac{1}{2}$

2 a $\frac{3}{2}$ or $1\frac{1}{2}$ b $\frac{7}{2}$ or $3\frac{1}{2}$ c $\frac{19}{2}$ or $9\frac{1}{2}$ d $\frac{11}{4}$ or $2\frac{3}{4}$
e $\frac{17}{4}$ or $4\frac{1}{4}$ f $\frac{31}{4}$ or $7\frac{3}{4}$ g $\frac{33}{2}$ or $16\frac{1}{2}$

3 a $1\frac{2}{5}$ b $1\frac{4}{5}$ c $2\frac{3}{5}$ d $2\frac{1}{5}$
e $1\frac{3}{5}$ f $2\frac{2}{5}$ g $2\frac{4}{5}$ h $3\frac{2}{5}$

4 a $\frac{6}{5}$ b $\frac{11}{5}$ c $\frac{17}{5}$ d $\frac{9}{5}$
e $\frac{11}{4}$ f $\frac{17}{6}$ g $\frac{14}{3}$ h $\frac{21}{8}$

1:33

1 a $\frac{2}{2}$ or 1, $\frac{3}{2}$ or $1\frac{1}{2}$. Rule: Add $\frac{1}{2}$
b $\frac{3}{4}$, $\frac{4}{4}$ or 1. Rule: Add $\frac{1}{4}$
c $\frac{2}{3}$, $1\frac{2}{3}$, 2, $2\frac{1}{3}$, $2\frac{2}{3}$, $3\frac{1}{3}$. Rule: Add $\frac{1}{3}$
d $\frac{1}{8}$, $\frac{2}{8}$ or $\frac{1}{4}$, $\frac{3}{8}$, $\frac{4}{8}$ or $\frac{2}{4}$ or $\frac{1}{2}$, $\frac{5}{8}$, $\frac{6}{8}$ or $\frac{3}{4}$, $\frac{7}{8}$.
Rule: Subtract $\frac{1}{8}$
e $\frac{2}{5}$, $\frac{5}{5}$ or 1, $\frac{7}{5}$ or $1\frac{2}{5}$, $\frac{8}{5}$ or $1\frac{3}{5}$, $\frac{9}{5}$ or $1\frac{4}{5}$, $\frac{10}{5}$ or $1\frac{5}{5}$ or 2.
Rule: Add $\frac{1}{5}$
f $\frac{1}{6}$, $\frac{2}{6}$ or $\frac{1}{3}$, $\frac{3}{6}$ or $\frac{1}{2}$, $\frac{4}{6}$ or $\frac{2}{3}$, $\frac{5}{6}$, $\frac{6}{6}$ or 1, $1\frac{1}{6}$.
Rule: Subtract $\frac{1}{6}$

1:34

1

	Millions	H Thous	T Thous	Thous	Hunds	Tens	Ones
a	0	6	2	8	4	7	2
b	0	9	6	5	7	2	3
c	0	4	8	0	9	3	5
d	0	8	4	9	6	7	0
e	0	3	1	6	9	0	2
f	1	0	0	0	0	0	0

2

2:01

1 a 5, 6, 7, 8, 9, 10, 11
b 10, 12, 14, 16, 18, 20, 22
c 15, 18, 21, 24, 27, 30, 33
d 20, 24, 28, 32, 36, 40, 44
e 25, 30, 35, 40, 45, 50, 55
f 30, 36, 42, 48, 54, 60, 66
g 35, 42, 49, 56, 63, 70, 77
h 40, 48, 56, 64, 72, 80, 88
i 45, 54, 63, 72, 81, 90, 99
j 50, 60, 70, 80, 90, 100, 110

2 a 12, 13 b 24, 26 c 36, 39 d 48, 52
e 60, 65 f 72, 78 g 84, 91 h 96, 104
i 108, 117 j 120, 130

3 a The last digits 2, 4, 6 8 and 0 repeat.
b The last digits 0, 4, 8, 2 and 6 repeat.
c The last digits 5 and 0 repeat.
d The last digits 0, 6, 2, 8 and 4 repeat.
e The last digits 0, 8, 6, 4 and 2 repeat.
f The last digit decreases by 1 for each term, i.e. 5, 4, 3, 2, 1, 0, 9. (Once it reaches 0 it goes back to 9.)
g The last digit is always 0.

4 Answers will vary.

2:02

1

×	0	1	2	3	4	5	6	7	8	9	10
1	0	1	2	3	4	5	6	7	8	9	10
2	0	2	4	6	8	10	12	14	16	18	20
3	0	3	6	9	12	15	18	21	24	27	30
5	0	5	10	15	20	25	30	35	40	45	50
10	0	10	20	30	40	50	60	70	80	90	100

2 a 12 b 20 c 0 d 10
e 30 f 9 g 0 h 25

3 a
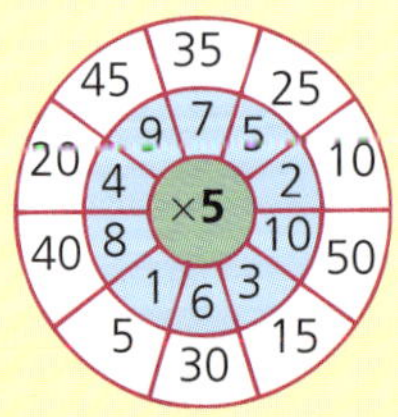

b

 • *AUSTRALIAN SIGNPOST MATHS 4* • ISBN 978 1 4886 2187 1

c 

d 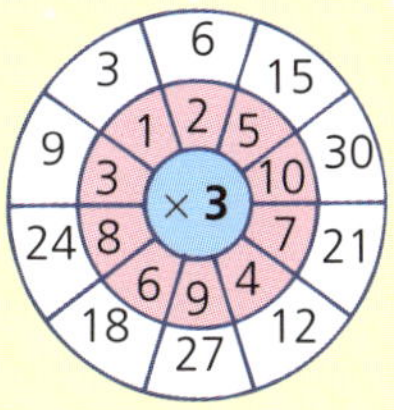

4 a 4 b 30 c 70 d 10 e 35
f 12 g 8 h 60 i 14 j 0

2:03

1 a 5 × 8 = 40 b 3 × 14 = 42 c 6 × 9 = 54
d 16 × 3 = 48 e 7 × 6 = 42 f 4 × 18 = 72
g 2 × 36 = 72 h 3 × 24 = 72

2 a 4 × 5 = 20 b 6 × 5 = 30 c 3 × 5 = 15
d 3 × 13 = 39 e 5 × 13 = 65 f 2 × 13 = 26
g 2 × 60 = 120 h 1 × 60 = 60 i 0 × 60 = 0
j 8 × 5 = 40 k 4 × 13 = 52 l 1 × 13 = 13

2:04

1 a 2 b 4 c 6 d 8 e 10
f 12 g 14 h 16 i 18 j 20
k 4 l 8 m 12 n 16 o 20

2 a 24 b 28 c 32 d 36 e 40

3 a

×	3	10	2	1	5	4	7	9	8	6
2	6	20	4	2	10	8	14	18	16	12

b

×	3	10	2	1	5	4	7	9	8	6
4	12	40	8	4	20	16	28	36	32	24

c 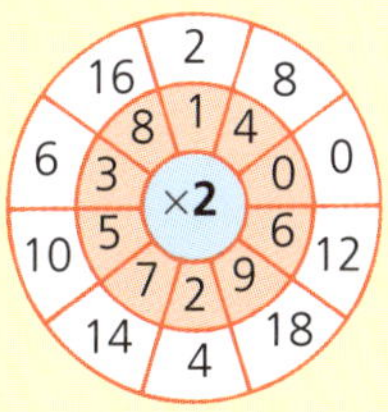

d 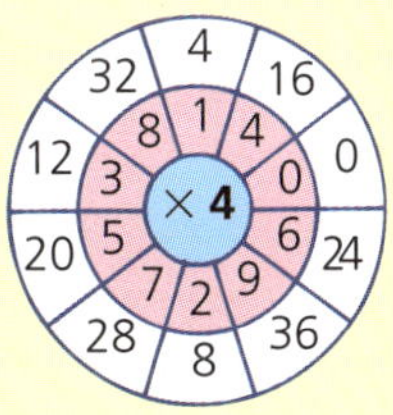

2:05

1 a 4, 8 b 8, 16 c 6, 12 d 12, 24

2 a 2, 4 b 14, 28 c 20, 40 d 16, 32

3

×	6	4	2	8	1	5	0	10	3	7	9
2	12	8	4	16	2	10	0	20	6	14	18
4	24	16	8	32	4	20	0	40	12	28	36

2:06

1 a 98 b 93 c 97 d 38 e 93 f 38
g 47 h 77 i $69 j $38 k $88 l $78

2 a 9 tens and 5 ones b 8 tens and 9 ones
c 8 tens and 8 ones d 4 tens and 6 ones
e 7 tens and 8 ones f 6 tens and 9 ones

3 a $79 b 27 c 46 d $78

2:07

1 a 88 candles b 69 lollies

2 a 69 b 75 c 76 d 69 e 76 f 78
g 69 h 88 i $87 j $88 k $77 l $99

2:08

1 a 41 b 90 c 73 d 84 e 43 f 82
g 90 h 95 i 92 j 90 k 51 l 84

2 a $71 b $62 c $66 d $62

2:09

1 a 94 b 90 c 97 d 74 e 99 f 86
g 85 h 92 i $93 j $71 k $76 l $71

2 a 30 emus b 83 kiwis c 54 metres d 40 birds

2:10

1 a 62 b 74 c 91 d 73 e 96 f 74

2 a 76 b 79 c 92 d 82 e $91 f $80

3 odd

2:11

1 a 61 + 88 = 149 b 94 + 53 = 147

2 a 118 b 138 c 157 d 106 e 159 f 128
g 155 h 127 i 129

2:12

1 a 8 b 16 c 24 d 32 e 40
f 48 g 56 h 64 i 72 j 40, 80
k 4, 8 l 8, 16 m 12, 24 n 24, 48 o 20, 40

2 a 24 b 48 c 40 d 32 e 56 f 64
g 8 h 72 i 16 j 0 k 80 l 88

3 a

×	4	0	2	8	3	7	5	9	1	6
2	8	0	4	16	6	14	10	18	2	12

b

×	5	1	9	8	2	10	6	4	0	7
3	15	3	27	24	6	30	18	12	0	21

c

×	4	0	2	8	3	7	5	9	1	6
4	16	0	8	32	12	28	20	36	4	24

d

×	4	2	8	10	5	9	6	1	7	3
5	20	10	40	50	25	45	30	5	35	15

e

×	4	0	2	8	3	7	5	9	1	6
8	32	0	16	64	24	56	40	72	8	48

f

×	4	2	8	10	5	9	6	1	7	3
10	40	20	80	100	50	90	60	100	70	30

2:13

1 a–c

1	2	3	4	5	6	7	8	9	10
11	12	13	14	15	16	17	18	19	20
21	22	23	24	25	26	27	28	29	30
31	32	33	34	35	36	37	38	39	40
41	42	43	44	45	46	47	48	49	50
51	52	53	54	55	56	57	58	59	60
61	62	63	64	65	66	67	68	69	70
71	72	73	74	75	76	77	78	79	80
81	82	83	84	85	86	87	88	89	90
91	92	93	94	95	96	97	98	99	100

d double or twice **e** double or twice
f yes (between 8 and 96)

2 **a** 12, 24, 48 **b** 4, 8, 16 **c** 6, 12, 24
d 10, 20, 40

3

×	10	3	2	7	0	6	1	5	4	9	8
2	20	6	4	14	0	12	2	10	8	18	16
4	40	12	8	28	0	24	4	20	16	36	32
8	80	24	16	56	0	48	8	40	32	72	64

4 **a** 32 **b** 2 **c** 18 **d** 20
e 56 **f** 0 **g** 28 **h** 16

2:14

1 **a** 38 – 7 = 31 **b** 36 – 4 = 32 **c** 47 – 35 = 12
d 74 – 61 = 13 **e** 56 – 34 = 22 **f** 35 – 22 = 13

2 **a** 23 **b** 35 **c** 32 **d** 37 **e** 1 **f** 26
g 33 **h** 10 **i** $4 **j** $30 **k** $67 **l** $44

2:15

1 **a** 26 **b** 36 **c** 37 **d** 44 **e** 19 **f** 18
g 73 **h** 47 **i** 38 **j** 18

2 **a** 52 **b** 42 **c** 57 **d** 37 **e** 71 **f** 81
g 99 **h** 71

2:16

1 **a**

Tens	Ones
4	10
~~5~~	~~0~~

b

Tens	Ones
6	10
~~7~~	~~0~~

c

Tens	Ones
3	10
~~4~~	~~0~~

d

Tens	Ones
8	10
~~9~~	~~0~~

2 **a** 56 **b** 73 **c** 44 **d** 82 **e** $14
f $13 **g** $45 **h** $48

3 **a** $\begin{array}{r} 40 \\ -\ 19 \\ \hline 21 \end{array}$ **b** $\begin{array}{r} 70 \\ -\ 43 \\ \hline 27 \end{array}$ **c** $\begin{array}{r} 60 \\ -\ 26 \\ \hline 34 \end{array}$

d $\begin{array}{r} 90 \\ -\ 41 \\ \hline 49 \end{array}$ **e** $\begin{array}{r} 80 \\ -\ 38 \\ \hline 42 \end{array}$ **f** $\begin{array}{r} 50 \\ -\ 23 \\ \hline 27 \end{array}$

g $\begin{array}{r} 80 \\ -\ 45 \\ \hline 35 \end{array}$ **h** $\begin{array}{r} 60 \\ -\ 17 \\ \hline 43 \end{array}$ **i** $\begin{array}{r} 50 \\ -\ 38 \\ \hline 12 \end{array}$

j $\begin{array}{r} 90 \\ -\ 61 \\ \hline 29 \end{array}$

2:17

1 **a** 6 **b** 7 **c** 8 **d** 7 **e** 18 **f** 16
g 19 **h** 15 **i** 16 **j** 19 **k** 38 **l** 44
m 28 **n** 29 **o** 19 **p** 55 **q** $26 **r** $18
s $37

2:18

1 **a** 52 **b** 46 **c** 83 **d** 24 **e** 9 **f** 35
g 18 **h** 7 **i** 16 **j** 31

2 **a** 79 **b** 28 **c** 58 **d** 45 **e** 55 **f** 19
g 25 **h** $17 **i** $17 **j** $35

3 **a** 18 **b** 48 **c** 36 **d** 35 **e** 28

2:19

1 **a** 49 **b** 64 **c** 83 **d** 91 **e** 51 **f** 37

2 **a** 77 **b** 45 **c** 87 **d** 91

2:20

1 **a** 6, 12 **b** 12, 24 **c** 9, 18 **d** 21, 42

2 **a** 3, 6 **b** 30, 60 **c** 24, 48 **d** 33, 66

3

×	1	2	3	4	5	6	7	8	9	10
3	3	6	9	12	15	18	21	24	27	30
6	6	12	18	24	30	36	42	48	54	60

2:21

1 3, 6, 9, 12, 15, 18, 21, 24, 27, 30

2 **a** 9 **b** 15 **c** 6 **d** 3 **e** 12 **f** 18
g 21 **h** 24 **i** 27 **j** 30

3 6, 12, 18, 24, 30, 36, 42, 48, 54, 60

4 **a** 18 **b** 30 **c** 12 **d** 6 **e** 24 **f** 36
g 42 **h** 48 **i** 54 **j** 60

5 **a** Each group of 6 is double each group of 3. (Answers will vary.)
b yes **c** yes

6 **a** 1 × 24 = 24, 2 × 12 = 24, 3 × 8 = 24, 4 × 6 = 24, 6 × 4 = 24, 8 × 3 = 24, 12 × 2 = 24, 24 × 1 = 24
b 1 × 15 = 15, 3 × 5 = 15, 5 × 3 = 15, 15 × 1 = 15
c 1 × 30 = 30, 2 × 15 = 30, 3 × 10 = 30, 5 × 6 = 30, 6 × 5 = 30, 10 × 3 = 30, 15 × 2 = 30, 30 × 1 = 30

 ISBN 978 1 4886 2187 1

2:22

1. 9, 18, 27, 36, 45, 54, 63, 72, 81, 90
2. a 9 b 18 c 27 d 36 e 45 f 54
   g 63 h 72 i 81 j 90
3. a 45 b 81 c 63 d 72 e 36 f 9
   g 27 h 54 i 18
4.

1	2	3	4	5	6	7	8	9	10
11	12	13	14	15	16	17	18	19	20
21	22	23	24	25	26	27	28	29	30
31	32	33	34	35	36	37	38	39	40
41	42	43	44	45	46	47	48	49	50
51	52	53	54	55	56	57	58	59	60
61	62	63	64	65	66	67	68	69	70
71	72	73	74	75	76	77	78	79	80
81	82	83	84	85	86	87	88	89	90
91	92	93	94	95	96	97	98	99	100

5.

×	0	1	2	3	4	5	6	7	8	9	10
0	0	0	0	0	0	0	0	0	0	0	0
1	0	1	2	3	4	5	6	7	8	9	10
2	0	2	4	6	8	10	12	14	16	18	20
3	0	3	6	9	12	15	18	21	24	27	30
4	0	4	8	12	16	20	24	28	32	36	40
5	0	5	10	15	20	25	30	35	40	45	50
6	0	6	12	18	24	30	36	42	48	54	60
7	0	7	14	21	28	35	42	49	56	63	70
8	0	8	16	24	32	40	48	56	64	72	80
9	0	9	18	27	36	45	54	63	72	81	90
10	0	10	20	30	40	50	60	70	80	90	100

2:23

1. a 9 b 20 − 2 =18, 9
   c 30 − 3 = 27, 9 d 40 − 4 = 36, 9
   e 50 − 5 = 45, 9 f 60 − 6 = 54, 9
   g 70 − 7 = 63, 9 h 80 − 8 = 72, 9
   i 90 − 9 = 81, 9 j 100 − 10 = 90, 9
2. a 1 b 1
   Conclusion: To add 9, add 10 then subtract 1.
   c 9 d yes (The digits add up to 9.)
   e no (The digits do not add up to 9.)
3. a 9 b 18 c 27 d 36 e 45
   f 54 g 63 h 72 i 81 j 90
   k yes l true m yes

2:24

1. a 21 b 21 c 28 d 28 e 35 f 35
   g 42 h 42 i 49 j 56 k 63 l 70
2. a 0 b 0 c 7 d 7 e 14 f 14
   g 21 h 21 i 40 j 72 k 48 l 64
3. a 28 b 42 c 35 d 70 e 63 f 56
   g 49 h 56 i 42 j 63
4. The table of sevens will be coloured and extended to 98.
5. The multiplication grid will be completed.

2:25

1. a

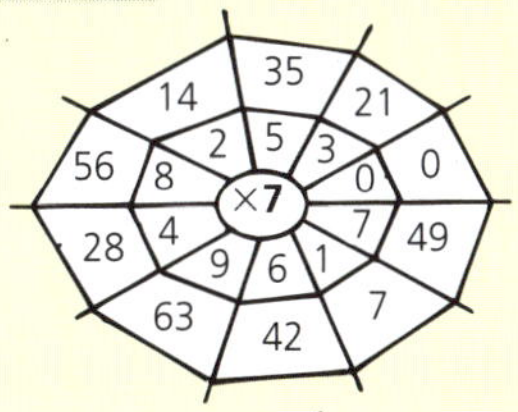

b

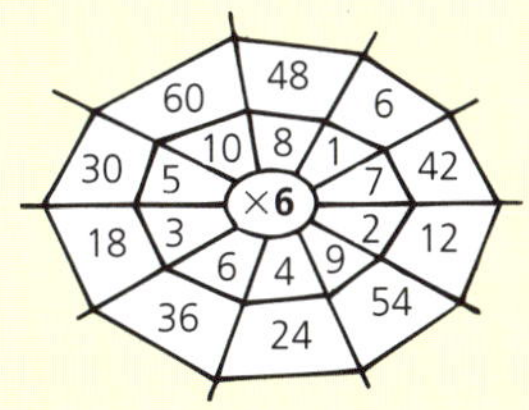

c

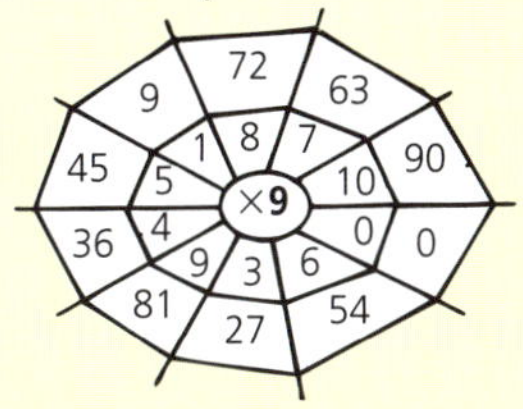

2. a 28 b 28 c 45 d 90 e 90 f 27
   g 54 h 54 i 36 j 72 k 63 l 81
   m 27 n 49 o 72 p 63 q 45 r 36
3. a 12 b 30 c 18 d 12 e 30 f 18
   g 6 h 24
4. a 42 b 0 c 21 d 45 e 36 f 48
5. a

×	3	7	4	5	2	10	6	9	8
4	12	28	16	20	8	40	24	36	32
8	24	56	32	40	16	80	48	72	64

b

×	3	7	4	5	2	10	6	9	8
6	18	42	24	30	12	60	36	54	48
7	21	49	28	35	14	70	42	63	56

2:26

1. a 154 b 117 c 108 d 283 e 790
   f 388 g 516 h 510 i 428 j 331
   k 514 l 733

2:27

1. a 328 b 601 c 815 d 713 e 561
   f 456 g 342 h 463 i 570 j 293
2. a 365 b 605 c 628 d 779 e 414
   f 819

2:28

1 a 334 b 662 c 361 d 432 e 615
f 531 g 792 h 657 i 727 j 762
k 321 l 731 m 805 n 640 o 962

2 a 432 minutes b 374 people
c 377 bags d 921 m^2

2:29

1 a 2 b 4 c 7 d 3 e 9
f 0 g 4 h 5 i 0 j 6
k 5 l 2 m 11 n 10 o 29

2 a 7 b 9 c 5 d 0 e 10
f 20 g 15 h 9 i 7 j 4
k 8 l 6 m 12 n 12 o 20
p 31

3 a 3 b 7 c 6 d 9 e 5
f 3 g 10 h 5 i 8 j 6
k 7 l 7 m 8 n 9 o 9
p 7

4 a 16 b 12 c 40 d 8 e 6
f 3 g 3 h 8 i 10 j 7

5 a 2 b 10 c 5 d 2 e 4
f 6 g 4 h 5 i 3

6 a 4 b 2 c 4 d 8

2:30

1 a 2 hundreds 4 tens 5 ones
b 6 hundreds 4 tens 5 ones
c 131 d 733 e 533 f 331
g 674 h 934 i 682 j 743
k 362 l 129 m 123 n 443
o 178 p 429 q 70 r 205

2:31

1 a 528 b 315 c 229 d 421 e 264
f 108 g 836 h 908 i 711 j 315
k 248 l 628 m 727 n 55 o 105
p 608 q 6 r 734 s 16

2:32

1 a 258 b 179 c 34 d 646 e 274
f 63 g 489 h 589 i 138 j 177
k 659 l 567 m 29 n 99 o 87
p 28 q 457 r 417 s 228

2:33

1 a 464 b 255 c 194 d 691 e 376
f 255 g 594 h 40 i 311 j 291
k 312 l 638 m 625 n 80 o 193
p 453 q 206 r 434 s 491

2:34

1 a 73 b 162 c 839 d 545 e 398
f 107 g 294 h 47

2 a 66 b 101 c 126 d 32

3 a 113 points b 419 albums c 253 books

2:35

1 a 63 b 252 c 829 d 231
e 215 f 63 g 79 h 307
i 254 j 329 k 551 l 344

2:36

1 a 3 b 2 c 4

2 a 3 groups, $15 \div 5 = 3$ b 6 groups, $30 \div 5 = 6$
c 5 groups, $15 \div 3 = 5$ d 6 groups, $36 \div 6 = 6$
e 10 groups, $30 \div 3 = 10$ f 9 groups, $36 \div 4 = 9$
g 6 groups, $24 \div 4 = 6$ h 8 groups, $48 \div 6 = 8$
i 4 groups, $36 \div 9 = 4$ j 3 groups, $24 \div 8 = 3$
k 6 groups, $48 \div 8 = 6$

3 a 7, $42 \div 6 = 7$ b 7, $35 \div 5 = 7$
c 8, $32 \div 4 = 8$

2:37

1 a 6 b 2 c 9 d 1

2 a 6 b 3 c 12 d 8 e 4 f 2

3 a 3 b 3 c 18 d 5 e 4 f 2

4 a $20 \div 4 = 5$, 5 b $16 \div 8 = 2$, 2
c $20 \div 2 = 10$, 10

2:38

1 a 4, 2 b 10, 2 c 9, 3

2 a 12, 4 b 16, 8 c 28, 4 d 27, 3 e 30, 5 f 15, 3

3 a 3, 4 b 8, 5 c 10, 9 d 6, 10 e 7, 8 f 5, 9

4 a 2, 2 b 6, 6 c 7, 7 d 6, 6 e 10, 10 f 3, 3

5 a 10 b 8 c 7

6 $34. Answers will vary.

2:39

1 a $24 \div 3 = 8$, $24 \div 8 = 3$ b $36 \div 4 = 9$, $36 \div 9 = 4$
c $30 \div 6 = 5$, $30 \div 5 = 6$

2 a 2 b 2 c 7 d 5 e 3 f 2
g 2 h 5 i 2

3 a 8 b 7 c 6 d 8 e 1 f 9
g 6 h 7 i 4 j 9 k 9 l 5

4 a

18	÷ 2	=	9
	÷ 3	=	6
	÷ 6	=	3
	÷ 9	=	2

b

24	÷ 3	=	8
	÷ 4	=	6
	÷ 6	=	4
	÷ 8	=	3

c

30	÷ 3	=	10
	÷ 5	=	6
	÷ 6	=	5
	÷ 10	=	3

2:40

1. **a** 10, 6 **b** 5, 8 **c** 6, 9 **d** 9, 8
   e 7, 6 **f** 4, 9
2. **a** 6 **b** 3 **c** 5 **d** 2 **e** 6
   f 8 **g** 1 **h** 6 **i** 7 **j** 8
   k 4 **l** 9
3. **a** 8 **b** 5 **c** 5 **d** 4 **e** 10
   f 6 **g** 5 **h** 10 **i** 3 **j** 8
   k 8 **l** 6 **m** 9 **n** 10 **o** 7
   p 6 **q** 9 **r** 8 **s** 9 **t** 9
   u 9 **v** 6 **w** 2 **x** 6

2:41

1. **a** 5 **b** 4 **c** 2 **d** 9 **e** 7
   f 1 **g** 4 **h** 6 **i** 8 **j** 9
   k 6 **l** 5 **m** 8 **n** 4 **o** 2
   p 9 **q** 3 **r** 6 **s** 5 **t** 9
   u 5 **v** 5 **w** 8 **x** 7
2. **a** 4 **b** 1 **c** 3 **d** 3 **e** 2
   f 3 **g** 8 **h** 3 **i** 5 **j** 4
   k 6 **l** 3 **m** 8 **n** 8 **o** 3
3. 2 → 6, 3, 18, 2, 20, 5, 50, 10, 80, 10 → 2

2:42

1. **a** \$50, \$5, \$2 **b** \$50, \$1, 50c
   c \$20, \$2, \$1 **d** \$50, \$10, 20c, 10c
   e \$100, \$50, \$20, \$10, \$5, \$2, \$1, 50c, 10c
   f \$50, \$20, \$20, \$5, \$1 50c, 20c, 20c, 5c
   g \$100, \$20, \$10, \$2, \$1, 50c, 20c, 10c
   h \$100, \$50, \$10, \$5, 50c, 20c, 5c
   i \$50, \$10, \$5, 20c, 10c, 5c
   j \$100, \$50, \$2, \$1, 20c, 10c, 5c
   k \$100, \$100, \$50, \$20, \$5, 20c, 20c
   l \$100, \$100, \$10, \$2, 50c, 20c, 10c
   m \$100, \$5, \$2, 20c, 5c
2. Answers will vary.

2:43

1. **a** 50c, \$1, \$2, \$20, \$20 = \$43.50
   b 10c, 20c, 50c, \$1, \$2, \$5 = \$8.80
   c 10c, 20c, \$1, \$20, \$20 = \$41.30
   d \$2, \$5, \$20, \$20, \$50 = \$97
   e \$36.50 … \$37 … \$39 … \$40 … \$50
   10c 50c \$2 \$1 \$10
   = \$13.60
   f \$67.40 … \$67.50 … \$68 … \$70 … \$90 … \$100
   20c 10c 50c \$2 \$20 \$10
   = \$32.80
2. Answers may vary.
   a 50c, \$1, \$2, \$10; change = \$13.50
   b 20c, 10c, 50c, \$20, \$20; change = \$40.80
   c 20c, 50c, \$1, \$20, \$10; change = \$31.70
3. Answers will vary.
   a 5c, \$2, \$1, \$5 **b** 5c, 50c, 50c, \$2, \$5
   c 5c, \$1, \$1, \$1, \$2, \$2, \$1

2:44

1. **a** 36 = 1 × 36 **b** 36 = 2 × 18
   c 36 = 3 × 12 **d** 36 = 6 × 6
   e 36 = 4 × 9 **f** 1, 36, 2, 18, 3, 12, 4, 9, 6
2. **a** 1, 10, 2, 5 **b** 1, 7
   c 1, 18, 2, 9, 3, 6 **d** 1, 21, 3, 7
3. **a** 10, 18 **b** 18, 21 **c** 18 **d** 7, 21
4. 1, 70, 2, 35, 5, 14, 7, 10
5. 1, 24, 2, 12, 3, 8, 4, 6

2:45

1. **a** 1, 2 **b** 1, 5
2. **a** 1, 8, 2, 4 **b** 1, 10, 2, 5
3. **a** 10 **b** 40 **c** 15 **d** 35
4.

Product	15	15	20	20	20	40	40	40	40	35	35
Factor	1	3	1	2	4	1	2	4	5	1	5
Factor	15	5	20	10	5	40	20	10	8	35	7

2:46

1. Find: number of books Emily has now
   Number sentence: 52 – 36 = ?, Answer: 16
2. Find: total cost of light bulbs
   Number sentence: 9 × \$3 = ?, Answer: \$27
3. Find: number of lollies each
   Number sentence: 24 ÷ 4 = ?, Answer: 6
4. Find: stars used altogether
   Number sentence: 8 × 25 = ?, Answer: 200
5. Find: stickers altogether
   Number sentence: 5 × 9 = ?, Answer: 45
6. Find: number of cupcakes each
   Number sentence: 36 ÷ 6 = ?, Answer: 6
7. Find: number of pizzas made
   Number sentence: 37 + 24 = ?, Answer: 61
8. Find: number of pencils each
   Number sentence: 32 ÷ 8 = ?, Answer: 4

2:47

1. 49
2. 72
3. 99
4. 110
5. 8 pens
6. 9 wagons, 3 wheels left over
7. 12
8. 56
9. B, 1 pin
10. 160 (nails)
11. 8

2:48

1. about 40
2. 6 × 7 = 42 and 7 × 6 = 42
3. 42 ÷ 6 = 7 and 42 ÷ 7 = 6

4 8 (children)

5 18

6 **a** 200, 100, 50, 25, $12\frac{1}{2}$ **b** 80, 40, 20, 10, 5
c 128, 64, 32, 16, 8, 4, 2, 1, $\frac{1}{2}$, $\frac{1}{4}$
d 6, 12, 24, 48, 96, 192, 384, 768
e 18, 36, 72, 144, 288, 576, 1 152, 2 304

7 **a** 48, 96, 192 **b** 36, 72, 144
c 126, 252, 504

8 9 ways (5 using one layer and 4 using two)

9 **a** 414 (For 9a and b you should have made an
b 213 estimate and shown working.)

2:49

1 **a** 40, 35, 30, 25 **b** 137, 147, 157, 167
c 995, 1 095, 1 195, 1 295 **d** 1 696, 2 696, 3 696, 4 696

2 **a** 77 **b** 99 **c** 400 **d** 88
e 1 543 **f** 500 **g** $3\frac{1}{2}$ **h** 2

3 **a** 40, 100, 120 **b** 499, 509, 529
c 2, $2\frac{1}{2}$ **d** 40, 56

4 **a**

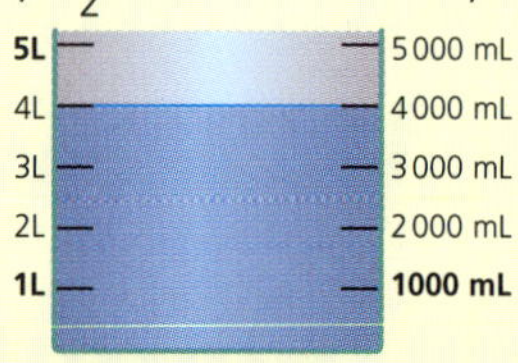

Volume of water = 4 L or 4 000 mL.

b

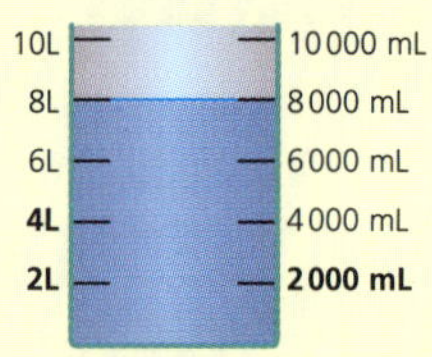

Volume of water = 8 L or 8 000 mL.

5 **a** 4 kg, 5 kg **b** 4 000 g, 5 000 g
c 8 km, 9 km **d** 8 000 m, 9 000 m
e 8 kg, 10 kg **f** 8 000 g, 10 000 g or 10 kg

2:50

1 **a** 20, 18, 184
28, 48, 162
32, 112, 188

even number plus even number = even

b 2, 0, 28
12, 66, 356
18, 190, 598

even number minus even number = even

c 24, 48, 8
32, 40, 36
24, 40, 16

even number times even number = even

d 24, 1, 40
10, 6, 5
3, 3, 6

even number divided by even number = odd or even

2

Number Types	+	−	×	÷
Even and Even	even	even	even	odd and even
Odd and Odd	even	even	odd	odd
Odd and Even	odd	odd	even	

3

6 248 + 396 = 6 645	✗	8 104 − 7 998 = 106		366 × 47 = 17 201	✗
6 107 ÷ 31 = 196	✗	7 319 + 2 997 = 10 313	✗	2 817 − 199 = 2 617	✗
283 × 654 = 28 353	✗	173 × 881 = 152 416	✗	68 019 ÷ 79 = 862	✗

2:51

1 **a** 85c **b** \$0.85 **c** \$1.50 **d** \$6.90
e \$3.25 **f** \$9.45 **g** \$8.90 **h** \$4.50
i \$7.45 **j** \$5.05 **k** \$0.60 **l** \$8.75
m \$7.15 **n** \$4.90 **o** \$0.40 **p** \$8.75
q \$3.55 **r** \$8.05

2 **a** \$11.75 **b** \$19.80 **c** \$60 **d** \$17.20

3

	Total	Total Rounded	Count on to Give Change	Change
a	\$11.66	\$11.65	5c \$11.70 30c \$12 \$8 \$20 \$30	\$38.35
b	\$20.33	\$20.35	5c \$20.40 60c \$21 \$9 \$30 \$20	\$29.65
c	\$21.59	\$21.60	40c \$22 \$8 \$30 \$20	\$28.40
d	\$10.02	\$10.00	\$10	\$10.00
e	\$2.77	\$2.75	25c \$3 \$7 \$10 \$10	\$17.25
f	\$30.51	\$30.50	50c \$31 \$9 \$40 \$60	\$69.50
g	\$53.68	\$53.70	30c \$54 \$6 \$60 \$40	\$46.30

2:52

1 **a** 8 **b** 13 **c** 14 **d** 24 **e** 1 **f** 125
g 96 **h** 30 **i** 10

2 **a** Add 2 **b** Add 4 **c** Subtract 2
d Multiply by 2 **e** Divide by 3 **f** Multiply by 5
g Subtract 1 **h** Add 7 **i** Divide by 2

3 **a** 7, 14, 21 **b** 21, 16, 11 **c** 4, 16, 64
d 44, 22, 11 **e** 34, 45, 56 **f** 38, 29, 20
g 12, 24, 48 **h** 9, 3, 1

4 **a** 4, 8, 12, 16, … Add 4 **b** 5, 9, 13, 17, … Add 4

2:53

1 **a** 24 **b** 36 **c** 48 **d** 60 **e** 72 **f** 84
g 96 **h** 108 **i** 120

2 **a** 42 **b** 70 **c** 112 **d** 26 **e** 65 **f** 52
g 60 **h** 135 **i** 45

3 **a** 320 **b** 288 **c** 192 **d** 336 **e** 384 **f** 448

3:01

1 **a** 5:05, 5 past 5 **b** 11:40, 20 to 12
c 4:35, 25 to 5 **d** 11:10, 10 past 11
e 6:50, 10 to 7 **f** 1:45, a quarter to 2 (or 15 to 2)
g 5:20, 20 past 5 **h** 8:35, 25 to 9
i 7:25, 25 past 7 **j** 2:55, 5 to 3

2 a  b  c 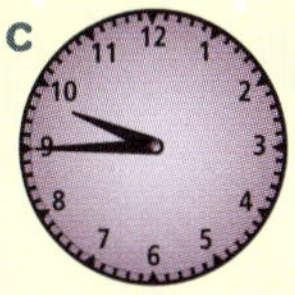
d  e  f 
g  h  i 
j 

3:02

1 a 5 to 2 b 25 to 6 c 10 past 10
d 20 to 5 e 20 past 12
f a quarter past 11 (or 15 past 11)
g a quarter to 7 (or 15 to 7)
h 25 past 2 i 10 to 8 j 5 past 3

2 a  b  c 
d  e 

3 a 10 past 8 b a quarter past 7 (or 15 past 7)
c 5 to 12 d 20 past 6
e a quarter to 8 (or 15 to 8)
f 10 to 3 g 25 past 3
h 20 to 3 i 9 o'clock

4 a 20 past 1 b 5 past 4 c 20 to 4
d a quarter to 5 (or 15 to 5)
e a quarter past 6 (or 15 past 6)
f half past 7 (or 30 past 7)
g 10 past 10 h 25 to 11 i 2 o'clock

3:03

1 a Friday b 3rd c 25th d four
e five f Tuesday g 29 h 28
2 a 2nd b 6th c 26th d 14th
3 a Thursday b Friday c Saturday d Monday
4 a 5th and 6th b 12th and 13th
c 19th and 20th d 2nd, 9th, 16th and 23rd

3:04

1 a January b October c May d September
e February f July g April h June
2 a 30 b 31 c 31 d 31
e 30 f 31 g 31 h 31
3 a Monday b Tuesday c Sunday
4 a 5th b 1st
5 a Thursday 19th b Wednesday 25th
6 a Friday b Friday c Saturday d Friday
e Thursday f Thursday

3:05

1 a 45 mm b 52 mm c 29 mm d 73 mm
e 82 mm f 16 mm
2 a 19 mm b 47 mm c 61 mm d 36 mm
e 55 mm f 72 mm g 28 mm h 93 mm
3 a 2 cm 5 mm b 6 cm 8 mm c 5 cm 1 mm
d 9 cm 2 mm e 4 cm 3 mm f 8 cm 7 mm

3:06

1 a 38 mm b 63 mm c 19 mm d 75 mm e 58 mm
f 41 mm
2 a 36 mm b 22 mm c 17 mm d 59 mm e 48 mm
3 a 32 mm b 81 mm c 19 mm d 53 mm e 74 mm
f 28 mm g 46 mm h 95 mm
4 a 3 cm 8 mm b 1 cm 9 mm c 4 cm 7 mm
d 8 cm 2 mm e 5 cm 1 mm f 2 cm 5 mm
g 6 cm 4 mm h 9 cm 3 mm i 7 cm 6 mm
5 a 12 cm or 120 mm b 14 cm or 140 mm
c 12 cm or 120 mm

3:07

1 a 256 cm b 316 cm c 540 cm d 781 cm
e 495 cm f 179 cm g 863 cm h 628 cm
2 a 1 m 19 cm b 8 m 53 cm c 5 m 82 cm
d 6 m 97 cm e 9 m 35 cm f 3 m 74 cm
3 a 2 cm 7 mm b 3 cm 6 mm c 5 cm 1 mm
d 1 cm 9 mm e 4 cm 2 mm f 8 cm 5 mm
4 a 54 mm b 69 mm c 11 mm d 87 mm
e 43 mm f 26 mm
5 a m b cm c m d mm
e cm f m g mm h mm

3:08

1 a 2 cm, 22 mm b 4 cm, 37 mm c 2 cm, 17 mm
d 3 cm, 33 mm e 1 cm, 8 mm f 5 cm, 46 mm
g 0 cm, 3 mm. A measurement of 0 cm means that the length is closer to 0 than to the next highest division on the scale.
2 a 8 kg b 15 kg c 1 kg d 11 kg
e 3 kg f 12 kg

3:09

1

The boiling point of water •	• is cool to warm weather.
35°C to 45°C •	• is cold weather.
25°C to 35°C •	• is 100°C
15°C to 25°C •	• is 0°C
5°C to 15°C •	• is very hot weather.
The freezing point of water •	• is warm to hot weather.

2 a 10°C b 90°C c 5°C d 40°C
e 80°C f 15°C g 45°C h 95°C

3 a b, e, h b a, c, f c d, g

3:10

1 a 25°C b 0°C c 18°C d 58°C
e 86°C f 37°C g 41°C h 73°C

2 a 40°C b 28°C c 36°C d 14°C

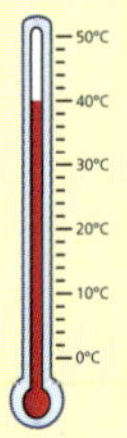 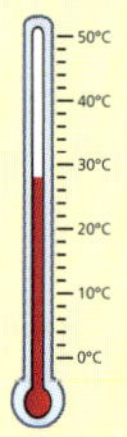 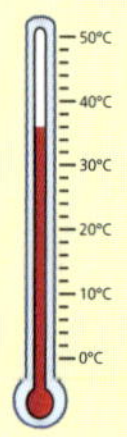 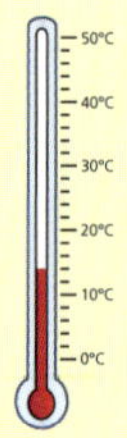

e 8°C f 33°C g 27°C h 15°C

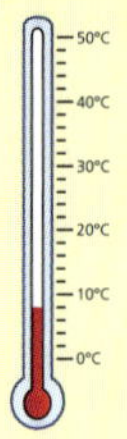 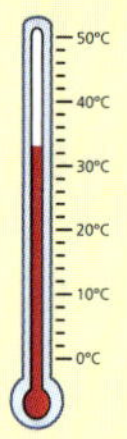 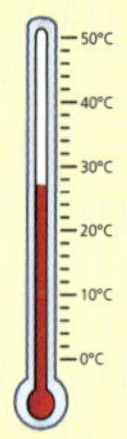 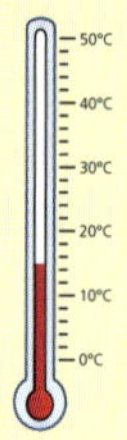

3 a 2°C b 40°C c 26°C d 70°C
e 96°C f 9°C g 100°C h 37°C
i 67°C j 17°C

3:11

1 a mL b mL c mL d L e mL
f mL g mL

2 a mL b L c mL d mL e L
f L g mL h L

3 a 1 L b 4 L c 2 L d 6 L e 5 L
f 7 L g 9 L h 3 L

3:12

1 a mL b L c mL d mL e L f mL

2 a 2000 mL b 7000 mL c 3000 mL d 5000 mL
e 8000 mL f 4000 mL g 6500 mL h 9500 mL

3 a 1 L b 7 L c 5 L d 3 L
e 6 L f 4 L g 2 L h 8 L

4 a 550 mL b 750 mL c 950 mL d 870 mL

3:13

1 a 8 b 16 c 32

2 a no b 1 book c 8
d Books are a good unit because they tessellate, however they would not be suitable to measure a small area.

3 Answers will vary. The square is considered to be the most useful shape because it tessellates and has 4 axes of symmetry.

3:14

1 A = 20 cm^2, B = 6 cm^2, C = 23 cm^2, D = 17 cm^2, E = 1 cm^2, F = 16 cm^2

2 a C b E c E d D

3:15

1 a 60 g b 1 kg c 350 g d 700 g e 500 g
f 3 kg g 9 kg h 900 g i 16 kg

2 a g b g c kg d g e g

3 a 1000 g b 3000 g c 5000 g d 9000 g

4 a 1 kg b 4 kg c 6 kg d 8 kg

3:16

1 a 1 kg b 450 g c 3 kg
d 935 g e 325 g f 750 g
g 1 kg 300 g h 1 kg 840 g i 1 kg 560 g
j 2 kg 125 g

2 a 3000 g b 5000 g c 9000 g d 7000 g
e 4000 g f 6000 g

3 a 1 kg b 8 kg c 2 kg d 4 kg e 6 kg
f 5 kg

4 a kg b kg c kg d g e g
f g g g h g i kg j g

3:17

1 a 4 cm^2 b 18 cm^2 c 20 cm^2 d 9 cm^2 e 15 cm^2
f 10 cm^2

2 a 12 cm^2 b 16 cm^2 c 24 cm^2

3:18

1 a 6 cm^2 b 10 cm^2 c 12 cm^2 d 14 cm^2 e 8 cm^2

2 a 15 cm^2 b 18 cm^2 c 3 cm^2 d 8 cm^2 e 24 cm^2

3:19

Answers will vary.

3:20

1 a 2 m^2 b 6 cm^2 c 5 m^2 d 12 cm^2 e 10 m^2
f 15 cm^2

2 a cm^2 b m^2 c cm^2 d m^2 e cm^2
f m^2 g m^2 h cm^2 i cm^2

3:21

1 a 12 cm^2 b 12 cm^2 c 10 cm^2 d 10 cm^2 e 16 cm^2 f 9 cm^2
2 a 9 cm^2 b 7 cm^2 c 9 cm^2 d 10 cm^2 e 15 cm^2 f 11 cm^2 g 15 cm^2

3:22

1 a 14 cm^2 b 13 cm^2 c 12 cm^2 d 9 cm^2 e 11 cm^2 f 10·5 cm^2 g 8 cm^2 h 7·5 cm^2
2 a 10·5 cm^2 b 13·5 cm^2 c 11·5 cm^2 d 10 cm^2

3:23

1 a 11:16, 16 past 11 b 7:24, 24 past 7 c 4:32, 32 past 4 d 5:08, 8 past 5 e 3:47, 47 past 3 f 12:37, 37 past 12 g 9:29, 29 past 9 h 6:53, 53 past 6 i 10:04, 4 past 10 j 8:58, 58 past 8 k 42 past 2 l 58 past 4 m 19 past 9 n 9 past 7 o 38 past 11
2 a 3:17 b 2:48 c 7:29 d 9:04 e 11:42 f 4:40 g 12:25 h 5:21 i 6:56
3 a 4:18 b 9:46 c 7:37 d 10:59 e 8:21 f 6:03

3:24

1 a

b

c
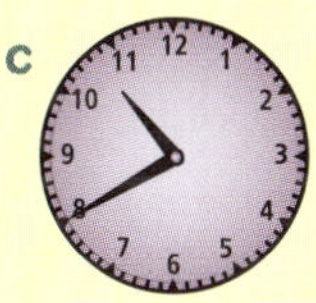
d

e

2 a 10 to 10 b 19 past 4 c a quarter to 8 (or 15 to 8) d 28 past 8 e 5 to 3
3 a 9, time is 9 to 7 b 22, time is 22 to 5 c 17, the time is 17 to 1 d 3, the time is 3 to 9 e 11, the time is 11 to 2
4 a 49 b 42 c 2 d 21 e 110 f 7 g 3 h 5 i 2
5 a 3, 1 b 1095 c 36 d 2 e 5 f 156 g 4

3:25

1 a 8 b 12 c 12
2 a 7 b 6 c 9

3:26

2 a 7 b 5 c 6 d 10 blocks e 9 blocks f 8 blocks
3 a B b D c G d F e A f E
4 C and H
5 B and F

3:27

1 a 3 b 5 c 4 d 8 e 7 f 6

3:28

1 a 6:10 b 8:45 c 11:40 d 7:50 e 10:05 f 1:15
2 a Richmond b Christmas Ck c Black Hill d Ironbark e Mulga f Wandong
3 a 8:10 b 1:40 c 3:15 d 12:05 e 2:30 f 10:45
4 a 12:00 b 8:15 c 11:10 d 12:45 e 5:40 f 9:35
5 a Black Hill b Mulga c Richmond d Wandong e Carbor f Mt Gordon
6
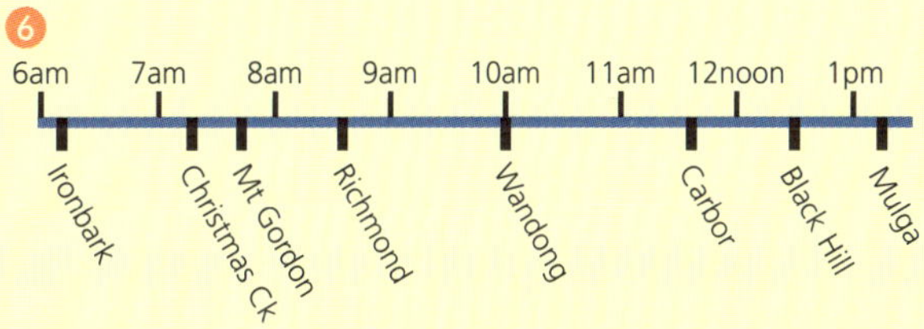

3:29

1 a 20 mm b 32 mm c 12 mm d 20 mm
2 a 36 mm b 32 mm c 69 mm d 64 mm e 8 mm f 48 mm g 53 mm
3 a 40 mm b 4 mm c 11 mm d 61 mm e 32 mm f 33 mm g 24 mm h 28 mm i 16 mm

3:30

Answers will vary.

3:31

1 a 26 past 7 b 6 to 7 c 19 past 12 d 2 past 10 e 18 to 12 f 13 to 2 g 12 past 12 h 21 to 4 i 4 to 8 j 9 past 6
2 a 27 past 6 b 16 past 7 c 10 past 8 d 4 past 12 e 26 past 4 f 25 to 6 g 20 to 2 h 5 to 5 i a quarter to 12 (or 15 to 12) j 10 to 7

3:32

1 a 5:14 pm b 3:32 am c 7:23 pm d 4:43 am e 10:09 pm f 1:17 pm g 12:49 am h 3:02 am i 8:53 pm j 3:59 pm

2 **a** after midday **b** before midday **c** before midday
d before midday **e** after midday **f** before midday
g after midday **h** after midday **i** before midday
j after midday

3:33

1

Millimetres	cm and mm	Rounded to the Nearest cm
39 mm	3 cm 9 mm	4 cm
63 mm	6 cm 3 mm	6 cm
56 mm	5 cm 6 mm	6 cm
72 mm	7 cm 2 mm	7 cm
81 mm	8 cm 1 mm	8 cm
45 mm	4 cm 5 mm	5 cm
94 mm	9 cm 4 mm	9 cm
26 mm	2 cm 6 mm	3 cm

2 **a** 35 mm **b** 29 mm **c** 46 mm
d 13 mm **e** 62 mm **f** 38 mm
g 81 mm **h** 56 mm
3 **a** 140 mm **b** 95 mm **c** 130 mm

3:34

1 **a** 1 500 mL **b** 1 250 mL **c** 1 750 mL **d** 1 600 mL
2 **a** 1 300 mL **b** 1 490 mL **c** 1 875 mL **d** 1 625 mL
e 1 750 mL **f** 2 500 mL
3 **a** 1 L 600 mL **b** 1 L 900 mL **c** 1 L 350 mL
d 1 L 250 mL **e** 1 L 425 mL **f** 2 L 750 mL
4 **a** L **b** mL **c** mL **d** mL **e** mL
f L **g** mL **h** mL
5 **a** L **b** mL **c** mL **d** mL **e** mL
f L **g** mL **h** L **i** mL

3:35

1 **a** 100 g **b** 600 g **c** 1 kg **d** 500 g **e** 3 kg
f 300 g **g** 10 kg **h** 956 g **i** 875 g
2 **a** 3 kg **b** 7 kg **c** 2 kg **d** 5 kg **e** 9 kg
f 4 kg
3 **a** 2 000 g **b** 1 200 g **c** 3 000 g **d** 1 600 g
e 5 000 g **f** 1 450 g **g** 1 980 g **h** 1 500 g
i 7 000 g
4 **a** 500 g **b** 750 g **c** 250 g **d** 1 500 g
e 1 750 g **f** 1 250 g **g** 2 500 g **h** 3 500 g
5 C, A, B, D

4:01

1 **a** C, E, F, I, J **b** B, E, G **c** A, D, H
2 E
3 **a** 1 **b** 2 **c** 1 **d** 0 **e** 1
f 1 **g** 0 **h** 0 **i** 1

4:02

1 **a** yes **b** no **c** no
2 **a** C and E **b** B, D and E **c** B and E
d 4, 4 **e** 3, 3 **f** 6, 6
3 **a** A and D **b** B **c** C **d** yes
4 Diagrams should look like this:
a 
b 
c 
d 
e 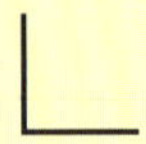
f 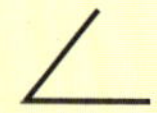
5 **a** B **b** C **c** A **d** A

4:03

1

My Angle (Right Angle)	Smaller	Same as My Angle	Larger
	E, C, G, I	A, F, D	B, H

2 **a** A **b** C **c** E **d** G
3

My Angle	Smaller	Same as My Angle	Larger
	C, E	G, I	A, F, D, B, H

4 **a** 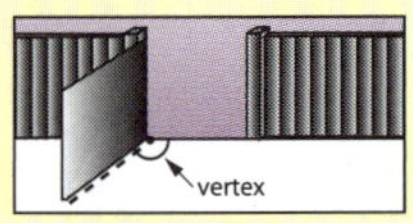

b 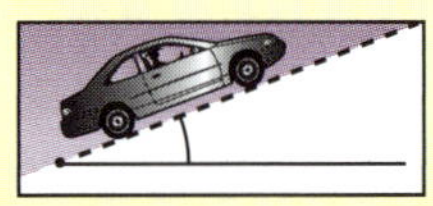
c 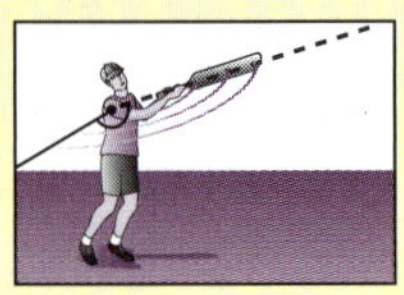

4:04

1 The prisms will be coloured red, the cylinders blue and the pyramids green.
2 **a** A, B, D, E **b** C, F
c A, D, E. The prisms will be coloured red.
d C. The cylinder will be coloured blue. **e** D
3 **a** A **b** D (or B) **c** E (or A) **d** B

4:05

1 **a** D, E, G, H **b** A, B, C, F **c** D, F
d A, E (also D, F) **e** C, G **f** B
2 **a** pyramid (square pyramid)
b prism (rectangular prism)
c pyramid (triangular pyramid)
d cube or prism (square prism)
e prism (triangular prism)

 ISBN 978 1 4886 2187 1

4:06

1. **a** A (prism) **b** B (pyramid) **c** A (prism) **d** B (pyramid)
2. **a** prism **b** pyramid **c** prism

4:07

1. Answers will vary.
2. **a** prism (triangular prism)
   b pyramid (triangular pyramid)
   c cube
   d pyramid (square pyramid)
   e prism (rectangular prism)
   f pyramid (pentagonal pyramid)

4:08

1. **a** A **b** C **c** C **d** A **e** D
2. **a**  **b** 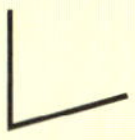 **c** Answers will vary.
3. A right angle is a quarter of a full turn. It is a square angle.
4. D, C, A, B

4:09

1. 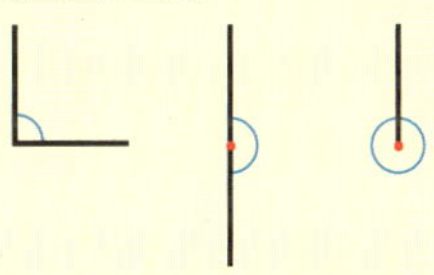
   Answers may vary.
2. **a** a half turn clockwise **b** a quarter turn clockwise
   c two quarter turns anticlockwise
3. **a** 4 **b** 4 **c** 2
4. **a** 12 **b** 3

4:10

1. **a** 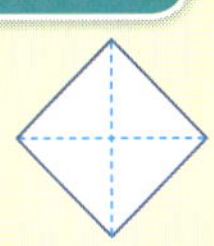 **b** 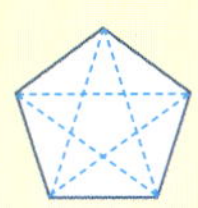 **c** 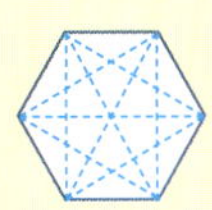
   d 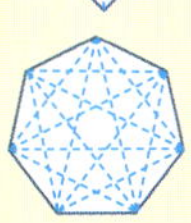 **e** 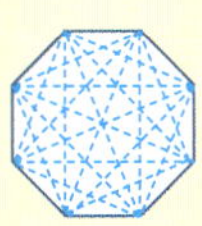
   f square, pentagon, hexagon
   g none (0) **h** the square **i** regular
2.

Regular Figure	Number of Sides	Number of Angles	Number of Diagonals
Triangle	3	3	none
Square	4	4	2
Pentagon	5	5	5
Hexagon	6	6	9
Octagon	8	8	20

3. **a** 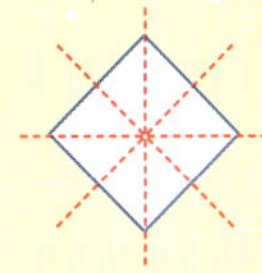 **b** 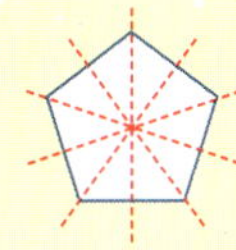 **c** 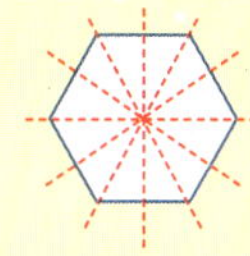
   d 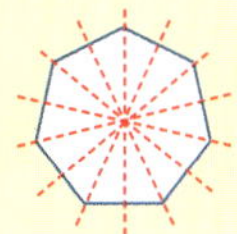 **e** 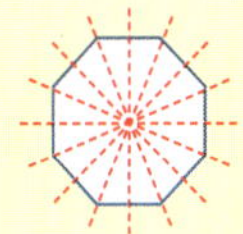
4. **a** 10 **b** 12 **c** 20

4:11

1. **a** 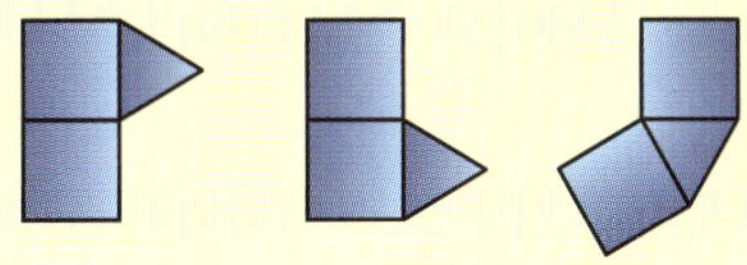
   b 2
2. **a** 1
   b Answers may vary.
   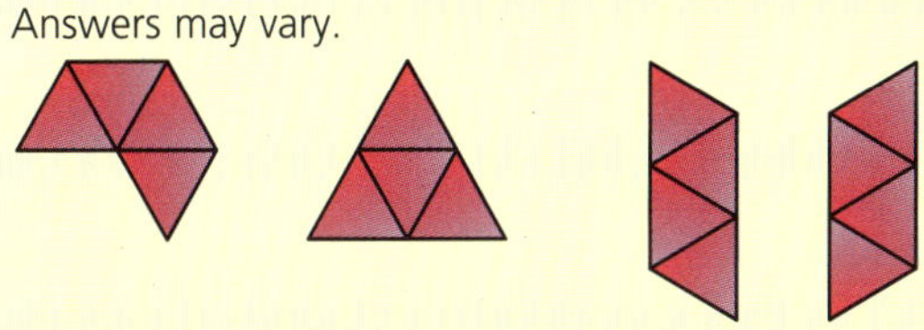
3. The complete set of tetrominoes is:
   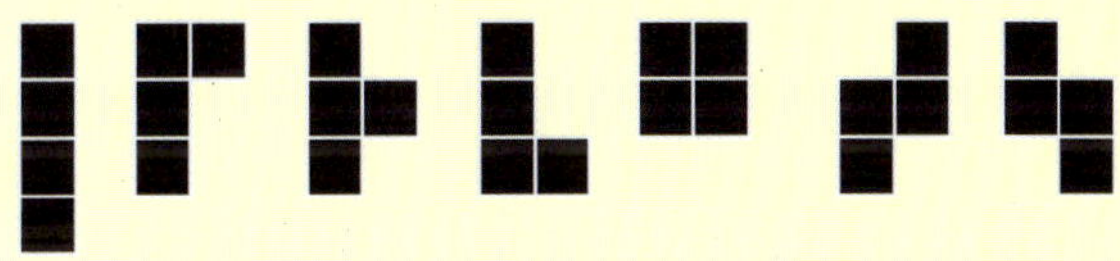
   a 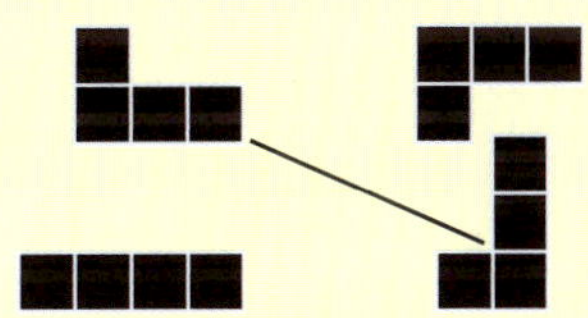 **b** 7

4:12

1. **a** Warn St **b** March St or Munn St **c** Joseph St
   d Meg St **e** school **f** hospital **g** police
   h traffic lights
2. **a** school **b** police **c** hospital
3. the intersection of March St and Munn St
4. **a** 2 km by road **b** 7 km by road

4:13

Answers will vary.

4:14

1.

Shape	Number of Surfaces	Number of Corners	Number of Edges
Cylinder	3	0	2
Cone	2	1	1
Sphere	1	0	0

2–4 Answers will vary.

5 A 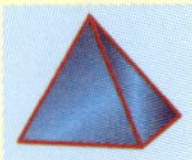
pyramid

B 
sphere

C 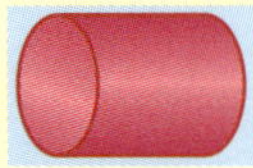
cylinder

D 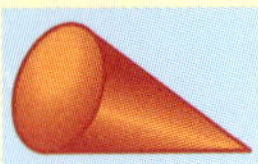
cone

E 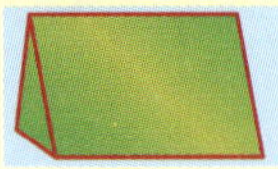
prism

a C, E
b A, D (also B, but B has no base)

4:15

1 Front 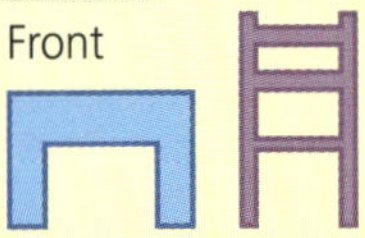

Top 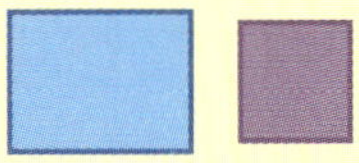

Side 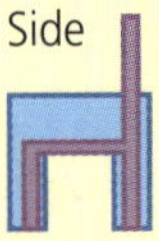

2 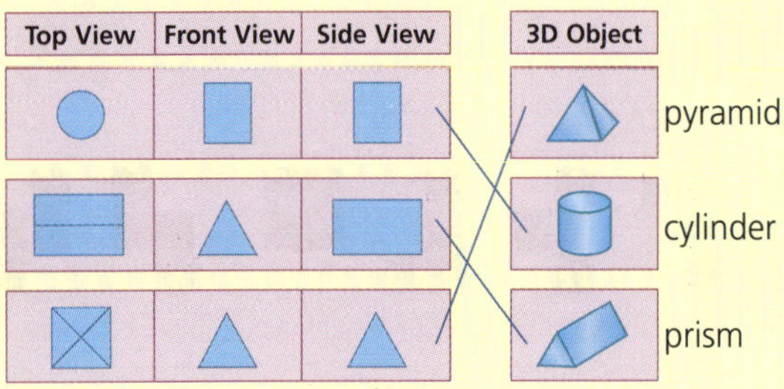

3 a 6 b 10 c 10 d 14

4 a  b  or 

c 4 d 10 e 8 f 0

4:16

1 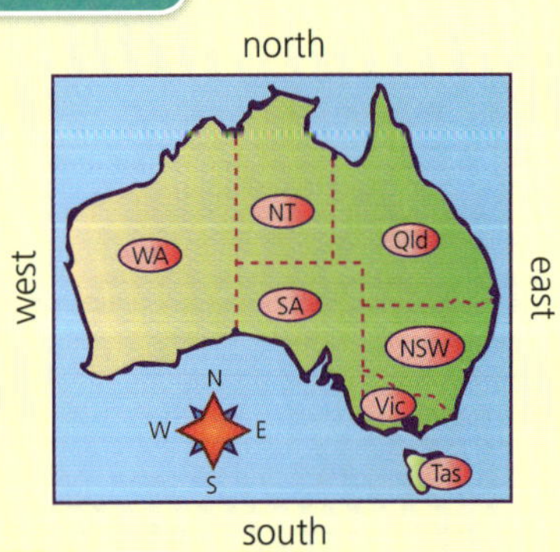

2 a Western Australia b Northern Territory
c South Australia

3 Queensland and New South Wales

4 a Jake's Grave b Bart's Cave c Queen Hill
d Land's End e Bart's Cave f Queen Hill
g Black Rock

5 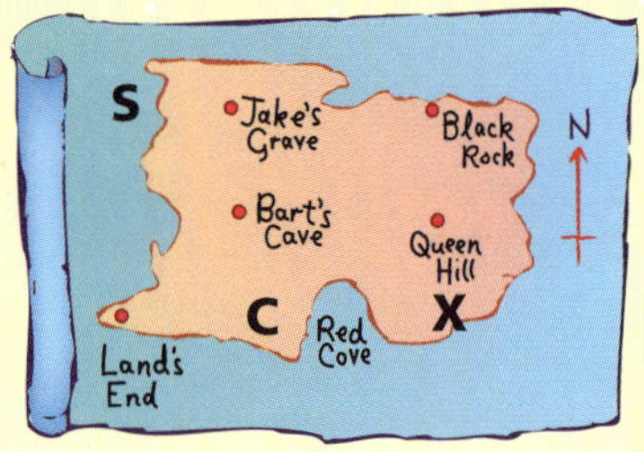

4:17

1 a prism (rectangular) b pyramid (triangular)
c prism (triangular) d pyramid (square or rectangular)
e sphere f cone g cube h cylinder

2 a museum b pool c park d station

3 a A b G c D d A e D

4:18

1 a cherries b carrot c apple
d banana e pear f strawberry
g left, middle h right, middle i left, bottom
j left, top k right, top l right, bottom

2 a Ruby b Rex c Tom d Chris
e Sophie f Ethan g Georgia h Oscar

3 a 2, A b 4, C c 2, B d 2, D e 1, B
f 1, A g 5, A h 1, C i 5, D j 5, C
k 3, B l 5, B m 3, A n 4, A

4:19

1 a Mildura b Lakes Entrance c Griffith
d Warrnambool e Horsham

2 a column 4, 2nd row b column 4, 3rd row
c column 3, 2nd row d column 3, 2nd row
e column 1, 1st row f column 5, 3rd row
g column 3, 3rd row h column 2, 4th row
i column 4, 1st row and column 5, 1st row
j Echuca k Albury l Mildura

3 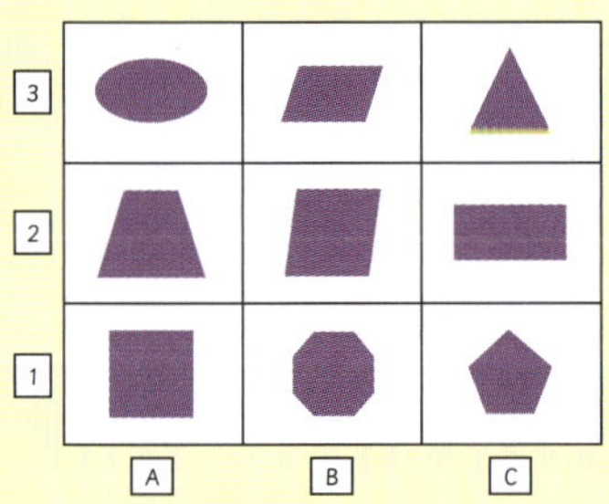

4:20

Answers will vary.

1 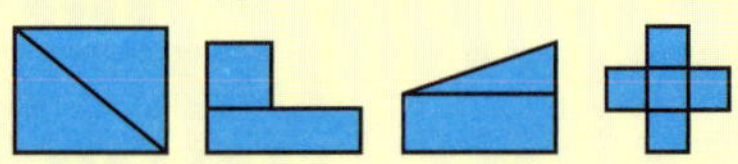

2 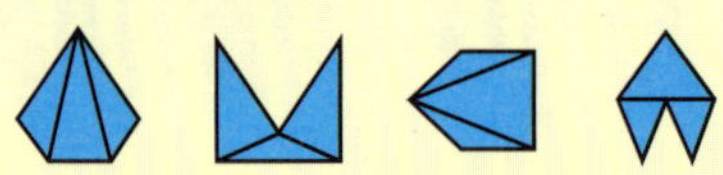

3 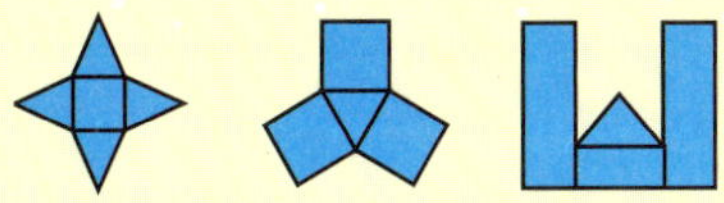

4 The shapes with line symmetry will be copied.

4:21

1 **a** A, B, G, H **b** C, D, J, K **c** E, F, I, L

2 **a** C, A and B **b** D, G, E **c** J, H, F
d K, B, L **e** K, C, G **f** G, F, L

3 **a** obtuse **b** right **c** acute
d obtuse **e** obtuse **f** acute
g acute **h** right

4 **a** B and E **b** B and C, or A and E
c A and D

4:22

1 **a** E, F, G, N, A, J, L, P **b** H, O, D, M **c** B, C, K, I

2 **a** Triangles are used to make rigid shapes in the structure, making it strong.
b O, D, M **c** P, N, F **d** L, E, G
e Rows of bricks, train lines etc.

3 **a** ABI IKP KLC CJB
b AHGO FHPM DEML DNOJ AJLP EFGN

4 **a** Yes, if the base is a square; no, if otherwise.
b H, O, F and N

4:23

1 **a** D, E, F, G, H
b 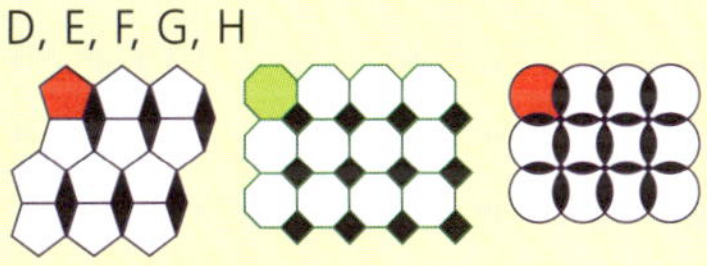

2 **a** 

Ten more tiles will be drawn on each.
b yes

4:24

Answers will vary.

5:01

1

Buttons			
	Red	Blue	Green
Large	5	8	6
Small	7	9	11

2

Buttons		
Red	Blue	Green
12	17	17

3

Buttons	
Large	19
Small	27

4 Graphs may vary.

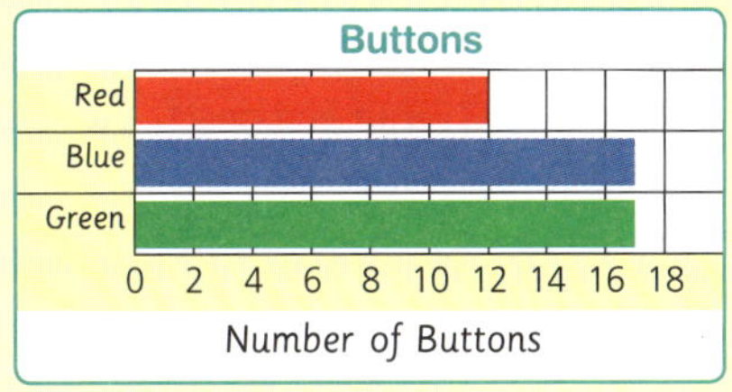

5 Answers will vary.

6

Number of Times Thrown					
⚀	⚁	⚂	⚃	⚄	⚅
4	4	4	3	3	3

5:02

1 **a** red **b** yes **c** no **d** 1 in 3 **e** 2 in 3

2 **a** red **b** yellow **c** red **d** zero **e** 4
f 3 **g** 1

3 Answers will vary.

4 Answers will vary.

5:03

1 E A D B C

2 Yes; answers will vary.

3 **a** impossible **b** not likely **c** certain **d** very likely
e certain

5:04

1 **a** 6 **b** Saturday **c** Thursday
d 10 **e** 48 **f** Everyone was at home.

2 

3 **a** 2 **b** 6 **c** 9

5:05

1 **a** i 8 ii 3 iii 0
b 14 **c** 16 **d** 66 **e** class discussion

2 **a** tennis **b** 29 **c** 3 **d** Sport Chosen by Class 4R

 • *AUSTRALIAN SIGNPOST MATHS 4* • ISBN 978 1 4886 2187 1

5:06

1 **a** unlikely **b** impossible **c** very likely
d certain **e** impossible **f** even chance
2 **a** two red **b** two blue **c** one red and one blue
3 **a** B and E **b** C and F **c** A and D
4 **a** no (they are just as likely).
b Two heads, two tails, or one head and one tail.

5:07

1 **a** 20 **b** **i** $\frac{9}{20}$ **ii** $\frac{11}{20}$
2 even chance
3 **a** likely **b** unlikely **c** impossible
d very unlikely **e** very unlikely
4 **a** Jack, Luke, Emma **b** Jack

5:08

1 **a** 1 **b** 10 **c** 2 **d** 14 **e** 28
f between 8 and 9 **g** 30; yes **h** Answers will vary.
2 Answers will vary.

5:09

1 **a**

Sport	Tally	Total
Tennis	卌 卌 卌 卌 卌 卌 卌 卌 卌	45
Athletics	卌 卌 卌 卌 卌 卌 卌	35
Swimming	卌 卌 卌 卌 卌 卌 卌 卌 III	43
Hockey	卌 卌 卌 卌 卌 III	28
Basketball	卌 卌 卌 卌 卌 卌 卌 III	38
Cricket	卌 卌 卌 卌 卌 卌 卌 卌 II	42
Football	卌 卌 卌 卌 卌 卌 卌 卌 卌 I	46
Other	卌 卌 卌 卌 IIII	24

b football **c** 8 **d** 80
e squash, netball, baseball, T-ball etc. (Two will be chosen.)

5:10

1 **a–g** Answers will vary.
h Yes, the more trials, the more accurate the result.
i A, B, C, D, E

5:11

1 **a** E **b** D **c** C **d** A **e** B
2 E, A, C, B, D
3 **a** D, C, B, A, E **b** B, D, C, E, A **c** E, D, A, B, C
4 Answers will vary.

5:12

Answers will vary.

5:13

1 **a** 30 **b** Tuesday **c** 100 **d** 100
e Answers will vary. The shop may have been closed on Thursday. **f** 10
2 **a** car **b** 100 **c** Approximately 200 students
d no
e

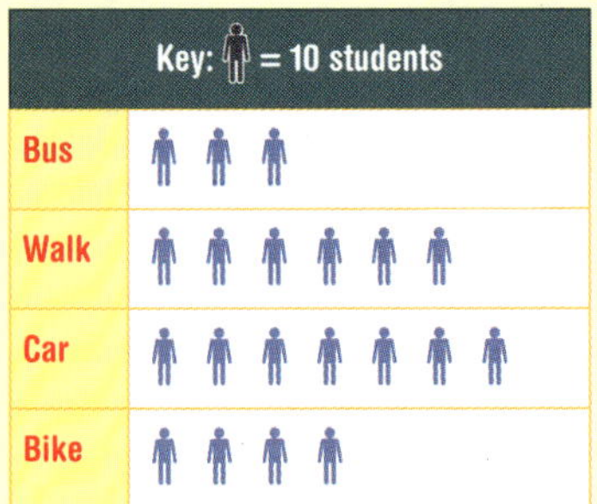

f Answers will vary. It is quicker to draw and count the number of students.
3 Answers will vary.

5:14

1–3 Answers will vary.
4 **a** E **b** N **c** E

5:15

Answers will vary.

5:16

1 **a** heads
b No, the results will vary each time the experiment is completed. There is an even chance that heads or tails will be tossed.
2 **a** 5 **b** Answers will vary.
c They are equally likely. **d** yes
e The more often an experiment is completed, the more accurate the results.
3 **a** 1 head and 1 tail
b–e Answers will vary.
f 2 heads and 2 tails have the same chance of occurring. 1 head and 1 tail is likely to occur twice as often as 2 heads or 2 tails.

ISBN 978 1 4886 2187 1

Diagnostic Test 1

Number

see page

1 Write the fraction for the shaded part. 1

a 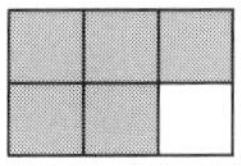 —

b 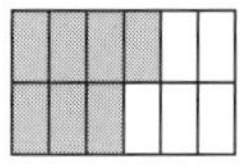 —

c 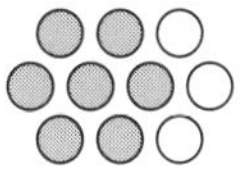 —

d 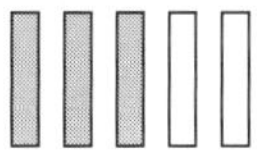 —

2 What part of each hundred square has been coloured? 2

a 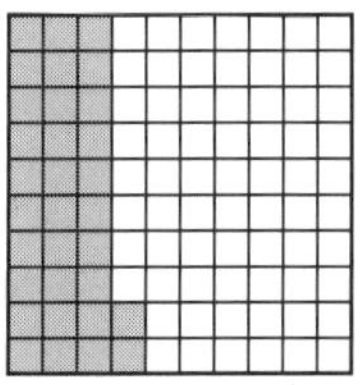 —

b 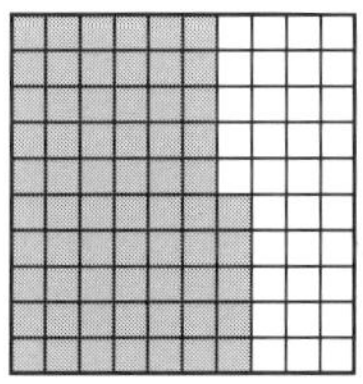 —

3 Colour the given fraction. 2

a 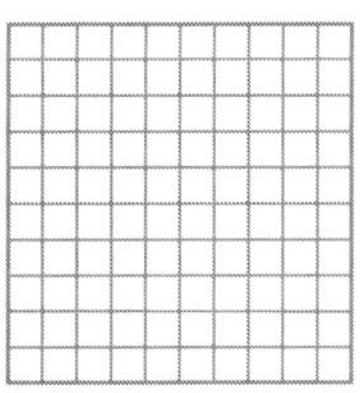 $\frac{16}{100}$

b 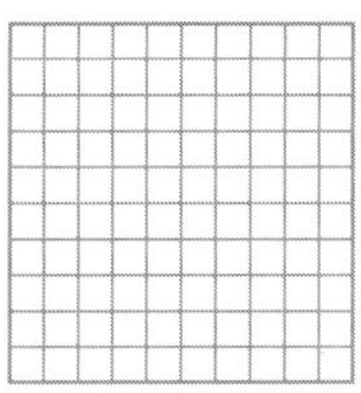 $\frac{49}{100}$

4 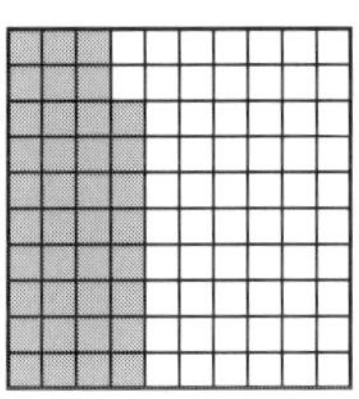 What decimal is shown? 0· 3

Operations

5 Use skip counting to continue the sequence. 35

7, 14, 21, 28, ☐, ☐, ☐, ☐, ☐, ☐, ☐

Explain the pattern made above.

see page

6 a $6 \times 2 =$ ☐ b $9 \times 10 =$ ☐ 36

c $7 \times 3 =$ ☐ d $5 \times 1 =$ ☐

e 6×5 f 8×2 g 4×10

7 Use this picture to find: 37

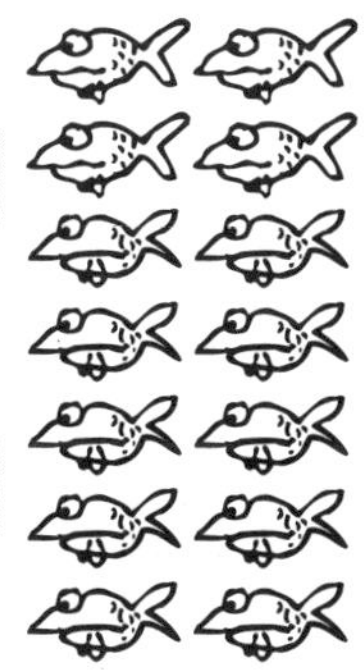

a 3 groups of 14 fish ☐

☐ × ☐ = ☐

b 4 groups of 14 fish ☐

☐ × ☐ = ☐

8 Write a word problem for the number sentence $9 \times 3 =$ ☐. 37

9 Complete the table. 38

×	3	10	2	1	5	4	7	9	8	6
2										
4										

Measurement

10 Show the time on each clock face. 88

a 

3:35

b 

10:15

see page

11 Write the analogue time shown. 89

a

☐ past ☐

b

☐ to ☐

c

☐ to ☐

d

☐ past ☐

12 Today is Wednesday, 4 April. 90
What is the date:

a next Saturday? ☐

b next Wednesday? ☐

c three weeks from now? ☐

Geometry

13 123

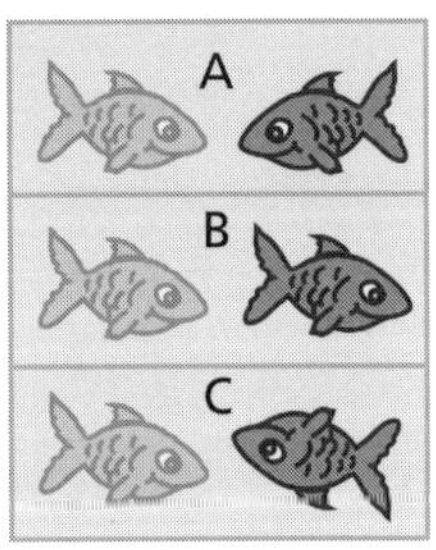

Which picture shows:

a a slide? ☐

b a turn? ☐

c a flip? ☐

d Continue this pattern.

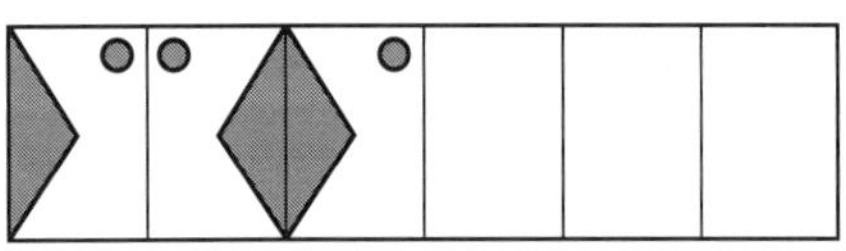

14 124

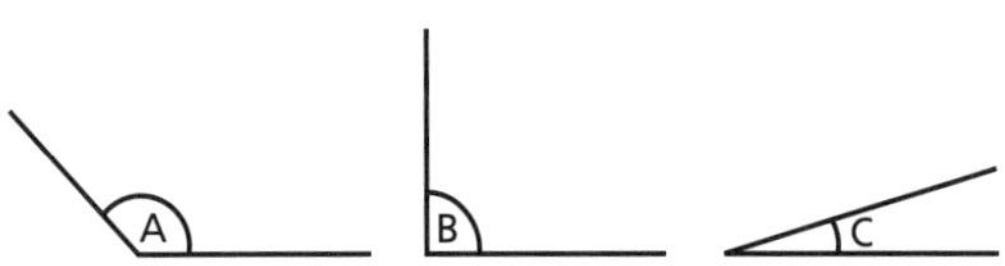

a Write the angles in order of size, from smallest to largest. ☐, ☐, ☐

b Which angle is a right angle? ☐

see page

15 124

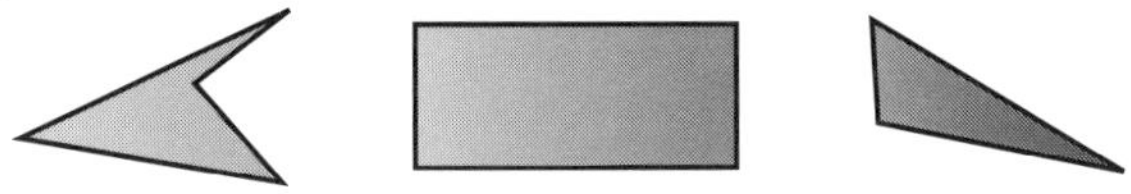

a Which of these shapes is not a quadrilateral? ☐

b Which has parallel sides? ☐

c Which shape has symmetry? ☐

d How many sides has shape **B**? ☐
How many angles? ☐

Statistics and Probability

16 147

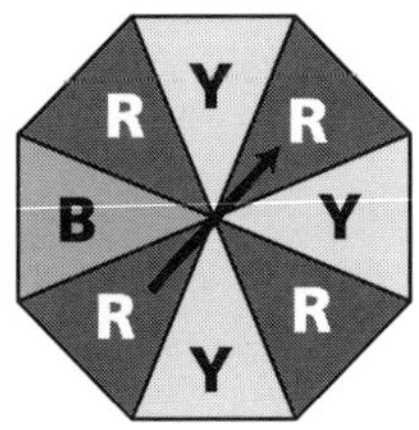

a The chance of landing on red (R) is ☐ out of 8.

b The chance of landing on blue (B) is ☐ out of 8.

c The chance of landing on yellow (Y) is ☐ out of 8.

17 Use the data to complete the table. 148

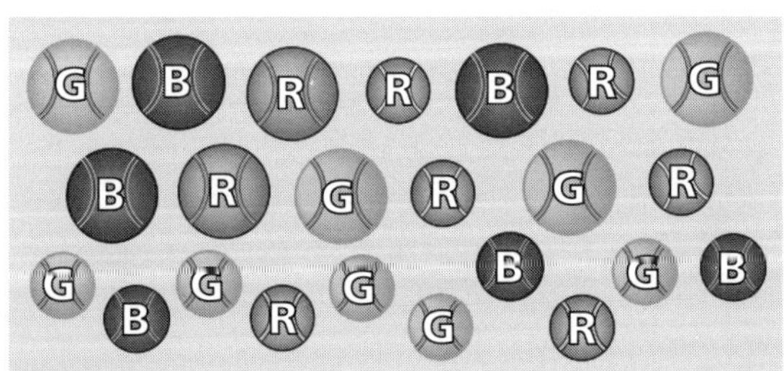

	Red (R)	Green (G)	Blue (B)
Small			
Large			

 • *AUSTRALIAN SIGNPOST MATHS 4* • ISBN 978 1 4886 2187 1

Diagnostic Test 2

Number

see page

1

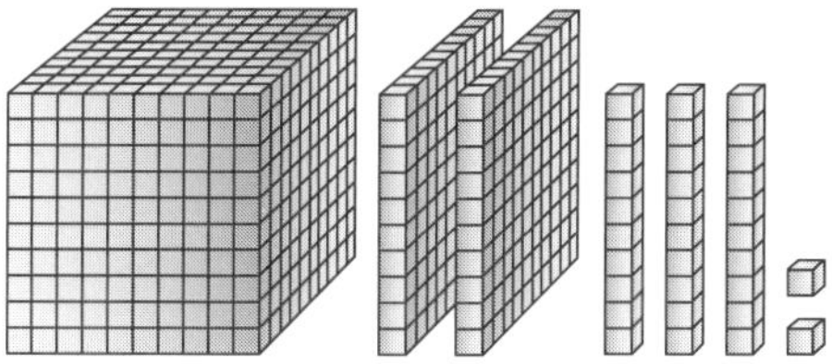

a Write the numeral. ☐ 4

Complete the numeral expander.

b Write 2063 in words.

2 How many digits in: 5

a 109? ☐ **b** 2365? ☐

Write the number that is:

c 6 more than 1236 ☐

d 10 less than 1645 ☐

3 Which is the larger number? ☐ 6

A 70000 + 6000 + 800 + 50 + 3

B 70000 + 9000 + 300 + 20 + 1

4 Which number has a 7 that stands for 700? ☐ 6

A 54762 **B** 37921 **C** 73524

5 Fill in the numeral expanders for 1436. 7

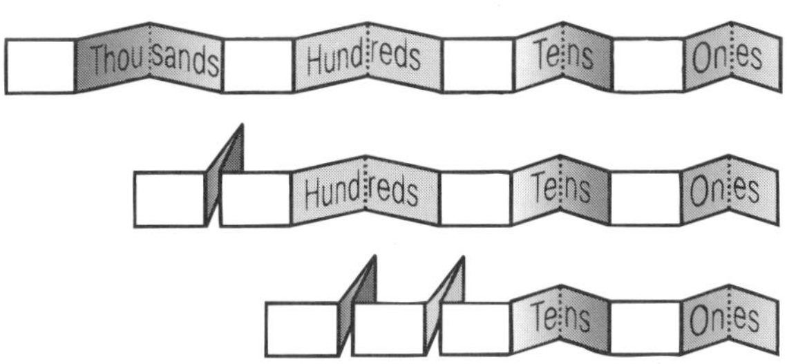

see page

6 How many hundreds in each number? 7

a 2875 ☐ **b** 6149 ☐

Operations

7 40 41

a

Tens	Ones
3	4
2	1
+ 4	3

b

Tens	Ones
$4	2
$1	6
+ $2	3

8 42 43

a

Tens	Ones
☐	☐
3	6
+	9

b

Tens	Ones
☐	☐
$4	9
+ $	7

c

Tens	Ones
☐	☐
3	6
+ 2	5

d

Tens	Ones
☐	☐
$5	7
+ $3	8

e 44

3	6
2	9
+ 1	7

f

$4	3
$2	8
+ $2	6

Measurement

9 Use the calendar to write the date for: 91

May

S	M	T	W	T	F	S
				1	2	3
4	5	6	7	8	9	10
11	12	13	14	15	16	17
18	19	20	21	22	23	24
25	26	27	28	29	30	31

a the first Wednesday ☐

b the last Thursday ☐

c the first weekend ☐

see page

10 Measure the length of each bar. 92

a 

☐ mm

b 

☐ mm

c 

☐ mm

11 Record each length shown. 93

a ☐ mm

b ☐ mm

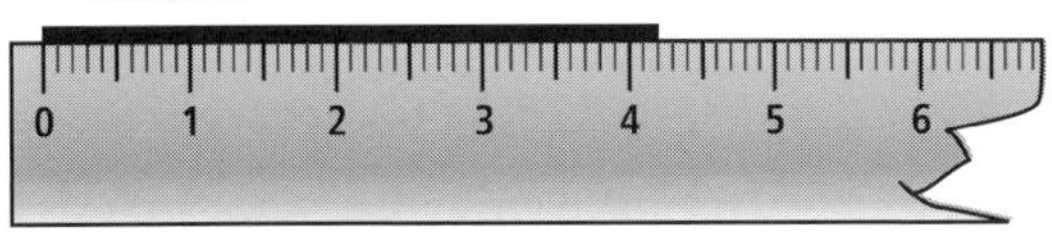

c Perimeter is ☐ cm.

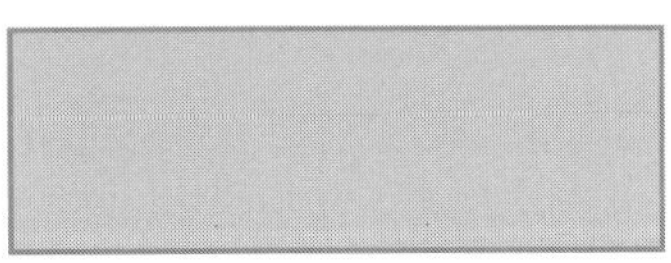

12 Write these as centimetres. 94

a 2 m 56 cm ☐

b 3 m 75 cm ☐

c 5 m 97 cm ☐

d 4 m 45 cm ☐

How many millimetres in:

e 5 cm 6 mm? ☐

f 9 cm 5 mm? ☐

g 8 cm 3 mm? ☐

Geometry

see page

13 125

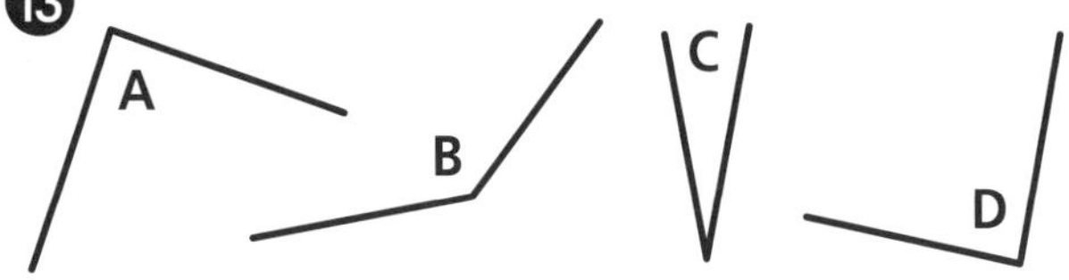

Which angle is smaller:

a **A** or **B**? ☐ **b** **C** or **D**? ☐

c **B** or **D**? ☐ **d** **A** or **C**? ☐

14 **a** Name this object. ☐ 126

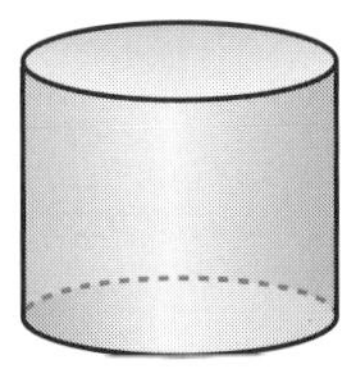

b Does it have any rectangular faces? ☐

c Is this a prism? ☐

Probability

15 Choose the best label. 149

A not likely **B** even chance

C likely

a You will visit the moon.

☐

b You will be in Year 5 next year.

☐

16 In a random selection, each number has the same ☐ of being chosen. 149

17 Does each number have the same chance of being chosen? ☐ 149

 ISBN 978 1 4886 2187 1

Diagnostic Test 3

Number

see page

❶ Write the numeral for: 9

a

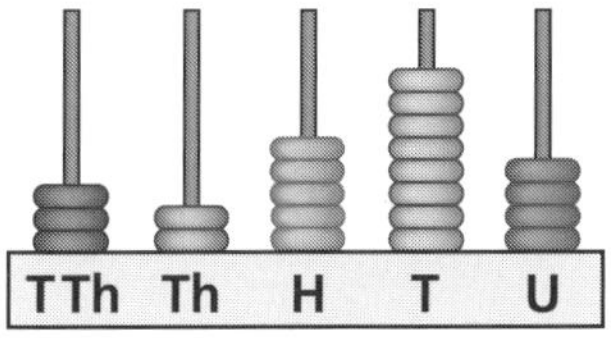

☐

b 50 000 + 900 + 30 + 7

☐

c thirty-four thousand three hundred and sixty ☐

❷ Round off to the nearest hundred. 8

a 34 967 ☐

b 73 431 ☐

Round off to the nearest thousand.

c 6259 ☐

d 91 530 ☐

❸ Name the part shaded. 10

a 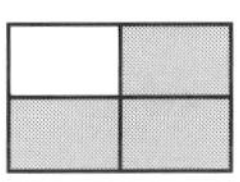 $\frac{\square}{\square}$ **b** 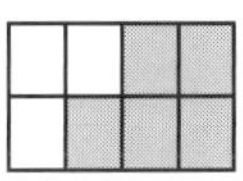 $\frac{\square}{\square}$

c 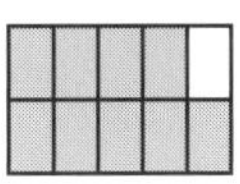 $\frac{\square}{\square}$ **d** 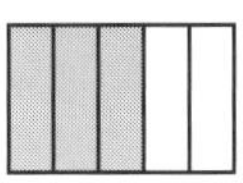 $\frac{\square}{\square}$

Write true or false.

e $\frac{4}{5} = 1$ ☐ **f** $1 = \frac{10}{10}$ ☐

g $\frac{4}{4} = 1$ ☐ **h** $\frac{8}{8} = 1$ ☐

Circle the larger fraction.

i $\frac{1}{2}$ or $\frac{1}{10}$ **j** $\frac{1}{4}$ or $\frac{1}{2}$

k $\frac{1}{10}$ or $\frac{1}{5}$ **l** $\frac{1}{5}$ or $\frac{1}{4}$

Use > or < to make each number sentence true.

m $\frac{1}{2}$ ☐ $\frac{1}{5}$ **n** $\frac{1}{10}$ ☐ $\frac{1}{4}$

see page

❹ Use the number lines to show equivalent fractions. 11

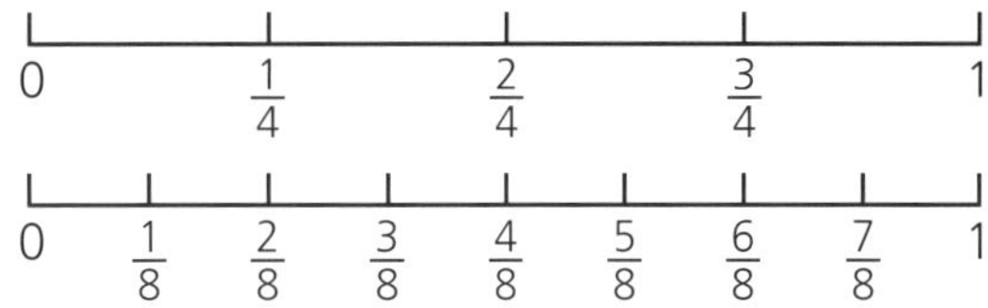

a $\frac{1}{4} = \frac{\square}{\square}$ **b** $\frac{6}{8} = \frac{\square}{\square}$

c $\frac{2}{4} = \frac{\square}{\square}$

Operations

❺ 45

Hund	Tens	Ones
☐	☐	☐
	9	6
+	7	3

❻ Complete this table. 46 47

×	4	1	5	7	10	3	8	2	9	6
2										
4										
8										

❼ **a** 35 buttons, 23 used. 48

How many buttons left?

☐ ☐ ☐ ☐ ☐

b

Tens	Ones
3	7
− 2	4

c

Tens	Ones
\$7	8
− \$5	6

d 49

Tens	Ones
☐	☐
9	4
− 3	7

e

Tens	Ones
☐	☐
8	2
− 6	9

Measurement

see page

8 Record the length of each interval. 95

a [ ] mm [ ] cm

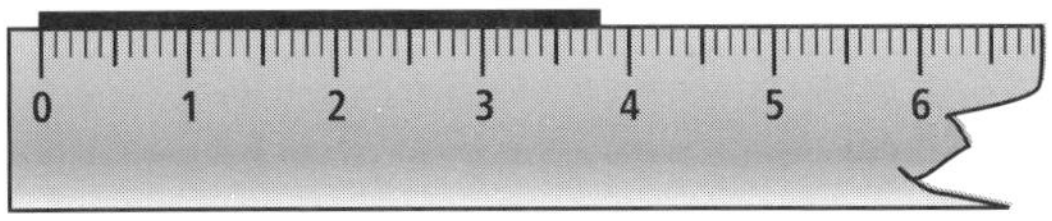

b [ ] mm [ ] cm

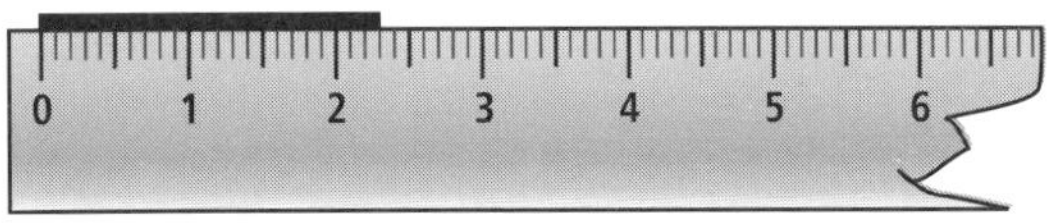

9 What temperature is shown on each thermometer? 96

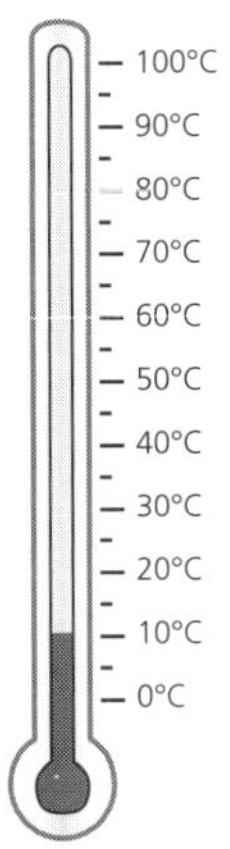

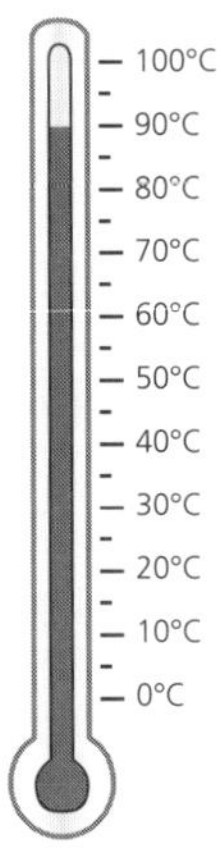

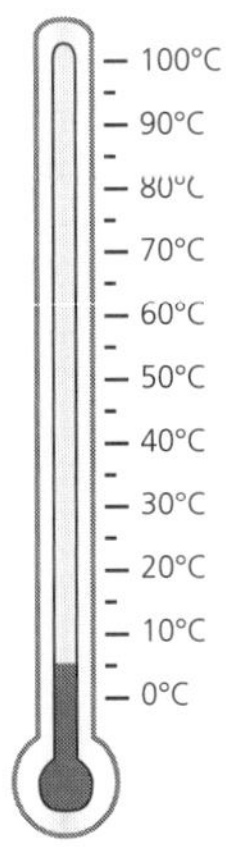

a [ ] **b** [ ] **c** [ ]

10 On each thermometer, colour the mercury to match the given temperature. 97

a 40°C **b** 28°C **c** 36°C

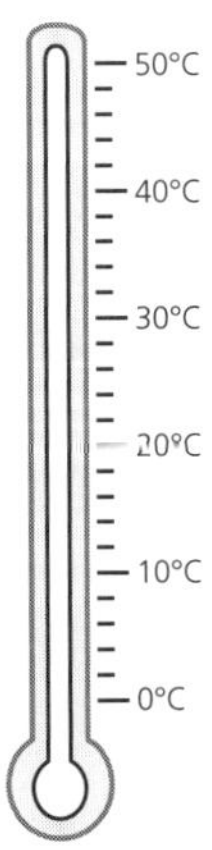

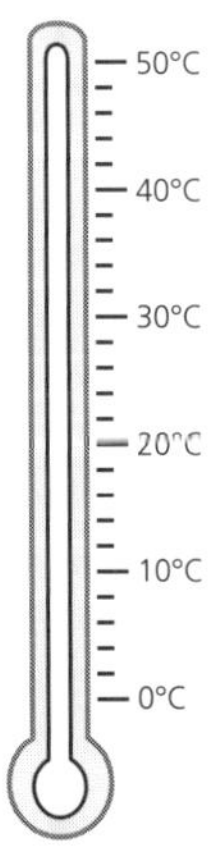

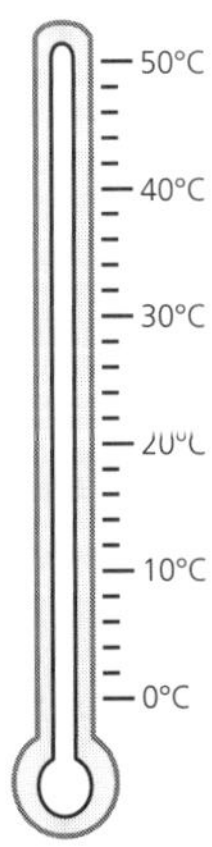

Geometry

see page

11 127

A 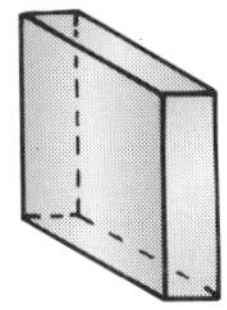 B 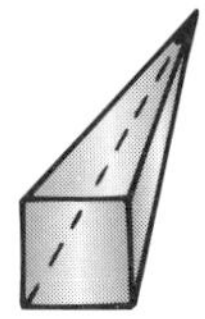 C 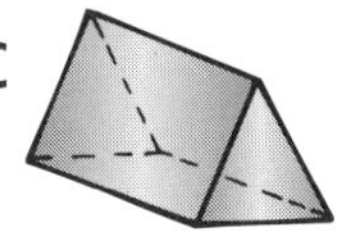

Is it a prism or a pyramid?

a A [ ]

b B [ ]

c C [ ]

12 How many faces has a: 128

a triangular prism? [ ]

b pentagonal prism? [ ]

How many faces are triangles on a:

c triangular prism? [ ]

d square pyramid? [ ]

Statistics

13 150

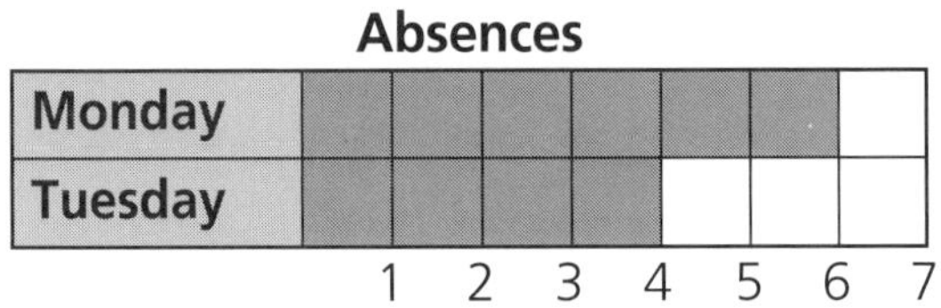

How many absences were there on:

a Tuesday? [ ] **b** Monday? [ ]

14 **a** Find the totals for the tallies. 150

Ford	𝍸 𝍸 𝍸 𝍸 𝍸 𝍸 𝍸 𝍷	[ ]
Holden	𝍸 𝍸 𝍸 𝍸 𝍸 𝍸 𝍸 𝍸 𝍷𝍷𝍷𝍷	[ ]
Toyota	𝍸 𝍸 𝍸 𝍸 𝍸 𝍸 𝍸 𝍸 𝍸 𝍷𝍷	[ ]

Which car was tallied:

b least? [ ]

c more than 45 times? [ ]

 • *AUSTRALIAN SIGNPOST MATHS 4* • ISBN 978 1 4886 2187 1

Diagnostic Test 4

Number

1 Name the shaded part. see page 12

a  ☐

b 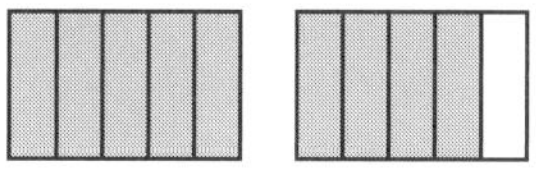 ☐

2 Use < or > to make each number sentence true. see page 12

a $\frac{1}{2}$ ☐ $\frac{3}{4}$ b $\frac{7}{8}$ ☐ $\frac{1}{4}$

c $\frac{3}{8}$ ☐ $\frac{3}{4}$ d $\frac{5}{8}$ ☐ $\frac{3}{4}$

3 Use the number lines to find equivalent fractions. see page 13

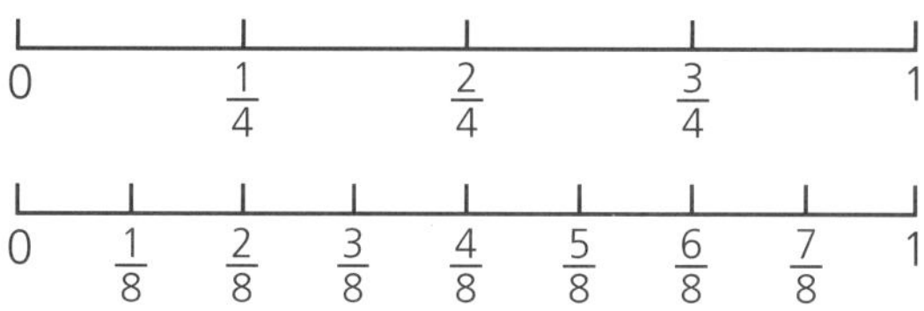

a $\frac{2}{8}$ = ☐ b $\frac{2}{4}$ = ☐

4 Complete the pattern. see page 13

a $\frac{1}{4}$, $\frac{2}{4}$, $\frac{3}{4}$, ☐, ☐, ☐, ☐

b $\frac{1}{8}$, $\frac{2}{8}$, $\frac{3}{8}$, ☐, ☐, ☐, ☐

5 Write the mixed number for: see page 14

a $\frac{7}{3}$ ☐ b $\frac{9}{5}$ ☐ c $\frac{11}{4}$ ☐

Write the improper fraction for:

d $2\frac{1}{3}$ ☐ e $2\frac{4}{5}$ ☐ f $3\frac{3}{4}$ ☐

Operations

6 see page 50

a Tens Ones
☐ ☐
  3 5
− 5

b Tens Ones
☐ ☐
  4 6
− 9

c Tens Ones
☐ ☐
  6 7
− 2 9

d Tens Ones
☐ ☐
  6 3
− 4 8

7 see page 51

a Tens Ones
☐ ☐
  \$4 5
− \$1 9

b Tens Ones
☐ ☐
  \$6 5
− \$2 7

see page 52

c Tens Ones
☐ ☐
  8 2
− 2 5

d Tens Ones
☐ ☐
  6 1
− 4 8

8 Complete this table. see page 54

×	4	1	5	7	10	3	8	2	9	6
3										
6										
9										

9 Why is × 6 double × 3? see page 55

Measurement

see page

10 How many litres in:

a 3000 mL? ☐

b 7000 mL? ☐

98

11 How many millilitres in?

a 1 L? ☐ **b** 7 L? ☐

c 5 L? ☐ **d** $1\frac{1}{2}$ L? ☐

99

12 How many tiles have been used to cover this area?

☐ tiles

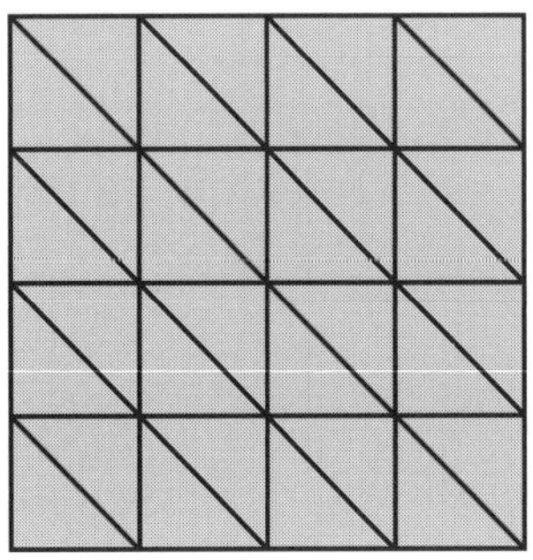

100

13

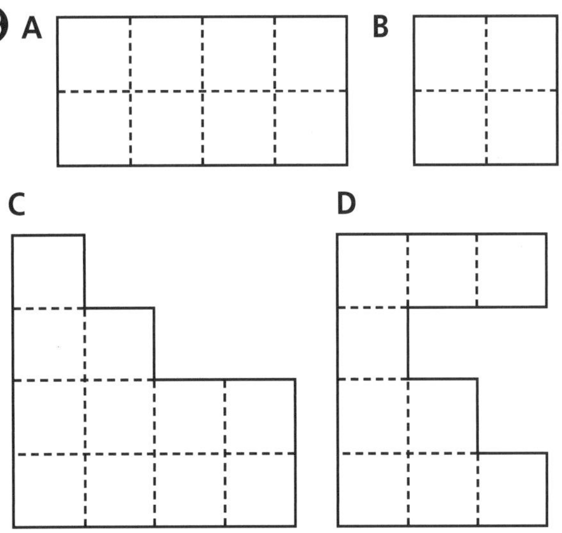

What is the area of:

a A? ☐ cm²   **b** B? ☐ cm²

c C? ☐ cm²   **d** D? ☐ cm²

e Which shape has the greatest area? ☐

f Which shape has the least area? ☐

101

Geometry

see page

14

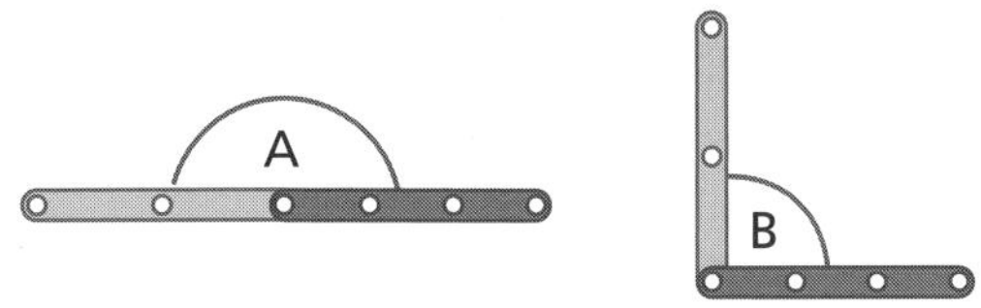

I used geostrips to make these angles.

a Which angle is larger? ☐

b What is the name of angle **B**?

☐

130

15

a Draw all the diagonals of this hexagon. How many can be drawn?

☐

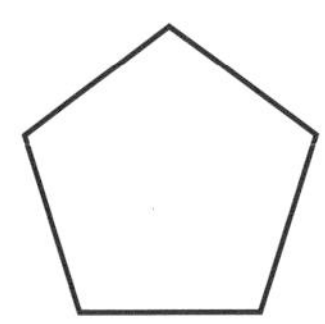

b Draw all the axes of symmetry of this pentagon. How many can be drawn? ☐

c How many angles has an octagon? ☐

132

Statistics and Probability

16 Tom has 14 soldiers, 8 cars and 10 balls. Complete the column graph to show the number of each toy.

Tom's Toys

Soldiers									
Cars									
Balls									

0 4 8 12 16

151

17 Choosing at random, which number is:

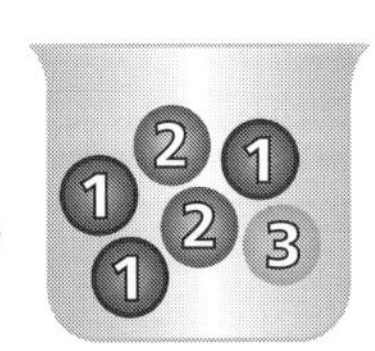

a most likely to be drawn?

☐

b least likely to be drawn? ☐

152

Diagnostic Test 5

Number

see page

1 Colour the circles to show the mixed number. 15

$1\frac{1}{2}$

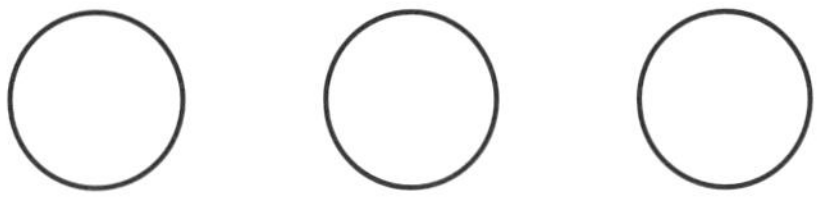

2 Write the numeral for: 16 17

a

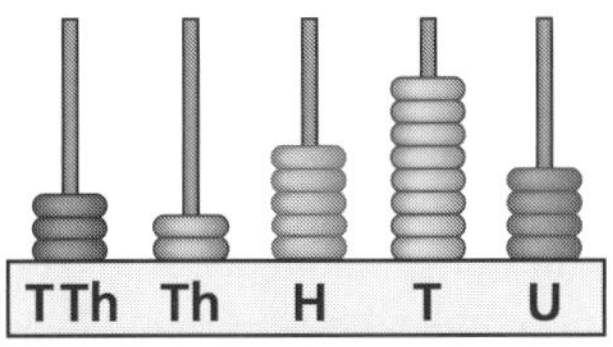

☐

b 50000 + 900 + 30 + 7 ☐

c thirty-four thousand three hundred and sixty ☐

3 Complete the number line. 18 19

0 $\frac{1}{8}$ ☐ $\frac{3}{8}$ ☐ ☐ $\frac{6}{8}$ $\frac{7}{8}$ 1

4 18 19

0 $\frac{1}{3}$ $\frac{2}{3}$ 1

0 $\frac{1}{6}$ $\frac{2}{6}$ $\frac{3}{6}$ $\frac{4}{6}$ $\frac{5}{6}$ 1

Use the number lines above to write a fraction equal to:

a $\frac{2}{6}$ ☐ **b** $\frac{2}{3}$ ☐

Operations

5 Complete the table. 57 58

×	3	10	2	1	5	4	7	9	8	6
7										
9										

see page

6 59

a 3 × 4 = ☐ **b** 2 × 5 = ☐

c 5 × 3 = ☐ **d** 5 × 6 = ☐

e 8 × 10 = ☐ **f** 7 × 8 = ☐

g 4 × 6 = ☐ **h** 6 × 2 = ☐

i The product of 3 and 5 is ☐.

j The product of 8 and 6 is ☐.

7 61

a

H	T	U
1	8	3
+ 1	0	9

b

H	T	U
6	2	5
+	9	7

62

c

H	T	U
4	7	5
+ 2	6	9

d

H	T	U
3	5	8
+ 4	7	5

8 Write the missing numbers. 63

a ☐ + 11 = 24

b 15 − ☐ = 8

c ☐ × 4 = 36

d 35 ÷ ☐ = 7

e ☐ squared = 25

Measurement

9 Use the short form to write: 102

a 60 grams ☐

b 379 grams ☐

c 16 kilograms ☐

see page

10 How many: 103

a grams in 2 kg? ☐

b grams in 5 kg? ☐

c kilograms in 7 000 g? ☐

11 Draw your own grid to find the area of each shape. 104 105

a 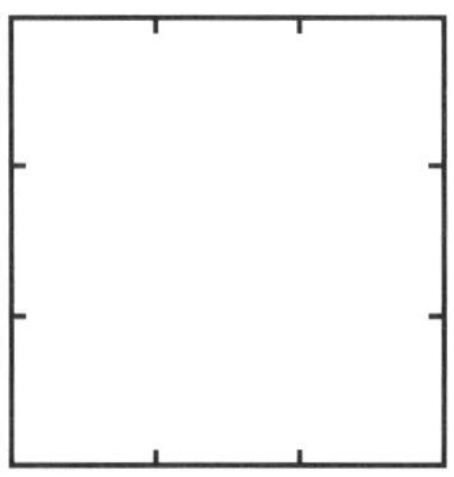 ☐

b 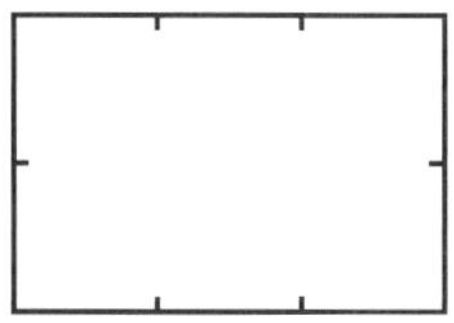 ☐

Geometry

12 We can combine shapes in different ways. Here a square and two triangles have been used. 133

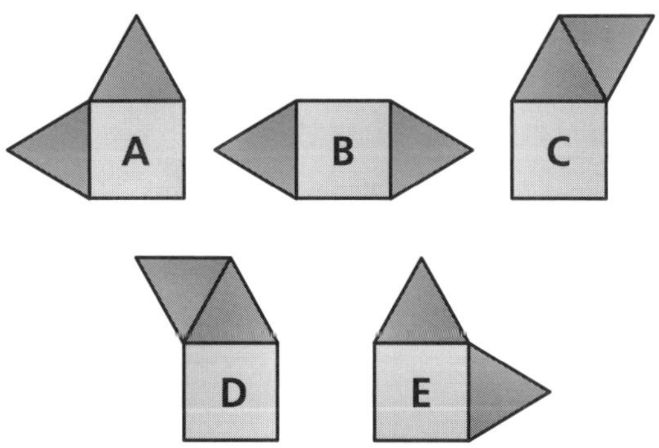

Which two shapes above are the same?

☐ and ☐

Why?

☐

☐

☐

see page

13 134

Scale: 1 cm = 1 km

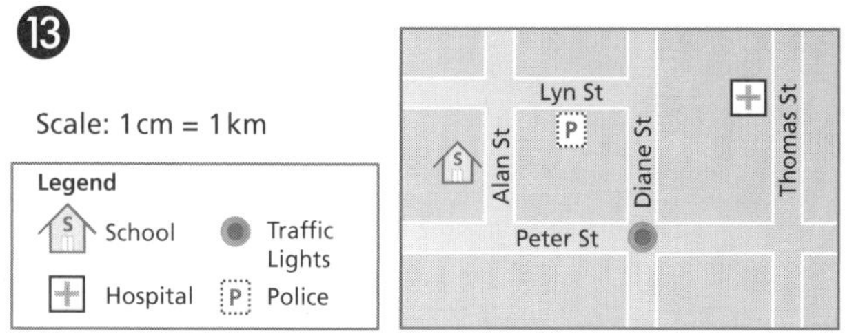

On which street on the map would you find the:

a school? ☐

b hospital? ☐

c police station? ☐

14 136

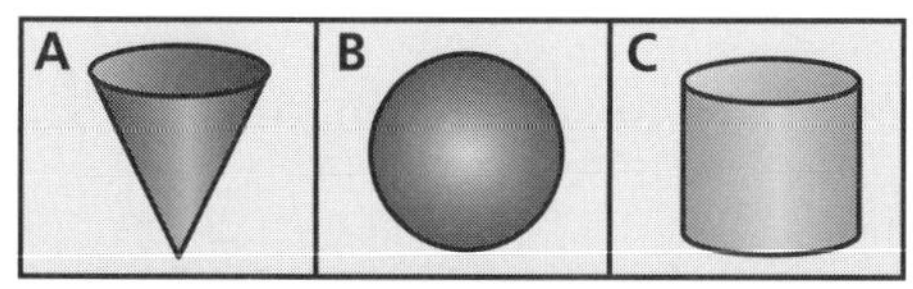

Write the name of:

a A ☐

b B ☐

c C ☐

Statistics

15 154 155

Year 4 Pets

Dogs	𝍸 𝍸 𝍸 |||
Cats	𝍸 𝍸 ||
Birds	𝍸 |
Others	𝍸 𝍸

How many Year 4 pets were:

a dogs? ☐ **b** birds? ☐

c cats? ☐ **d** others? ☐

e not birds? ☐

f either dogs or cats? ☐

g there altogether? ☐

Diagnostic Test 6

Number

see page

1 20

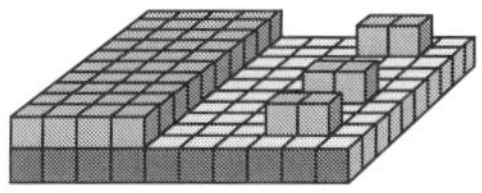

What fraction is shown? $\frac{\square}{\square}$

2 Complete the numeral expanders with 0·58. 21

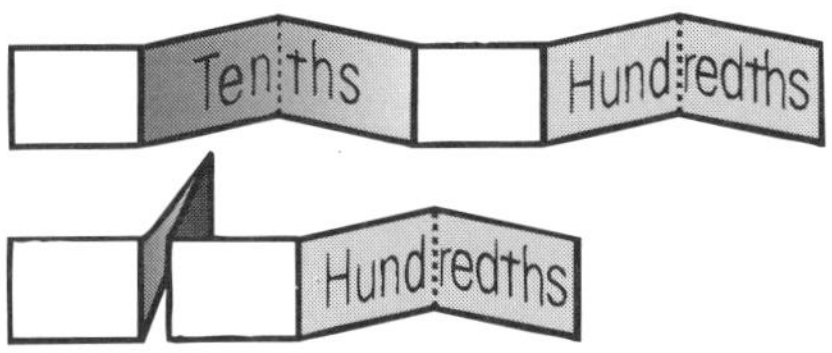

3 Write the decimal for: 22

a $\frac{7}{10}$ ☐ **b** $\frac{9}{10}$ ☐

c $\frac{35}{100}$ ☐ **d** $\frac{67}{100}$ ☐

e 5 tenths ☐

f zero point eight ☐

4 22

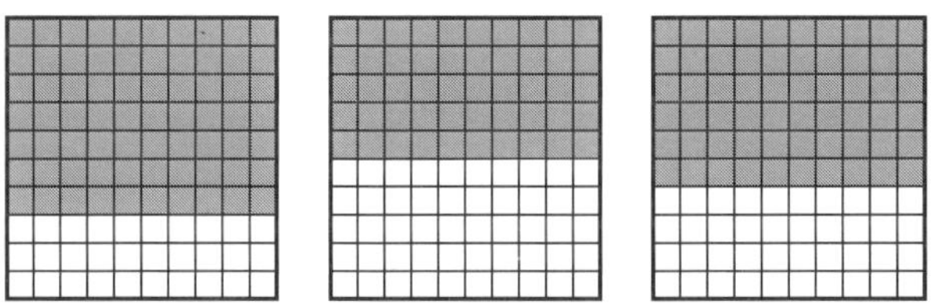

Which shaded area is smaller:

a **A** or **B**? ☐ **b** **A** or **C**? ☐

5 Colour part of each square to match the given decimal. 23

a 0·25 **b** 0·4 **c** 0·06

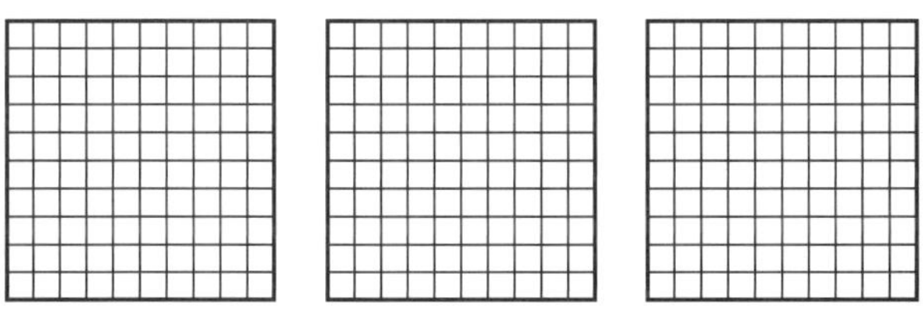

6 Circle the larger number. 24

a 5963 or 5693

b 46 319 or 49 631

see page

7 Write the numeral for: 24

a 20 000 + 9000 + 500 + 60 + 3 ☐

b 70 000 + 8000 + 300 + 2 ☐

c sixty-four thousand five hundred and twenty-nine ☐

Operations

8 64 65

a Hund Tens Ones

5	4	9
− 3	2	6

b

6	3	4
− 4	2	7

66 68

c

4	2	6
− 1	7	8

d

7	0	0
− 3	6	5

66 68

e

4	6	7
− 2	6	9

f

8	1	6
− 3	0	7

9 69

7 9 9 + 1

~~8 0 0~~

− 5 4 6

+ 1 = ☐

10 How many groups of 5 fish are in 15? 70

☐ groups. So, 15 ÷ 5 = ☐

11 71

a 8 × 9 = 72 **b** 7 × 6 = 42

72 ÷ 8 = ☐ 42 ÷ 7 = ☐

72 ÷ 9 = ☐ 42 ÷ 6 = ☐

 • *AUSTRALIAN SIGNPOST MATHS 4* • ISBN 978 1 4886 2187 1

Diagnostic Test 6 (continued)

Measurement

see page

12 Use the short form to write: 107

a 4 square metres ☐

b 10 square metres ☐

13 Find the area of each coloured shape. 109

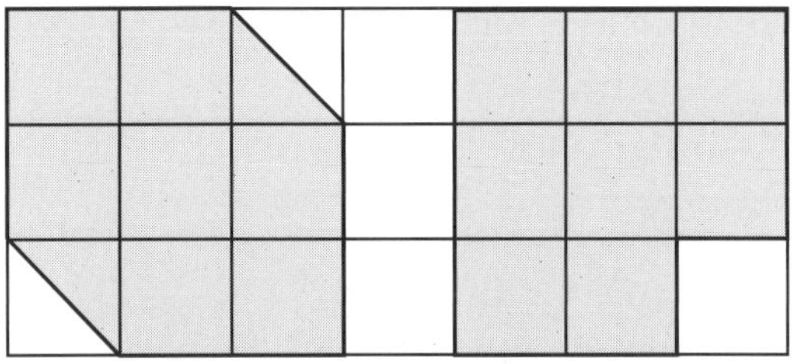

a ☐ **b** ☐

14 Write the volume of each model. 112

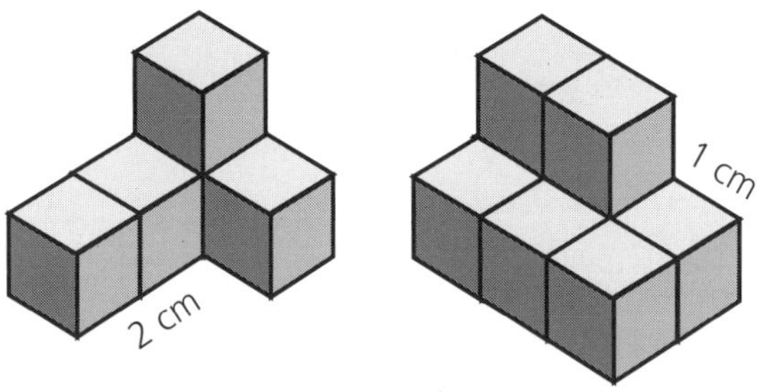

a ☐ cm^3 **b** ☐ cm^3

15 Write each time. 110

a ☐ **b** ☐

c 7:45 is ☐ to ☐. 111

d 3:29 is ☐ past ☐.

Geometry

see page

16 137

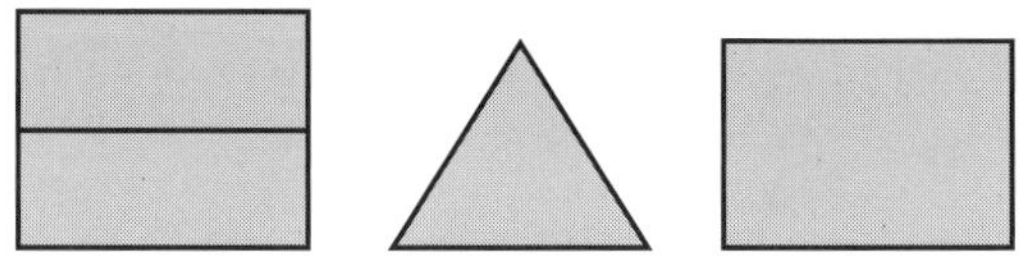

Are these views of a pyramid or prism?

☐

17 What is north of: 138

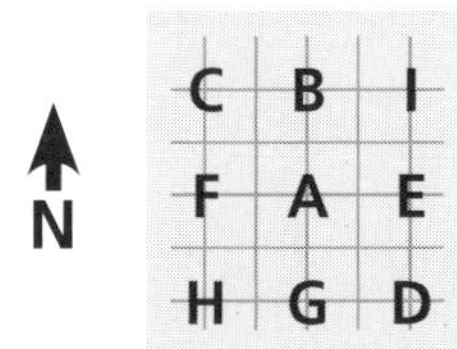

a E? ☐

b A? ☐

What is east of:

c G? ☐ **d** A? ☐

What is south of:

e F? ☐ **f** E? ☐

What is west of:

g B? ☐ **h** G? ☐

18 Label the directions on the compass. 139

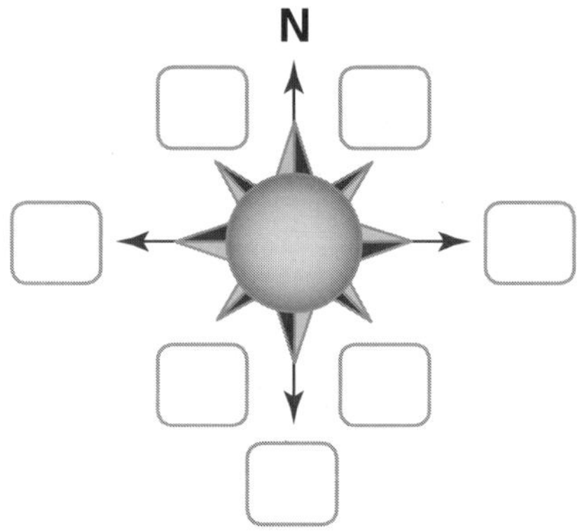

Probability

19 Put the spinners in order, from least likely to most likely chance of spinning grey. 157

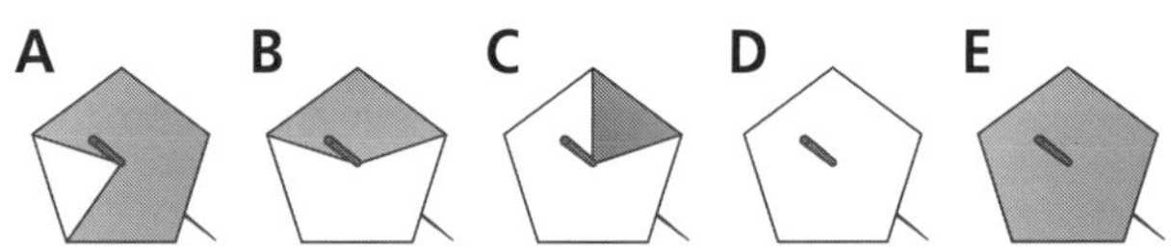

☐

 • *AUSTRALIAN SIGNPOST MATHS 4* • ISBN 978 1 4886 2187 1

Diagnostic Test 7

Number

see page

1 Use < or > to show the larger number. 25

a 32 604 ☐ 35 971

b 63 423 ☐ 63 243

2 Write: 26

a one point nine seven ☐

b

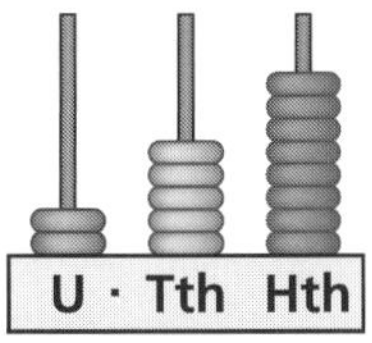

c

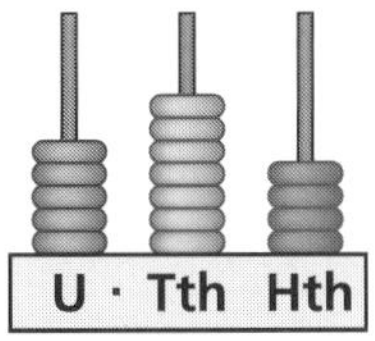

d

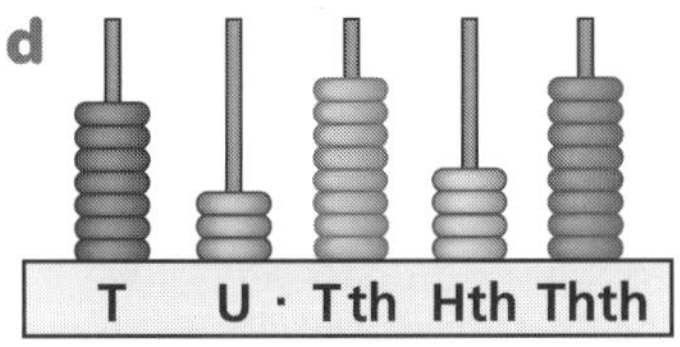

3 Is 3·619 < 3·916 true? ☐ 27

4 28

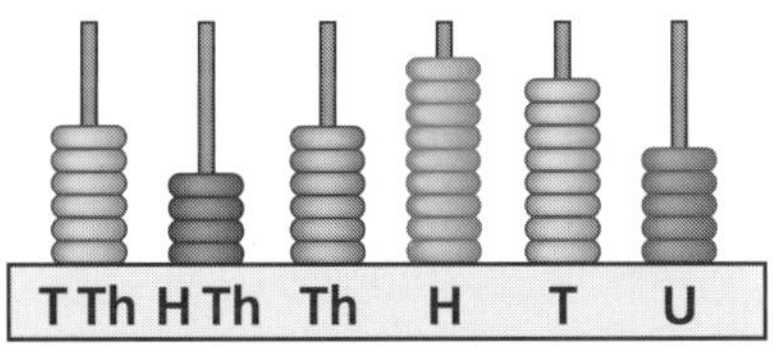

5 Write the numeral for: 28

a 300 000 + 60 000 + 9000 + 600 + 3

b five hundred and eight thousand six hundred and twenty-two

c Use the expanded form to write 80 703.

Operations

see page

6 **a** ☐ ÷ 5 = 7 **b** 15 ÷ ☐ = 5 72

c ☐ ÷ 4 = 5 **d** 30 ÷ ☐ = 6

e If 3 × 6 = 18,

6 × 3 = ☐ 18 ÷ 6 = ☐

18 ÷ 3 = ☐

7 Complete the table. 73

30	÷ 10	=	
	÷ 5	=	
	÷ 6	=	
	÷ 3	=	

8 **a** 12 ÷ 2 = ☐ **b** 15 ÷ 3 = ☐ 74

c 42 ÷ 6 = ☐ **d** 28 ÷ 7 = ☐

9 75

a 8)32 **b** 9)45 **c** 7)56

d 6)24 **e** 4)28 **f** 7)35

10 Make each amount using the highest value of notes and coins possible. 76

a $57 ☐ ☐ ☐

b $60.20 ☐ ☐ ☐

11 Find the change from $50, if $36.50 is spent. 77

Change = ☐

 • *AUSTRALIAN SIGNPOST MATHS 4* • ISBN 978 1 4886 2187 1

Measurement

see page

12 A B

113

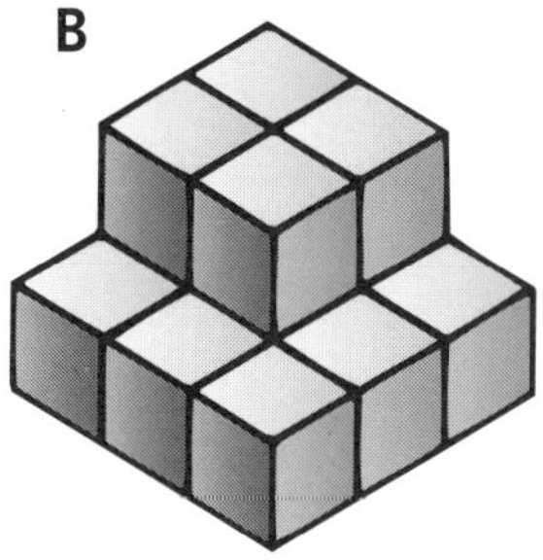

What is the volume of model:

a A? [ ] blocks

b B? [ ] blocks

13 What is the volume?

113
114

a

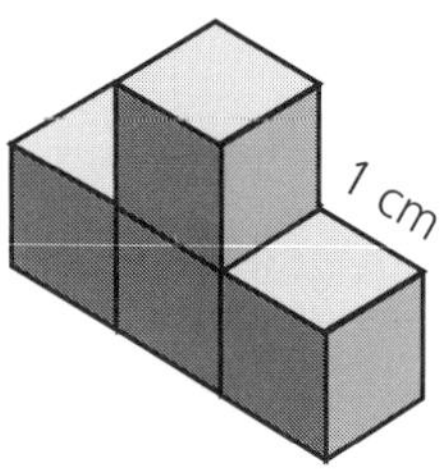

b

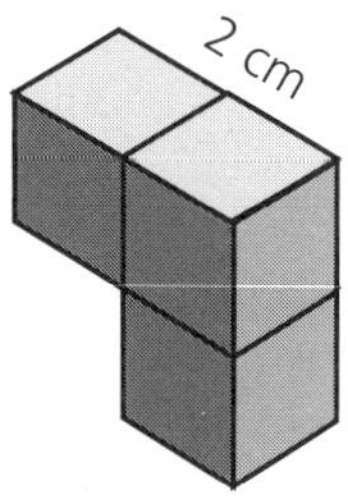

[ ] cm³ [ ] cm³

14 Write the time half an hour after:

115

a 3:20 [ ] **b** 5:50 [ ]

c 9:05 [ ] **d** 11:35 [ ]

15 Record each length shown.

116

a [ ] mm

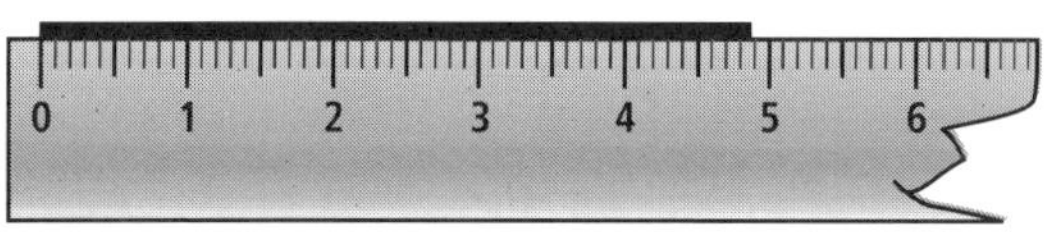

b [ ] mm

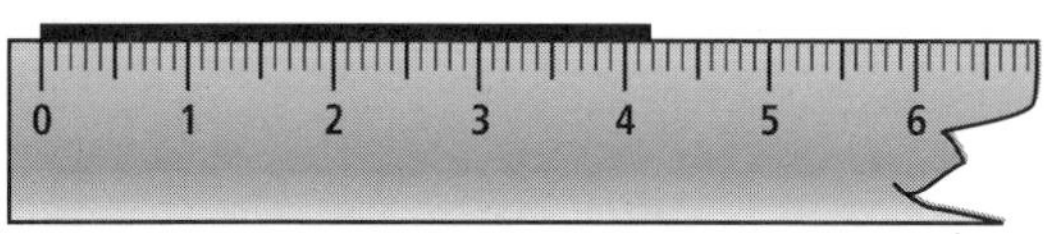

Geometry

see page

16

140

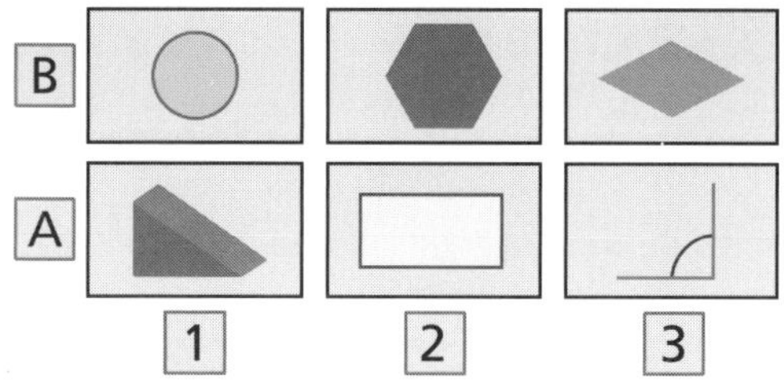

Give a number and letter for the:

a circle [ ]

b rectangle [ ]

c right angle [ ]

17

141

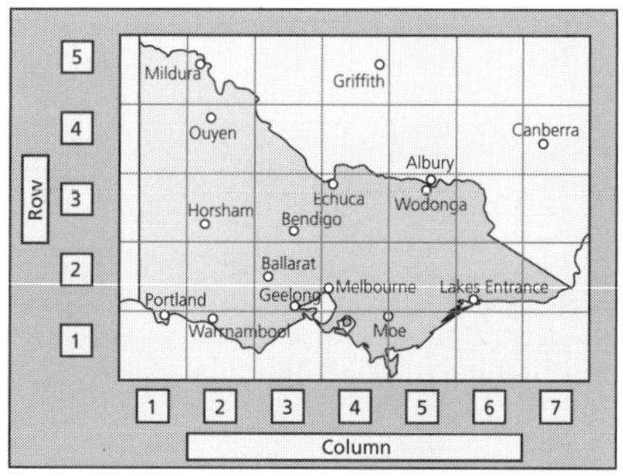

What place can be found in:

a column 2, row 3? [ ]

b column 4, row 5? [ ]

Statistics

18 20 girls were asked to choose the colour they liked best.

158

3 chose red, 6 chose green, 4 chose blue and 7 chose gold.

Use these choices to finish this graph.

Colour Chosen

Red							
Green							
Blue							
Gold							

2 4 6

Diagnostic Test 8

Number

see page

❶ Write the numeral for: 29

a (6 × 10 000) + (4 × 1 000) + (9 × 100) + (7 × 10)

b 100 000 + 70 000 + 6 000 + 300 + 6

❷ Write the value of each bold digit. 29

a 37**6**19

b 91**3**683

❸ Write in words: 29

503 277

❹ Write the mass shown on the scale. 30

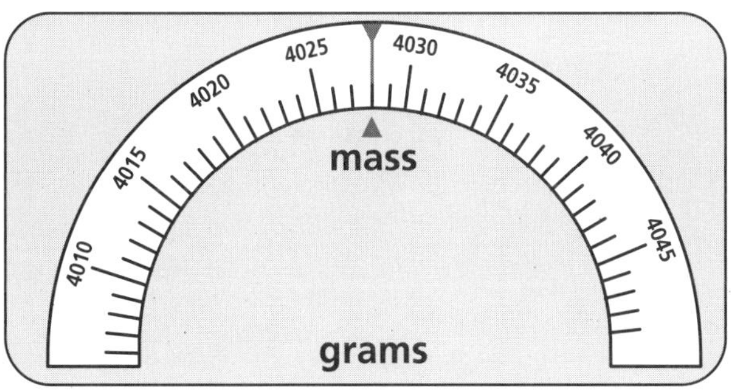

❺ Round off each number to the nearest thousand. 31

a 879 835

b 642 747

Operations

see page

❻ Write the first six multiples of: 78

a 6 ___, ___, ___, ___, ___, ___

b 9 ___, ___, ___, ___, ___, ___

❼ What is the product of: 79

a 5 and 4?

b 6 and 5?

c 9 and 7?

❽ **a** Greg made 37 pizzas in May and 24 more in June. How many did he make? 80

b Scott shared 32 stickers among 8 children. How many each?

❾ When a number is added to 28, the answer is the same as 57 minus 19. 81

What is the number?

❿ Halve the number and keep halving to fill the boxes. 82

a 200 ___, ___, ___, ___

b 80 ___, ___, ___, ___

⓫ Use the fact given, and doubling, to answer the questions. 82

a 2 × 12 = 24 4 × 12 =

b 6 × 7 = 42 12 × 7 =

Measurement

see page 118

12 Complete each label.

a [ ] past [ ] **b** [ ] to [ ]

c [ ] past [ ] **d** [ ] to [ ]

119

13 Write a digital label for each.

a

morning

[ ] : [ ]

b

evening

[ ] : [ ]

120

14 Measure the length of each bar.

a [ ] mm

b [ ] mm

c [ ] mm

Geometry

144

15 In this picture, colour the horizontal lines red and the vertical lines green.

see page 143

16 Name each angle as acute, right or obtuse.

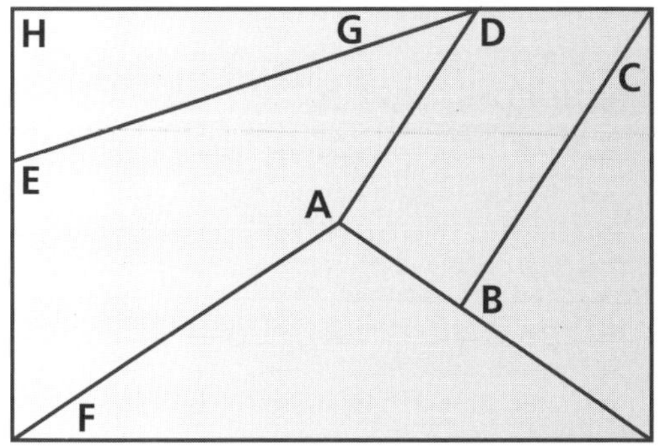

a A [ ]

b C [ ]

c E [ ]

d B [ ]

Statistics

159

17 **Children in School**

Year 3	卌 卌 卌 III
Year 4	卌 卌 II
Year 5	卌 卌 IIII
Year 6	卌 卌 卌 I

Use the data to complete the picture graph below.

Children in School

Key 웃 = 2 children

Year 3	
Year 4	
Year 5	
Year 6	